World Small-Scale Fisheries
Contemporary Visions

World Small-Scale Fisheries
Contemporary Visions

Edited by
Ratana Chuenpagdee

Eburon Delft
2011

ISBN 978-90-5972-539-3

Published by Eburon Academic Publishers,
P.O. Box 2867, 2601 CW Delft, The Netherlands

On the cover: Small-scale marine 'group' fishing in Thua Thien Hue, Vietnam. Photograph by Anthony Charles.

Cover design: Studio Hermkens, Amsterdam
Graphic design: Textcetera, The Hague

CONTENTS

PREFACE

Small-scale fishing is an occupation, a source of livelihood and a way of life for millions of people around the world. It also contributes food security, economic growth and development to communities and nations. However, its conditions are not always favorable. The sector needs better governance, designed to ensure a healthy ecosystem while enhancing fishers' well-being. Small-scale fishers and their families must have an active role in this process

Several efforts have been made in the last decades by intergovernmental organizations like the Food and Agriculture Organization of the United Nations (FAO), fishers' associations and community groups – notably the International Collective in Support of Fishworkers (ICSF) – funding agencies, and scientific communities to support and protect the interests of small-scale fisheries. Yet, more can be done to improve our understanding about this important sector and to elevate its status on the policy agenda.

As this volume illustrates, fisheries are lived experience, and thus need to be examined in their own contexts, namely through in-depth analysis often conducted in social science research. Together, and individually, the 25 essays make a case for why we must pay more attention to small-scale fisheries. The essays cover a broad range of topics that illustrate the diversity, complexity and dynamics of small-scale fisheries in several places around the world, emphasizing their vulnerability and their strengths, their contributions and their limitations, as well as reflecting on their future. The unifying message from these essays is that small-scale fisheries are too big and too important to ignore. This volume is also a testimony of how social science research contributes to de-marginalizing this sector, as well as the role of this discipline in fisheries policy.

The essays are organized into seven parts. In *Chapter 1* (**Section 1**: Introduction) I recount how small- and large-scale fisheries have coexisted, describe their dynamic relationships in the shared space, and discuss the conditions needed to support the viability of small-scale fisheries. **Section 2** (Challenges and Issues) contains four essays illustrating the struggles of small-scale fisheries in maintaining their existence and livelihoods. Learning from their experience working for a community-based organization, Lokuge and Munas (*Chapter 2*) give a full account of how small-scale fishers struggle to make a living in post-war Sri Lanka. Their situation is worsened with the small-scale trawler fleet from Palk Bay, India, competing for resources, as explained by Sathyapalan and colleagues (*Chapter 3*). The Palk Bay trawl fishery has its own problem of over-capitalization and inefficient measures to reduce capacity, due also to ineffective enforcement. In both cases, the authors offer their thoughts about what needs to be done to address the problems. In *Chapter 4*, Khattabi and Jobbins present another major threat facing small-scale fisheries – climate change. Drawing from an adaptation research project in Morocco, they explain fishing communities' vulnerability and what can be done to enhance their adaptive capacity and strengthen their resilience. The last essay in this section by Walsh (*Chapter 5*) brings us to Newfoundland, Canada where the collapse of the cod fishery resulted in major hardship to small-scale fishing people. Recovery and rebuilding efforts have not worked and the latest recommendations to restructure and rationalize the industry created more debate rather than resolution.

Section 3 (Livelihoods and Communities) includes four essays that bring us to the community level of analysis about small-scale fisheries. Starting with Frangoudes (*Chapter 6*), we learn about the prominent role of women in income and livelihood diversification, and how legislation can facilitate or inhibit their contribution in European small-scale fisheries. Lowitt follows in *Chapter 7*, with a colorful description of fish, especially cod, as a dietary staple and an important part of the traditional 'foodways' of the people in Newfoundland, and how present-day small-scale fisheries may be able to take advantage of the rise of culinary tourism to improve their food security. In *Chapter 8*, Felt and Natcher turn our attention to aboriginal communities in northern Canada, and explore challenges related to governance, integration of traditional knowledge and science, and impact of environmental change. Although we refer mostly to capture fisheries in our deliberation, we are reminded by Belton and Little (*Chapter 9*) about the importance of small-scale aquaculture in supporting livelihoods and alleviating poverty. Similar to capture fisheries, this sector faces challenges related to globalized trade, production regulation and climate variability that threaten its future. In all cases, direct policy interventions are required to promote and sustain small-scale fishing communities.

There is no shortage of tools and approaches employed to manage fisheries around the world. As exemplified in **Section 4** (Instruments and Reforms), not all of them are suitable in small-scale fisheries contexts. Property rights are among the most contentious ones, and its consequences in Icelandic fisheries are often discussed. In *Chapter 10*, Sabau offers her views on the formulation of the individual transferable quota (ITQ) in Iceland and discusses its consequences, in the short- and long-term, especially on small fishing vessels. She ends her essay with a suggestion that management systems such as ITQs are not as irreversible as they may seem, but it does require more art than science to get it straight. Discussion about rights continues with the essay by Gasalla (*Chapter 11*) that examines fishing rights and the trade-offs with marine conservation, using territorial concepts. Based on her research on Brazil's community-based and territorial initiatives, she argues for an environmental stewardship program that considers fishers' perspectives and respects local territorial rights.

Another well-known territorial institutional arrangement is marine protected areas (MPAs). As illustrated by Pascual-Fernández and De la Cruz-Modino in their essay about MPAs in Spain (*Chapter 12*), application of this tool is not without challenges. MPAs may fuel existing conflicts or may help alleviate them. Given amenable conditions, they may provide small-scale fishers with increased control over their traditional fishing areas. Some governance arrangements can, however, undermine small-scale fisher rights. Isaacs (*Chapter 13*) provides a good example of this in her description of how the national legislation to legalize small-scale fishers in South Africa generated conflict with the practice of their livelihoods. She also calls for a new small-scale fisheries policy with clear rules about how resources should be regulated, who should participate in the harvesting, and how benefits from the fisheries should be distributed.

The other two essays in this section reinforce the importance of fishers' participation in resource management. Donda (*Chapter 14*) describes the shift from government-centered, conservation-based fisheries management to participatory fisheries management in Malawi in the early 1990s. Encouraged by the success of this management scheme, the Mbenje fishing community followed with their own initiatives, and unique achievement stories. Although public involvement in resource management is not new in Indonesia, legislation enacted in 2007 provided impetus for local governments to establish a new system of integrated and decentralized coastal management planning, focusing on marine conservation, poverty alleviation, alternative income generation and business development. The effectiveness of this program, Ginting (*Chapter 15*) submits, depends on the ability of government agencies to coordinate activities in coastal areas and to broaden coastal community empowerment.

For these various institutional arrangements to work, careful deliberation about the fundamentals of governance is needed. Five essays in **Section 5** (Governance and Principles) offer thoughtful perspectives on how to investigate them. Mahon and McConney (*Chapter 16*) provide a much needed examination of the structure and function of national fisheries departments and their suitability to manage small-scale fisheries. They submit that an appropriate mix of technical and organizational skills needs to be developed, especially in small or developing country fisheries departments, to deal with the prevailing management issues facing small-scale fisheries. McConney and colleagues continue the discussion in *Chapter 17* calling for changes in the institutional arrangements for fisheries that aim at, among other things, creating enabling policy environments and facilitating self-organization. Drawing on the emerging concept of fisheries networks and their experience in the Caribbean, they argue that a shift in perspective is urgently needed, and is possible by translating the concepts of adaptive social-ecological systems into everyday thinking.

But as Charles suggests in *Chapter 18*, there are no universal solutions in small-scale fisheries governance. He identifies a set of 'good practices,' which include adopting a rights-based approach to fisheries, securing access rights, enhancing legal space and organizational capacity, adopting integrated system thinking and sustainable development, promoting food sovereignty and community well-being, expanding policy and livelihood options beyond fisheries, and ensuring effective communication. Whereas Charles suggests that progress can be made through international instruments on small-scale fisheries, Juntarashote and Chuenpagdee (*Chapter 19*) turn our attention to basic devices closer to home. Inspired by the concept of 'sufficiency economy,' which touches at the core values that people have about their lives and prospects, these authors examine how small-scale fishers in Thailand were empowered to create thriving communities, while protecting the marine ecosystem. The notion that small-scale fishers are able, and in many cases in a better position, to take control of their own future is further reflected in Bavinck and Jentoft's essay (*Chapter 20*). Following the subsidiarity principle, they argue that when small-scale fishers are able to fish with their technology, they, rather than large-scale fisheries, should be given the priority to access the resources. Likewise, if they are able to do the job of governing the resources just as well or better, they should be given the privilege. The essays in this section similarly call for more discussion about the roles and responsibilities of governments, at all levels, to facilitate governance transformation, recognizing that it essentially requires innovative thinking and imagination.

The final main section in this volume (**Section 6**: Reflections and Visions) offers us an opportunity to reflect on lessons learned in our efforts to understand and contribute to making small-scale fisheries viable and prosperous. In *Chapter 21*, Kurien reminds us that the theory and the practice of governance are two different processes. Using his experience in fisheries development and negotiating governance arrangements in three different contexts – India, Cambodia and Indonesia – he emphasizes how we must be aware and ready to address possible dilemmas generated by proposed governance transformation. Challenges in fisheries governance go beyond the divide between theories and practices. As noted by young scholars like Song and Khan (*Chapter 22*), the current, and at times ferocious, scientific debates in fisheries create confusion in policy arenas and with the general public. Perplexed by the inconsistencies appearing in the media nowadays, they raise questions about future research directions for small-scale fisheries. There are unfortunately no easy answers to their enquiry, but further probing and alternative perspectives are valuable. Angerbrandt and colleagues (*Chapter 23*), for instance, urge us to revisit the conceptualization of the term community and what it is good for. While not dismissing the importance of community and locally-driven initiatives for fisheries sustainability, they submit that embedding communities in society and redressing any asymmetrical community relationships is necessary to substantiate such claims. Finally, Jentoft (*Chapter 24*) challenges the traditional fisheries paradigm based on Hardin's "Tragedy of the Commons," and evokes Sen's "Development as Freedom" as a contrasting image to address fisheries problems. Illustrated using some historical events in Norwegian fisheries, he argues that governance outcomes would have been very different, had it been Sen's idea that inspired fisheries policies.

As Pauly reiterates in his foreword, for small-scale fisheries to realize their potential of becoming the fisheries of the future, they must be seen as a viable part of the solution, rather than a problem. In the last essay (*Chapter 25,* **Section 7**: Way Forward), I present the global research network for the future of small-scale fisheries, "Too Big To Ignore," as one possible way to move in this direction. The network is initiated to facilitate interdisciplinary and participatory, action-oriented research that addresses current and emerging issues facing small-scale fishing people. It also encourages information and knowledge sharing, through a database system and online curriculum, and ongoing dialogue between fishing communities, scientists, policymakers, environmental groups and funding agencies about the sustainable future of the world's fisheries, with small-scale fisheries at its core.

This book could not have happened without the hard work and dedication of 35 other authors from around the globe. The essays embody the experiences and viewpoints of fisheries scientists, practitioners, post-doctoral fellows, and graduate students and, most of all, of the small-scale fishers around the world who inform our research. I sincerely thank all the authors for telling their stories and sharing with us their enthusiasm and optimism for small-scale fisheries. Technical assistance was provided by Theresa Heath, Ian Ivany, Jennifer Dyke and Andrew Song. I extend my gratitude to Professor Daniel Pauly, not only for his encouraging words in the foreword, but also for his unyielding belief in the value of good science. I continue to learn from him.

I greatly owe the participants of the first World Small-Scale Fisheries Congress (WSFC) and many partner organizations, including Memorial University of Newfoundland (MUN). My special thanks go to the three conference co-hosts: Department of Fisheries, Government of Thailand; the Southeast Asian Fisheries Development Center (SEAFDEC); and the Coastal Development Centre (CDC), its members and the Director, Dr. Kungwan Juntarashote, for the logistical organization of the congress. Several people played significant roles at the congress, prominent among them were Andrew Song of MUN, Shettapong Meksumpun and Sangtien Ajjimangkul of CDC, and Phattareeya Suanrattanachai and Namfon Imsamran of SEAFDEC. Funding for the congress came from various sources, most notably the Norwegian-funded PovFish Project and the Social Sciences and Humanities Research Council of Canada.

We have already taken major steps towards improving knowledge, renewing interests, and contemplating visions in small-scale fisheries with the launch of the WSFC and with this volume. I invite readers and everyone interested in small-scale fisheries of the world to join us in making them *Too Big To Ignore.*

Ratana Chuenpagdee
St. John's, Canada

FOREWORD

Daniel Pauly
University of British Columbia, Canada

Small-scale fishing is the practice by which we began to utilize the coastal resources of our East African home, some 120,000 years ago. Gradually, along with the conquest of new continents, fishing effort increased, and some pre-industrial fisheries became rather substantial, involving large sailing fleets targeting herring in the North Sea or cod in Newfoundland.

But the most significant transition in fishing, and the start of the dichotomy which colors each and every one of its various aspects, was the deployment of the first steam-powered trawlers along the English coast in the late 1880s. This marked the transition to a mode of fishing using non-renewable but abundant fossil energy to exploit a renewable resource, itself ultimately limited by the rate at which the sun's energy is converted into edible fish.

In the short-term, industrial fisheries simply had to 'win' in their competition with small-scale fisheries: they could deploy major capital and fishing power to obtain massive catches – even if they consisted of the accumulated biomass of old fish – which helped to fuel population and fish consumption growth everywhere. And their fishing power expanded to cover the whole world when, after two world wars and the cold war, a vast array of new fish finding and processing technologies became available to them. From the 1950s to the late 1980s, the fishing world was consumed by the exploits of industrial fishing which, at the end of this period, had engulfed the globe.

And so, almost everyone forgot small-scale fisheries – or they were seen as a relict of the past, soon to disappear, just like steam engines and Apple II computers. Only a few maritime anthropologists and fisheries sociologists soldiered on, and their accounts often emphasized quaint features of the lives of fisher folks, rather than their core activity. Fisheries scientists also tended to avoid small-scale fisheries, whose largely undocumented catch, when known, contained little of the detailed per-species information that was required for application of a succession of population dynamics models then fashionable, whose relevance to the management of small-scale fisheries was, moreover, dubious at best.

In many countries, especially so in developing countries, the urban elites who were recruited to staff the bureaucracy could not be bothered with the logistical and administrative nightmare that monitoring and reporting on small-scale fisheries often represents, and thus mostly ignored them. And, since they did not report on them to the United Nations Fisheries and Agriculture Organization (FAO), the coverage of small-scale fisheries in the FAO's global database of fisheries statistics vastly underestimates the contribution of small-scale fisheries to global fisheries catches. As a result, the army of academic economists and other consultants hurriedly writing reports for the World Bank and other international development agencies, and even environmental NGOs, bought into this alternative reality, and accepted that small-scale fisheries are, well, small, and best left to those dealing with folklore or caritative projects.

But then, throughout the 1990s, and increasingly in the first decade of the 21st century, many realized that the party may be soon over for industrial fisheries. Some of them are managed well, in Alaska and a few other places, with strong harvest control rules. But in most of the rest of the world, the very size of industrial (particularly distant-water) fleets translates into political power often exceeding that of coastal states, and they abuse this power and operate, literally, out of control. And, other than destroying habitats, as in the case of trawlers, they also regularly crash the stocks they depend on. Then they continue, further offshore, into deeper waters and further south because, at least in the rich countries of the global north, they get government subsidies to do so, and so escape the economic consequence of their activities – a market failure if there ever was one. However, these fleets now have less room – if any – to expand, and they are squeezed by increased fuel costs which cannot always be absorbed by additional subsidies; hence the widespread crisis of fisheries.

Thus, it is high time to recall that there is an alternative, and that it is staring us in the face: the small-scale fisheries. They survived, albeit with difficulties, while industrial fisheries grew in the absence of checks or balances. Some of these difficulties resulted from the application of brute force against them, such as when coastal fixed gears are destroyed by trawlers operating – illegally – close inshore, or when politically connected industrial fleet owners redirect to themselves funds that governments or international agencies allocated to 'fisheries development'.

But a major problem has always been the conceptual swamp which fisheries policies are stuck in, and in which small-scale fisheries are always seen as a problem, rather than a viable part of the solution of the crisis of fisheries. It is true that small-scale fisheries can be destructive (e.g., when they use dynamite, insecticide or other poison as fishing gear), and they often serve as a social 'safety valve' for landless farmers or cattle-less pastoralists, with the decline of coastal fish resources as an all too frequent result.

However, because small-scale fishers invariably live close to their fishing grounds, and depend on the resource therein and on no others, they can be (re-)connected, if need be, with the idea of caring for the resources; which the hired hands on board of industrial vessels cannot afford to do. Combined with the obvious advantage of adjacency (short sailing time, and hence limited fuel consumption, if fuel is used), this is a major reason why small-scale fisheries have the potential of becoming the fisheries of the future.

It is therefore very opportune that, finally, a 'World Small-Scale Fisheries Congress' (WSFC) was held in October 2010. This volume features selected papers presented at this event and others, and they span the entire gamut of small-scale fisheries operations and management, showing that they are indeed, too big to ignore. Also, these essays show that there are solutions to problems in small-scale fisheries. I congratulate the editor of this volume – also the initiator of the WSFC – for her initiative and enthusiasm which, I am convinced, will help validate the systematic, quantitative and comparative study of small-scale fisheries worldwide.

Section 1:

Introduction

CHAPTER I

A matter of scale: prospects in small-scale fisheries

Ratana Chuenpagdee[1]

Small- and large-scale fisheries have long coexisted. Their relationship has been increasingly uneven and strenuous since the major modernization of fishing gears and technology in the beginning of the 20th century. In many parts of the world, large-scale fisheries dominate in terms of catches and revenues. Yet, the majority of fishers are small-scale. The significant contribution of small-scale fisheries to food security and livelihoods of millions of people is generally acknowledged, but information about this sector is relatively poor. This essay examines the relationship between small- and large-scale fisheries, and the conditions needed for their coexistence. Specifically, it asks what natural and social environments and governance arrangements are required to support the viability of small-scale fisheries, and make it possible for them to coexist with large-scale fisheries.

Shifted existence

In current fisheries discourse, we refer to fishing and fisheries mostly in the modern time, without due attention to the fact that fishing is a form of primary production that is as old as humankind (Gabriel *et al.*, 2005). Major changes have taken place throughout the history of fishing, affecting aquatic environments, fishing people, and society at large. The distressing state of world fisheries today is likely a reflection of disconnect between fisheries policy and present reality, and the history of their development (McGoodwin, 1990). Without a proper examination of the changes, also of fisheries institutions and governance systems, and how they affect the existence of,

1 Corresponding author: Memorial University of Newfoundland, Canada and Coastal Development Centre, Thailand, ratanac@mun.ca

and relationship between, small- and large-scale fishing sectors, we may be endorsing policies that inhibit the sustainability of small-scale fisheries, as well as limit their ability to adapt to environmental and economic changes.

According to Gabriel *et al.* (2005), fishing and hunting may be traced to the same origin, although there is no consensus in terms of which came first. Several examples suggest a blurred distinction between these two primary production activities. For instance, it is not clear whether the spear was designed and used for fishing, hunting, fighting or for ceremonial purposes. Drawings in Egyptian tombs also reveal that the use of netting in hunting may be older than that used in fishing (Gabriel *et al.*, 2005). Kusnetzow (1971, cited in Gabriel *et al.*, 2005) referred to hunting as a 'violent occupation of living creatures,' by which hunters were seen as 'experienced warriors' with more prestige and higher social status than fishers. Others argue that, while fishing may be a relatively 'peaceful' occupation, fishers face grave danger at sea, and thus only the experienced and courageous survive.

The first large-scale fisheries were established in the Middle Ages in Europe with bulk demands for salted cod and herring and whale oil (Gabriel *et al.*, 2005). It was not until the late 1700s, however, that introduction of the steam engine enabled fishing at greater depths, further offshore, with bigger and heavier gear, thus marking a major transformation in fisheries, which later spread to other parts of the world (Smith, 2000). The ocean was then considered an inexhaustible source of wealth that could easily handle the 'blue revolution,' shown for instance in tropical fisheries (Bailey, 1985), the United States of America (McEvoy, 1986), India (Bavinck, 2001), and Southeast Asia (Butcher, 2004). By the time studies suggesting overfishing and unsustainable fishing practices started to emerge, it was much too late. Large-scale, industrialized fisheries have, literally and figuratively speaking, gained important grounds. As shown in Pauly *et al.* (2003), fishing areas have expanded over the past 50 years, farther from shore and into deeper waters. With the high export values they bring to fishing nations in the north and south, large-scale fisheries have garnered attention from policymakers at all levels. Considered to be of less importance from the national economic point of view, small-scale fisheries have therefore been largely ignored.

In *The Closing of the Frontier,* Butcher (2004) provided a vivid and comprehensive description of how fisheries in Southeast Asia went through a series of changes. As in other areas of the world, the abundance of resources in the sea was the main impetus for the intensification of fish harvesting and major developments in fishing technology, which led to the 'great fish race,' and the eventual closing of the frontier. But even before the mechanization of fishing boats and technological advancement in fishing methods, fishing areas were expanded to increase catch levels, thus supporting the view that small-scale fisheries can also cause resource degradation, if equipped with effective and non-selective fishing gears. The most drastic change in the South China Sea area was caused by the introduction of trawls from the 1950s to the late 1970s; first in the Philippines, then later in Thailand, Malaysia and Indonesia, which contributed to over-exploitation of fisheries in the region. In a paper about the Gulf of Thailand fisheries, the development of the trawl fishery was referred to as an 'unplanned experiment,' which resulted in major change in the marine ecosystem, induced by alteration of the seafloor, massive removal of fish, and switching of target species (Pauly and Chuenpagdee, 2003). Similar consequences were observed in other areas with the heavy exploitation of fisheries using bottom-touching, towed gears such as trawls, and highly efficient gears like purse seines. Other changes associated with the modernization of fishing technology are increased investment in shipbuilding, processing and other post-harvest industries which, once made, are difficult, if not impossible, to suspend. In the case of Thailand, the government was reluctant to put controls on expansion due to concerns about unemployment. The only option was to continue to fish, at the peril of the resources, leading to further inappropriate policy. Instead of curtailing fishing effort after the collapse of demersal fisheries in the Gulf of Thailand, the government went on to promote fishing in distant seas, which gave rise to conflicts and disputes with neighboring countries. In an attempt to continue reaping benefits from fisheries resources exploitation beyond domestic waters, Thailand was the last country in the region to comply with the International Law of the Sea in declaring the Exclusive Economic Zone (EEZ) in 1980.

There is considerably less documentation of what happened to traditional small-scale fishers and their communities, who dominated fisheries prior to the blue revolution. They were undoubtedly affected by the changes, which were not necessarily welcome, as shown in the case of Galicia when purse seining was introduced to catch sardines at the turn of the 20th century (Pascual-Fernández and De la Cruz-Modino, this volume). These major fisheries developments resulted in an enormous increase of catches, and contributed to improving food security for the millions of people who rely on fishing employment (Bavinck, 2011). While many small-scale fishers with access to capital may have benefited from it, others may have been pushed to the brink of poverty, as competition for resources and markets intensified. In Europe and North America, some displacement occurred, which resulted in decline of the fishing population. Pascual-Fernández and De la Cruz-Modino; and Walsh (this volume) have illustrated similar situations in Spain and Canada (Newfoundland), respectively. This could happen because fishing is no longer viable, compared to other economic opportunities, or because certain fishery policies (like licences and permits) or property arrangements (like quotas) make it difficult for some fishers to stay in the industry (Bavinck, 2011; Sabau, this volume).

A look at different fishing gears used to capture fish tells a story about the unequal existence of small- and large-scale fisheries. Information from the *Sea Around Us Project* (SAUP) database[2] shows that three of the five most 'popular' fishing gears employed in the world in 2004 were bottom trawls, purse seines and mid-water trawls. These gears are mostly operated in large-scale, industrialized fisheries. Together, they contributed about 72% of world catches, and were used by more than 170 countries, in the case of bottom trawls and purse seines, while almost 150 fishing nations and territories used mid-water trawls. The other two gears in the top five – gillnets and hooks or gorges – were also used by over 170 countries and, like most gears, can be operated as part of large-scale, mechanized fisheries, or at a much smaller scale. They accounted for an additional 13% of total catches. The other 15% of global fisheries production came from at least 29 different gears reportedly employed in fishing nations around the world in 2004. These gears are mostly small-scale, like lift nets, cast nets, raking, bag nets, diving, and grasping by hands.

National fisheries statistics can be used complementarily with these catch by gear data to provide a further picture about the relationship between large- and small-scale fisheries. In 2008, the Food and Agriculture Organization of the United Nations (FAO) reported China, Indonesia, India, Philippines and Myanmar among the top ten capture fishing nations, with combined production of about 32% of the world's total capture fisheries from marine and inland systems (about 90 million tonnes in total; 80 million of which were marine) (FAO, 2010). Given that all of these are developing countries, both small- and large-scale fisheries are likely to play important social and economic roles. The other top producers – Peru, USA, Japan, Chile, and Russia – on the contrary, are likely dominated by large-scale, industrialized fisheries, which contribute to the high volume of catches of Peruvian anchoveta, Alaska pollock, skipjack tuna, Atlantic herring and Chilean jack mackerel. Together, they contribute another 23%, for a total of

2 www.seaaroundus.org; The SAUP database is compiled from existing sources such as the Food and Agriculture Organization of the United Nations (FAO), the Northwest Atlantic Fisheries Organization (NAFO), and the International Council for the Exploration of the Seas (ICES).

58% of world fisheries capture production from the top ten countries. Yet, as noted by the FAO, one-quarter of the total catch comes from 82 low income 'food-deficit' countries, half of which are in Africa and another 30% in Asia (including China). For these countries, it is not unreasonable to suspect that small-scale fisheries outweigh the large-scale sector in terms of fishing employment, livelihood dependency and importance to food security.

These statistics show a mixture of small- and large-scale fisheries worldwide. This applies to capture fisheries (discussed here), as well as to culture fisheries (discussed in Belton and Little, this volume). Although large-scale fisheries came into existence much later and have managed to take over the small-scale counterpart in many cases, small-scale fisheries are still very important, especially in developing and less-developed countries. As reported by the FAO (2010), only about 2% of fishers and fish farmers are in Europe and North America, but their average annual production per capita is about 24 tonnes in Europe and 18 tonnes in North America. In contrast, Asia has about 85% of the world's fishers and fish farmers, whose individual production averages about one-tenth of those in the North. Put simply, large-scale, industrial fisheries dominate in developed countries, while small-scale fisheries hold prominence in other parts of the world.

Compromised condition

Small- and large-scale fisheries have long coexisted, but their relationship has never been straightforward. In a few places where fisheries consist mainly of either industrialized fishing enterprises or traditional subsistence fisheries, sector interaction (in harvest and post-harvest activities) is minimized. For the most part, mid- to large-size fisheries and small-scale, commercial fishing sectors coexist in a complex relationship since they share fishing grounds, use similar gears that differ in power and efficiency, often target the same species, and sometimes compete in the same markets. In other words, they interact in the physical surroundings, on land and at sea, in their operations, and in their post-harvest activities. Inescapably, conflicts arise, despite efforts to minimize them. For instance, many nations have regulations that prohibit the large-scale fishing sector from operating within certain inshore areas reserved mainly for small-scale fisheries. But weak enforcement and other flaws in the governance system often render these rules and regulations ineffective. Other privileges may also be given to small-scale fisheries, as when those using non-destructive fishing practices are allowed to operate in marine protected areas (MPAs). Still, by and large, small-scale fisheries are at a disadvantage, being at the receiving end of the ecological and economic impacts induced partly by large-scale fisheries, in addition to the demise of their own.

In the book, *Fish for Life* (Kooiman *et al.*, 2005), we identified ecosystem health as one of the key concerns facing world fisheries. This issue well illustrates how small-scale fisheries are compromised by their large-scale counterpart. Large-scale, industrialized fisheries cause great ecological impacts, with the high volume of fish they remove from the sea, and severe habitat damages and high percentage of bycatch and discards when certain gears are employed (Chuenpagdee *et al.*, 2003). Although thousands of small-scale fishing units may cause equivalent damage to a few large-scale fishing vessels, the latter still dominate world fisheries, taking in over 70% of the total production. The poor health of the oceans creates more hardship to small-scale fishers due to their high dependency on fisheries resources, and their limited options to improve their efficiency or alter their fishing activities, like fishing further offshore. In some areas, small-scale fishers and their household members, especially women, as noted by Frangoudes (this volume), have employment in other sectors, like farming, small-business, and tourism, whether or not they are desirable or suitable alternatives.

In Newfoundland, since the eighteenth century when fishing enterprises were small-scale, family-based activities, fishers hunted, trapped, and had gardens as part of their occupational pluralism and traditional foodways (Lowitt, this volume). The seasonal small-scale fishing of cod using handlines was replaced by other gears, like traps and trawls, that could be used year-round. Fishers also became professionalized, which enable them to receive full benefit from government policies, such as subsidies to the commercial fishing sector. Fishing turned into a single occupation, instead of a part of a livelihood portfolio. Such professionalization may work for some small-scale, commercial fishers, who are capable of operating in a competitive, profit-driven economy. Otherwise, they stand to lose under these government policies, which make it harder, if not impossible, for them to maintain their preferred lifestyle (Cadigan, 2009). The reverse situation is reported in Brazil where a process of 're-artisanalization' of fisheries activities, supported by the government through the development of marine 'extractive' reserves, resulted in reoccupation of coastal waters by small-scale fishers (Gasalla, this volume).

When resources become scarce, and other livelihood options are no longer available or viable, small-scale fishers are under pressure to modernize their gears, increase the power of their vessels, take high interest loans, sell at low prices, or employ illegal and/or destructive gears. These are short-term solutions at best since, in the long run, they contribute to pushing small-scale fishers into the vicious circle of poverty, rather than improving their well-being. As illustrated in Chuenpagdee and Jentoft (2011), there are several conditions throughout the fish chain, i.e., in the aquatic environment, harvest and post-harvest activities, that make small-scale fishers vulnerable to poverty. These include natural disasters, unfavorable environmental conditions, perturbations affecting fisheries productivity caused mainly by large-scale fishing gears, as well as war like the one experienced in Sri Lanka (Lokuge and Munas, this volume), and climate change, as in the case of Morocco (Khattabi and Jobbins, this volume). It is notable that the scale of the issues affecting the fish chain is not confined to the specific area where the problems occur, and some of them are transnational. As demonstrated in Sathyapalan *et al.* (this volume), the unrest in Sri Lanka created problems for Palk Bay fisheries in India.

Other issues related to safety at sea, cost of fishing, market accessibility, inappropriate post-harvest handling, and poor processing facilities worsen the situation of small-scale fishers. Moreover, it is far too common in developing countries that small-scale fishing people are at the mercy of moneylenders, who have full control over the prices and marketing of their products. The difference between small- and large-scale fisheries is obvious with regards to access to major domestic and international fish and seafood markets. Only in some niche markets, such as those catering to cultural-based tourism supporting local harvests, are small-scale fisheries well positioned to benefit.

Pauly (1997; 2006) has long argued that small-scale fisheries fare much better than the large-scale sector in several key areas, such as contribution to job and food security, fuel consumption, and environmental sustainability. Yet, they are largely marginalized, not only socially and economically, but also institutionally, as suggested by Mahon and McConney (this volume). If the FAO recognizes that small-scale fisheries contribute more than half of the world's fisheries catch and that the majority of the world's fishers are small-scale, we must ask why there is no comparable data collection system to that of large-scale fisheries, and thus no official statistics that properly differentiate the two sectors. The general omission of the roles of women in fisheries is even more disturbing (Frangoudes, this volume). The reasons may be related to the characteristics of small-scale fisheries, that they are highly diverse and numerous, are often located in remote areas, with poor infrastructure, and that records of their fishing activities and marketing do not always exist. Nevertheless, these reasons are not sufficient to abandon any attempt to verify this work. Some efforts have indeed been made to systematically document and examine the extent of small-scale fisheries (see, for instance, the global database of small-scale fisheries by Chuenpagdee *et al.*, 2006), but more can be done, especially by intergovernmental organizations like the FAO, to help provide policy balance to these fisheries sectors.

In all fairness, several recent FAO initiatives are directed at small-scale fisheries. For instance, many articles in the FAO Code of Conduct for Responsible Fisheries acknowledge the importance of local traditional knowledge and the role of small-scale fishers in decision-making (FAO, 1995). In 2003, the FAO established a Working Party on Small-Scale Fisheries to evaluate the importance of this sector, as well as to set a research agenda for small-scale fisheries (FAO/RAP/FIPL, 2004). Another major effort by the FAO, in collaboration with the WorldFish Center, was to develop a framework for an integrated assessment of small-scale fisheries, partly as a means to operationalize the Code of Conduct and other management approaches (Garcia *et al.*, 2008). In 2008, the FAO organized a global conference on small-scale fisheries in Bangkok, often dubbed the 'Peoples Conference,' to recognize the conscious effort to bring fishers' voices to the world stage. One of the aims of the conference was to discuss and explore issues related to ensuring sustainability and securing the 'rights' of fishing people in the broadest sense, including rights of access, and social, economic and human rights (see Charles, this volume). The conference certainly helped to elevate small-scale fisheries issues to the global level, leading also to several follow-up initiatives, including the World Small-Scale Fisheries Congress that provided the backdrop for this book (See Chuenpagdee, Chapter 25, this volume).

Several fishers' organizations and community groups, notably the International Collective in Support of Fishworkers (ICSF), play important roles in protecting the interests of small-scale fisheries and empowering them to participate in fisheries in a decent, equitable, gender-just, self-reliant and sustainable manner. Formed in 1986, ICSF is an international, non-governmental organization, aimed mainly at monitoring and disseminating information on issues concerning fishworkers around the world, and helping create space and momentum for the development of alternatives in the small-scale fisheries sector. Their mandate was drawn from the international conference of fishworkers and their supporters, held in Rome in 1984 in parallel with a meeting organized by the FAO.

On the research side, fisheries scientists from both natural and social science disciplines have been paying attention to small-scale fisheries. At the 2004 World Fisheries Congress, about 60% of more than 200 papers presented addressed subsistence and small-scale fisheries as well as aboriginal and recreational fisheries, which may also be considered small-scale (Chuenpagdee and Bundy, 2005). A literature search of scholarly articles about small-scale fisheries shows a major increase from about 70 articles published during the 20 years prior to 1990, to more than 1,600 articles, twenty years later (see Song and Khan, this volume). Numerous books and reports focusing on small-scale fisheries have also been published, the latest of which are two edited volumes by Pomeroy and Andrews (2011), and Jentoft and Eide (2011).

Restoring balance

"Fishing is a living occupation" (Gabriel *et al.*, 2005, p. 2). This statement emphasizes how changes take place throughout history. Although it may not alter how fish is caught, as in some traditional small-scale fisheries, the efficiency of catching and handling fish has greatly improved, particularly in the commercialized and industrialized sector, both large and small. This also implies that what dominates fisheries economies today, like trawl fishing, may lose significance in the future with increasing oil prices and other policies to reduce carbon emissions. Concurrently, small-scale fishing gears and devices employed in some parts of the world are becoming more sophisticated with the use of global positioning systems (GPS), and almost everywhere information technology, especially mobile phones, has enabled efficient marketing of small-scale fisheries catches. In the same way that small-scale fishers are affected by different conditions throughout the fish chain, which may result in increasing their marginality, restoring balance in fisheries requires an understanding of the whole fisheries spectrum, as well as anticipated changes, for effective governance interventions.

Some governance arrangements, regardless of their original intention, serve in favor of large-scale fisheries. Government subsidies for modernization of fishing vessels and gears and for offsetting fishing costs (e.g., fuel subsidies) have encouraged the growth and expansion of domestic and distant water fleets. Despite implementation challenges, removal of subsidies and reducing fishing capacities has already been suggested as the way forward (see Swan and Gréboval, 2004; Sumaila *et al.*, 2008). Channelling these subsidies to support sustainable fisheries, whether large or small, is another option. The challenge is in determining what sustainable fisheries, or the contrary, look like. The interactive governance theory posits that such exercise requires a careful examination of the underlying values, principles and images held by public and private actors in the state, market and society (Kooiman *et al.*, 2005). This departs from the existing approach of applying indicators to assess various dimensions of sustainability, in that it encourages interaction among actors in the deliberation of what these meta-elements may be, and how best to negotiate them when they do not align. In other words, interactive governance focuses on process more than tools, and outcomes more than outputs. Similar to the network concept presented in McConney *et al.* (this volume), it also pays close attention to interactions among fisheries stakeholders and between stakeholders and the natural environment.

Of these three elements, principles and values appear regularly in fisheries governance discourse. While principles such as those related to precaution apply to all fishing sectors, Jentoft *et al.* (2011) argue that certain principles are more pertinent for small-scale fisheries, because of their attention to local contexts and the problems that small-scale fishing communities face. They refer, for instance, to the "dexterity principle," which calls for governance initiatives to be sensitive to details and the diversity of local contexts. The "subsidiarity principle," which states that decision-making about fisheries should be vested with the lowest possible organization is also emphasized (see Bavinck and Jentoft, this volume). The transition from top-down, centralized approaches to co-management and decentralized systems, which has taken place in many fisheries around the world, suggests an uptake of these small-scale fisheries principles in governance. It also shows that well organized small-scale fishing communities are able to contribute to, and possibly take the leadership in, the governance of their own resources. Donda (this volume) presents a good example of how Mbenje fishers in Lake Malawi initiated co-management as a way to protect fisheries resources, considered as their 'gold mine.' A similar situation related to MPAs in the Canary Islands and Galicia was referred to by Pascual-Fernández and De la Cruz-Modino (this volume). These examples do not imply that large-scale fisheries should bear no responsibility in fisheries management. Rather, they should be held accountable if their fishing operations result in collateral damages.

Like principles, values have been considered in fisheries governance, but with a major caveat that they are mostly market-based. This reflects how fish are mainly treated as a commodity, traded in markets for revenue generation and export earnings. Food security, nutrition, and cultural and spiritual values of fish come second to the economic values. The under-appreciation of fish, fisheries and the aquatic ecosystem has led to policies that undermine the existence of small-scale fishing communities, who value these aspects of resources. When governments sell fishing agreements to foreign fleets, promote clear-cutting of mangrove forests, allow industrial development to take place in areas where small-scale fisheries operate, or when they support certification and labelling of seafood products designed with the characteristics of large-scale fisheries in mind, they diminish the values of small-scale fisheries and their actual and potential contributions to address global concerns. There is no shortage of approaches to elicit values of fisheries, although some are more appropriate than others in capturing non-market values, such as those using non-monetary metrics like damage schedules (Chuenpagdee et al., 2001) or other qualitative measurements, including the narrative valuation method (Satterfield et al., 2000). If values are well understood in the context of small- and large-scale fisheries, interventions can be made to strengthen the viability and sustainability of both sectors. For instance, environmental education and awareness programs can be promoted to enhance local stewardship. Governments can support the establishment of cooperatives for small-scale fisheries to help develop local markets and secure reasonable prices for their products.

The final element, 'images', is often referred to in fisheries governance, at least metaphorically, but without proper examination. Jentoft *et al.* (2010) defines images as representation of ideas about what is and what should be; they provide meaning and direction, as well as viewpoints about the relationship between humans and their environment. They can be expressed in the form of narratives, perspectives, and metaphors. Two powerful examples of influential images in the history of fisheries include Thomas Huxley's 1881 view that "*all the great sea fisheries are inexhaustible*" (cited in Gordon, 1991), and Garrett Hardin's more recent 'tragedy of the commons' image based on the premise that "*freedom in a commons brings ruin to all*" (Hardin, 1968). These two images were associated with different historical contexts and to some degree motivated different policy frameworks, as elaborated by Jentoft (this volume). The latter image, in particular, became the root metaphor for policy prescriptions and management designs in fisheries, especially after the declaration of the EEZ, which led to the empowerment of the nation state in fisheries (Jentoft *et al.*, 2010). Another contrasting image of small-scale fisheries is about how they '*rhyme with poverty,*' which suggests that there are problems with the sector (Béné, 2003). The alternate image is how small-scale fishing is a '*way of life,*' which offers solutions. Unfortunately, the former image dominates fisheries policy and development efforts, which tend to focus on promoting non-fishing livelihood options. Studies show that these alternatives do not always align well with the culture and traditions, or the preferences, of fishing communities.

Making values, principles and images explicit is essential in dealing with fisheries issues because of the broad range of stakeholders with divergent, if not conflicting, interests. It helps provide transparency to management and decision-making, thus encouraging stakeholder participation in the process. Governance is obviously not only about these meta-elements. Fisheries managers are often occupied with first-order (daily actions and problem-solving), and occasionally with second-order (design and setting of institutions) governance issues. Spending time discussing meta-order governance may be considered too much of a luxury, especially when crisis management is a permanent feature. The argument for moving governance discussion to the meta-level is based on the premise that knowledge about values, principles and images can help make fisheries more governable, thus easing the governing tasks. Put differently, knowing what matters to small-scale fishers and large-scale fishing industries can help reduce tension and misunderstanding between them, and build legitimacy into management decisions. Discussion about meta-order governance and fisheries knowledge takes on an added dimension in aboriginal fisheries, as shown in Felt and Natcher (this volume), due to the dynamics between the knowledge system and the evolving and changing science.

Re-arranging governance

What should a governance arrangement that fosters the coexistence and viability of small- and large-scale fisheries look like? In saying that fisheries governance is a wicked problem (Jentoft and Chuenpagdee, 2009), we suggest that there is no simple answer to this question since there is no one size fits all solution or technical quick fix. But, talking about governance as a holistic concept that goes beyond management does not imply that governments have no roles to play. Instead, it calls for small- and large-scale fishers, fishing communities, industries, environmental organizations and governments to take part in governance, playing different roles and having various levels of engagement, according to the conditions, demands and challenges that they face (see also Kurien, this volume). As shown in Isaacs (this volume) in the case of South African fisheries, the power differential between these governing actors needs to be recognized and adjusted to enable marginalized groups to fully participate in a democratic, transparent and accountable process. This implies that governance may not yield outcomes that are equally acceptable to all involved. Further argument can be made that governance performance cannot be judged as good or bad. Measurement instruments useful in examining governance are those that illuminate both potentials and limitations of governance arrangements, and suggest innovative ways to enhance its performance. The importance of exploring other livelihood opportunities to either replace or supplement fisheries means that a broad framework like integrated coastal management needs to be drawn upon (Ginting, this volume).

The governance assessment begins at understanding the historical existence of small- and large-scale fishers and identifying what factors in the fish chain foster or inhibit their ability to maintain and develop their fishing activities. Each fishing sector needs to be examined on its own, as well as in terms of its relationship to the other, taking into account the dynamics of the natural, social, economic and political systems. An analysis of past and existing institutions, including rules, norms, and cultural-cognitive elements, what influences how they are set up and designed, and how they affect fishing sectors, is needed (see, for example, de la Torre-Castro and Lindström, 2010). Also, as suggested by Angerbrandt *et al.* (this volume), an understanding of how community and society are perceived and treated is required, if they are to play a greater role in fisheries management and development, as they should.

The next step is to inspect what underlies fishers' values, what constitutes their images, and what principles serve as their foundation. These values, images and principles are compared and analyzed to detect similarity and/or disparity. Some of these are illustrated in the essays included in this volume by Bavinck and Jentoft (on subsidiarity) and Juntarashote and Chuenpagdee (on sufficiency), as well as that by Belton and Little in their discussion of 'risk-averse' subsistence ethics among fishers/farmers in Bangladesh. Once these components are mapped, spatially and temporally, it will be possible to discern dominating governance patterns, and the causal relationship that they have with small- and large-scale fisheries.

Other examples in this volume, such as fisheries reform in South Africa (Issacs), bottom-up co-management in Malawi (Donda), and the broader concept of integrated management in Indonesia (Ginting), all suggest that innovative governance arrangements are possible in making life better for small-scale fishers. This implies that, in addition to the 'soft' approaches required in illuminating underlying values, images and principles, we also need 'hard' approaches from governments at local, national and global levels, to directly intervene and correct social justice and power imbalance issues, and to design institutions and governance mechanisms that foster the viability of small-scale fisheries and enable them to successfully coexist with large-scale fisheries.

References

Bailey, C. (1985) Blue revolution: the impact of technological innovation on Third World fisheries. *The Rural Sociologist*, 5(4), 259-266.

Bavinck, M. (2011) The megaengineering of ocean fisheries: a century of expansion and rapidly closing frontiers. In Brunn, S. D., ed., *Engineering earth: the impacts of megaengineering projects*. Dordrecht: Kluwer, pp. 257-273.

Béné, C. (2003) When fishery rhymes with poverty: a first step beyond the old paradigm on poverty in small-scale fisheries. *World Development*, 31, 949-975.

Butcher, J. G. (2004) The closing of the frontier – a history of the marine fisheries of Southeast Asia c. 1850-2000. Leiden: KITLV Press.

Cadigan, S. T. (2009) *Newfoundland and Labrador: A History*. Toronto: University of Toronto Press.

Chuenpagdee, R. and Bundy A., eds. (2005) *Innovation and Outlook in Fisheries: An Assessment of Research Presented at the 4th World Fisheries Congress*. Fisheries Centre Research Report, 13(2). Vancouver, Canada: Fisheries Centre, University of British Columbia.

Chuenpagdee, R. and Jentoft, S. (2011) Situating poverty: a chain analysis of small-scale fisheries. In Jentoft, S. and Eide, A., eds., *Poverty Mosaics: Realities and Prospects in Small-Scale Fisheries*. Amsterdam: Springer.

Chuenpagdee, R., Knetsch, J. L. and Brown, T. C. (2001) Environmental damage schedules: community judgments of importance and assessment of losses. *Land Economics*, 77(1), 1-11.

Chuenpagdee, R., Liguori, L., Palomares, M. L. D., and Pauly, D. (2006) Bottom-up, global estimates of small-scale fisheries catches. Fisheries Centre Research Report, 14(8). http://www.fisheries.ubc.ca/publications/

Chuenpagdee, R., Morgan, L. E., Maxwell, S. M., Norse, E. A., and Pauly, D. (2003) Shifting gears: assessing collateral impacts of fishing methods in the U.S. waters. *Frontiers in Ecology and the Environment*, 1(10), 517-524.

de la Torre-Castro, M. and Lindström, L. (2010) Fishing institutions: addressing regulative, normative and cultural-cognitive elements to enhance fisheries management. *Marine Policy*, 34, 77-84.

FAO (2010) The State of World Fisheries and Aquaculture 2008. Rome: Food and Agriculture Organization.

FAO (1995) Code of Conduct for Responsible Fisheries. Rome: Food and Agriculture Organization.

FAO/RAP/FIPL (2004) A research agenda for small-scale fisheries. Bangkok, Thailand: FAO Regional Office for Asia and the Pacific.

Gabriel, O., Lange, K., Dahm, E., and Wendt, T. (2005) *Fish catching methods of the world*. Oxford: Blackwell Publishing.

Garcia, S. M., Allison, E. H., Andrew, N. J., Béné, C., Bianchi, G., de Graaf, G. J.,

Kalikoski, D., Mahon. R., and Orensanz, J. M. (2008) Towards integrated assessment and advice in small-scale fisheries: principles and processes. FAO Fisheries and Aquaculture Technical Paper No. 515. Rome: Food and Agriculture Organization.

Gordon, H. S. (1991) The economic theory of common-property resource: the fishery. *Bulletin of Mathematical Biology*, 53(1/2), 231-252.

Hardin, G. (1968) The tragedy of the commons. *Science*, 162, 1243-1248.

Jentoft, S. and Chuenpagdee, R. (2009) Fisheries and coastal governance as wicked problems. *Marine Policy*, 33, 553-560.

Jentoft, S. and Eide, A., eds. (2011) Poverty Mosaics: Realities and Prospects in Small-Scale Fisheries. Amsterdam: Springer.

Jentoft, S., Chuenpagdee, R., Bundy, A., and Mahon, R. (2010) Pyramids and roses: alternative images for the governance of fisheries systems. *Marine Policy*, 34, 1315-1321.

Jentoft, S., Eide, A., Bavinck, M., Chuenpagdee, R., and Raakjær, J. (2011) A better future: prospects for small-scale fishing people. In Jentoft, S. and Eide, A., eds., *Poverty Mosaics: Realities and Prospects in Small-Scale Fisheries*. Amsterdam: Springer.

Kooiman, J., Bavinck, M., Jentoft, S. and Pullin, R. S. V., eds. (2005) *Fish for Life: Interactive Governance for Fisheries*. Amsterdam: Amsterdam University Press.

McEvoy, A. F. (1986) The Fisherman's Problem – Ecology and Law in the California Fisheries 1850-1980. Cambridge: Cambridge University Press.

McGoodwin, J. R. (1990) Crisis in the World's Fisheries: People, Problems, and Policies. Stanford: Stanford University Press.

Pauly, D. (1997) Small-scale fisheries in the tropics: marginality, marginalization and some implications for fisheries management. In Pikitch, E. K., Huppert, D. D. and Sissenwine, M. P., eds., *Global trends: fisheries management*. Bethesda: American Fisheries Society Symposium.

Pauly, D. (2006) Major trends in small-scale marine fisheries, with emphasis on developing countries, and some implications for the social sciences. *Maritime Studies*, 4(2), 7-22.

Pauly, D., Alder, J., Bennett, E. and Christensen, V. (2003) The future for fisheries. *Science*, 302, 1359-1361.

Pauly, D. and Chuenpagdee, R. (2003) Development of fisheries in the Gulf of Thailand large marine ecosystem: analysis of an unplanned experiment. In Hempel, G. and Sherman, K., eds., *Large Marine Ecosystems of the World: Trends in Exploitation, Protection, and Research.* Elsevier Science, pp. 337-354.

Pomeroy, R. and Andrew, N., eds. (2011) Small-scale Fisheries Management: Frameworks and Approaches for the Developing World. Cambridge: CABI.

Satterfield, T., Slovic, P. and Gregory, R. (2000) Narrative valuation in a policy judgment context. *Ecological Economics*, 34, 315-331.

Smith, H. D. (2000) The industrialization of the world ocean. *Ocean & Coastal Management*, 43, 11-28.

Sumaila, U. R., Teh, L., Watson, R., Tyedmers, P. and Pauly, D. (2008) Fuel prices, subsidies, overcapacity, and resource sustainability. *ICES Journal of Marine Science*, 65(6), 832-840.

Swan, J. and Gréboval, D., eds. (2004) Report and documentation of the International Workshop on the Implementation of International Fisheries Instruments and Factors of Unsustainability and Overexploitation in Fisheries, Mauritius, February 3-7, 2003. FAO Fisheries Report No. 700. Rome: Food and Agriculture Organization.

Section 2:

Challenges and issues

CHAPTER 2

Survival of small-scale fisheries in the post-war context in Sri Lanka

Gayathri Lokuge[1] and Mohamed Munas

The post-war era has afforded Sri Lanka, especially those people in the North and the East, a chance to re-start their lives and livelihoods that were disrupted for 30 years. One of the main industries affected was the fisheries. Within this sector, there is a prevalence of poverty among small-scale fishers as well as others along the value chain. With the end to the armed conflict, the fisheries sector has been identified by both state and non-state actors as an important source of income generation for the people affected by war. Amidst this renewed interest in the fisheries sector, this essay discusses the current practices and types of fisheries that poor people are engaged in, the opportunities and constraints they face, and proposes ways of increasing the sustainability of small-scale fisher livelihoods.

Introduction

Although there is no standard definition of small-scale fisheries, various classifications of fisheries include: small-scale versus large-scale, subsistence versus commercial, artisanal versus industrial, inshore versus offshore; or fisheries may be classified according to vessel size (Panayotou, 1985). The literature on small-scale fisheries indicates that the definition differs depending on the context of the country, as those considered small-scale in one context could be considered large-scale in another. Thus, ranges or rough categorizations of the technical and socioeconomic characteristics of fishers' activities are generally used.

In the Sri Lankan context, a range of small- and large-scale fishing can be identified based on characteristics such as the size of the catch and the type of species caught, the technology – type of craft and equipment

1 Corresponding author: Centre for Poverty Analysis, Sri Lanka, gayathri@cepa.lk

used, the amount of capital (human and physical) invested, and the returns to the producer. Apart from those who engage in capture fishing, there are culture fish producers – both small- and large-scale – and a complex web of men and women who engage in fishing-related activities along the value chain. For the purpose of this discussion, we consider fishers who engage in capture fishing in the coastal and inland waters using traditional methods, equipment and crafts as small-scale fishers.

The fisheries sector in Sri Lanka is divided into three main sub-sectors, depending on the location where the fishing activities take place: coastal fisheries, offshore fisheries and inland/aquaculture fisheries. Coastal fishing occurs within the continental shelf and is undertaken by fishing crafts in single-day operations and beach seine[2] activities. Offshore fishing involves deep sea fishing outside the continental shelf, extending up to the edge of the exclusive economic zone, carried out by multi-day boats. Inland and aquaculture mainly involves fishing in fresh water, brackish water and fish culture. Sea fishing in Sri Lanka is naturally controlled by the two monsoons that affect the country; the main fishing season in the Southern and Western coastal belt is from September to April and on the Eastern coast from March to September (focus group discussion, Trincomalee).

This essay draws upon information and field experience gathered by the Centre for Poverty Analysis (CEPA) on different assessments and studies conducted in the Northern and Eastern provinces and in Puttalam district in the North Western province of the country. The studies and assessments looked at livelihoods in general in the conflict and post-war context of the country, and we draw specifically on the fisheries related information. The first part of this essay provides an overview of the definitions and concepts related to small-scale and other fisheries in the Sri Lankan context. This is followed by details of the current status of small-scale fishers, highlighting the opportunities and constraints, drawing upon the personal stories and experiences of fishers affected by the war. The essay goes on to discuss the current policy environment and action plans related to the small-scale fisheries sector, followed by potential solutions for the challenges identified.

2 Beach seines in Sri Lanka are called *'maadal'*. The average beach seine measures about 2-3km in extent with the length of the rope varying from 1-1.5km. The body and the bag (or the cod end) measures around 7m. The ropes are made of coir fibres whereas wings are made of coir rope meshes and Kuralon meshes. The bag is totally made of Kuralon. The net is set in water with the use of a wooden craft (8m in length) known as *'maadal paaruwa'*. The beach seine craft is paddled by four oars, two on either side, and each oar is paddled by two men. In addition, there are two men in the craft to set the net, and a leader to give directions and coordinate activities.

Overview of the fisheries in Sri Lanka

Coastal fishing involves different types of techniques to capture fish within the continental shelf, as single-day operations and beach seines, using traditional methods such as rod fishing and stilt fishing. For those engaged in this sub-sector, fishing is their sole income source. Their dependency on one livelihood makes them vulnerable to the uncertainty and risks involved as well as lack of income or savings in the offseason. The net income distribution of the coastal sub-sector compared to the offshore sub-sector is low. The number of people engaged in a given coastal fishing activity is high and the harvest is low compared to the offshore sub-sector. For example, a beach seine is estimated to support about 40 families, but the harvest is lower compared to that of multi-day boats (Munas *et al.*, 2009).

Coastal marine resources offer direct and indirect income generation opportunities for a cross section of coastal society; men, women and even children, after school and during school holidays. While the actual capture of fish is the man's domain, women and children boost household income by engaging in post-harvest and support activities such as cleaning fish, preparing fish for drying, and mending fishing nets.

According to the Ministry of Fisheries, Sri Lanka's fishing fleet is mainly mid- to small-scale, with very little large-scale fishing due to lack of technology, boats and capital. Therefore, Sri Lanka has not yet capitalized on offshore fisheries, and new policy directives are looking to improve the offshore sector. Even though the offshore sub-sector has the highest level of productivity in the North and the East, there is a marked lack of fishing equipment and crafts suitable for offshore fishing. Thus, this sub-sector will likely take time to develop due to the lack of capital that is essential for purchasing multi-day boats.

In the recent past, inland fishing and aquaculture has become important to the fisheries sector for various reasons. Some fishers engage in inland fishing as the primary source of income while for others it is a secondary form of income combined with other non-fisheries related activities, such as agriculture, which helps them maintain a mixed livelihood portfolio. This is a coping mechanism that can mitigate the impact of external shocks to income. During the offseason in the marine sub-sector, certain fishers engage in inland activities in the brackish waters as a secondary livelihood option.

The total fisheries production from the three sub-sectors has increased, but with high fluctuations, from about 187,500 tonnes in 1980 to about 320,000 tonnes in 2008. In the early period, almost all production came from the coastal sub-sector. While more than half of the total production still comes from this sub-sector, the offshore fishing sector is gaining a significant part of the catch share (about 33% of the total in 2008). A similar trend is observed in the inland and aquaculture sub-sector, which currently contributes about 12% of total production.

Role of small-scale fishing as an income generation activity in the post-war context

Agriculture and fisheries are the mainstays of Sri Lanka's North and East economy. They were important livelihood strategies before the war, and are vital for the economic recovery of these conflict-affected areas. The growth rate of the agriculture sector (which includes fisheries) is higher than the industry and service sectors in the North and East. During the ceasefire of 2002-2005, which is closest to 'normalcy' that the North and East provinces experienced prior to the end of the war, agriculture and fisheries showed the highest percentage contribution to the provincial GDP. Also, between 2002 and 2003, this sector expanded by an average of 32% per year in Northern Province and by 19% in Eastern province (Anonymous, 2007). This shows the potential of the agriculture and fisheries sector in the two conflict-affected regions.

About 60% of the coastal area of Sri Lanka is in the North and East Provinces. The geography of these provinces makes them suitable for coastal and inland fisheries. They have an abundance of freshwater and brackish water sources and are very rich in coastal, offshore and deep-sea marine resources. As a consequence, 60% of fishers in the country are from these regions. Even though the North and East Provinces contributed 64% of the total fish production in the country in 1980, it dramatically declined to 32% in 2002 as a result of the civil war. In contrast, Eastern province saw an increasing trend in fish production due to relative accessibility; as opposed to the North where there was limited access to the sea.

The Northern and Eastern provinces also employ the highest number of people in the fishing sector, which comprised 60% of the total fishing population in the country in 2009. They are mostly in the coastal sub-sector and lagoons that fall under small-scale fisheries. In contrast, the offshore sub-sector shows the highest productivity but employs lower numbers of people compared to the other two sub-sectors. According to the available data, Trincomalee and Batticaloa districts in the Eastern province have the highest numbers of traditional crafts throughout the island. These are used by the small-scale coastal fisheries communities.

During the conflict, security restrictions on time, distance, access and type of craft used in these regions severely affected the fishery sub-sector. The lack of skill, capital and capacity available in the region, loss of livelihood-based assets due to multiple displacements of the fishing communities, damaged infrastructure from landing centers, storage facilities to market infrastructure, financial establishments such as banks and road networks added to further deterioration of the situation.

The end of the armed conflict has made it possible for long-term development interventions to take place in these provinces. The fisheries sector contributes 70% of the nation's animal protein intake, and so is a vital contributing factor to the health of the population. The policies are also looking at promoting alternative livelihoods and pushing people to move towards offshore or culture fishing mainly targeting the export market. There are also implications of overfishing and resource depletion of Sri Lanka's fish stocks. This has spurred a renewed call for tighter regulations and limits of licenses for fisheries, especially on coastal/near-shore fishing. These changes have implications for small-scale fishers; testing their ability to adjust and benefit from post-war development.

Experiences of small-scale fishers

During the war from the 1980s to 2009 and in the immediate post-war context, communities lost family members, property and livelihood assets; experienced multiple displacements; and faced fear and uncertainty, restricted physical mobility and limited access to livelihood opportunities. Some resettled communities in the post-war North and East provinces of Sri Lanka engage in small-scale fishing because of their pre-displacement experience. Also, the relatively low capital requirement in engaging in small-scale fishing using traditional crafts and equipment means that they are in a position to restart these activities in the fragile time of immediate return to their original communities. Some small-scale fishers were able to take their fishing gear (such as nets/rods) with them during their displacement so this enabled them to restart a livelihood with relatively low capital cost.

Some experiences of fishers in the East, immediately following the end of the armed conflict and the resettlement process, highlight some of the challenges on the ground.

Vicious cycle of debt: Is there a way out for small-scale fishers?
Marine fishing in Sri Lanka is seasonal with the peak season from February to July in the Eastern and North-Eastern coastal belt. Small-scale fishers with their non-motorized one-day boats – unlike the offshore/multi-day boats – cannot engage in fishing activities during the off peak season. A secondary form of livelihood – either related to fishing or not – which can cushion them during the offseason is not available to them. This leaves them with no other alternative but to take credit at high interest rates from individual money lenders, trapping them in a cycle of debt. Consequently, fishers' indebtedness binds them to sell the catch to these money lenders, who are also the collectors/buyers (*Mudalalis*) during the fishing season.

> Mudalali (local collector/buyer) is the one who decides the price of the catch we sell. We don't have a say in the price because we are indebted to him. We are in debt to him because during the offseason, we survive on the credit that we get from him. If we can supply the product to a different buyer then we can get a better price.
> (Focus Group Discussion, Trincomalee)

Lessons for offseason survival can be learned through resourceful fishers from the same area. As an alternative to engaging in marine fishing during the offseason, some fishers switch to lagoon and brackish water systems to supplement their income. Some look for employment in the informal agriculture sector on a daily wage basis. They also try to diversify their household portfolio with members taking on different income generation activities.

> I have one son and one daughter. Son in grade nine. My husband died in 1996 during the war. My daughter who is in grade five also engages in fishing after school, and my son is a member of the fisheries society. So he can use a small boat to fish in the village tank. I fish using a rod and undertake home gardening. I have enough land of about ½ acre for home gardening, so in Yala (rainy) season I cultivate chilli, long beans, tomato, brinjals, and bitter gourd. (Female fisher, Batticaloa)

There is a clear need to break away from the cycle of debt that traps the fishers and tightens around them during the offseason. Finding ways to reduce the monopoly of the buyers and their control of prices is one option. Better management of their spending patterns during the peak season when most of them earn a sizable income is another. Finally, diversifying income sources to tide them over the off peak season should be encouraged.

A future for beach seines?

Beach seine fishing is an important aspect of small-scale fishing, especially in the Eastern coastal belt. However, many beach seines in this area are owned by people from the Northwest coast. Local people can obtain employment on the beach seines either as temporary workers (where they are given fish as payment) or as permanent workers with cash wages. However, most of the permanent workers on these beach seines are from outside the district and they have been brought to the East on a contract basis. Some women from the area engage in the last phase of the beach seine activity, “the illippu”, and are paid in kind with two or three baskets of fish.

> I support myself by taking part in the beach seine related activities. Women are usually needed to bring in the last part of the beach seine and we are usually paid in kind with fish. We usually get about four fish per net. But this depends on the amount of fish that is caught in the net. Sometimes there aren’t any fish caught in the nets at all and we come off empty handed. We can sell the fish that we get to the buyers that come to collect the fish and take cash. (Female villager, Trincomalee)

However, being employed in beach seine activities is seen to bring less return, as most of the time they are paid in kind. Fishers thus prefer to go out to sea in their traditional crafts. Hence, this type of fishing is undertaken by the most vulnerable fishers of the community because it does not require any capital cost.

Based on the experience of the Southern coastal belt, beach seine activities are not encouraged by policymakers due to resource depletion in the near-shore area. One way of controlling beach seine activities is by not issuing the license that is necessary to operate a beach seine in the East. However, each beach seine supports about 40 families and, given that a large proportion of these families belongs to the most vulnerable group, sufficient attention should be paid to including them in the post-war development phase.

Emerging sub-sector: Freshwater capture and culture fishing

Freshwater capture fishing has been identified as a potential area for support by both state and non-state entities in the conflict-affected areas of the country. The quick returns of this sub-sector can enhance the livelihoods of the rural poor as well as add nutritional value as a protein supplement in the diet. Fishing provides the communities with a secondary source of income which helps them maintain a mixed livelihood portfolio to mitigate the impact of external shocks to their income.

The communities lack the initial capital and the necessary technology to engage in culture fishing. Since this is a relatively new livelihood option for most of these communities, market linkages and maintaining quality production levels will be a challenge at the initial stage. External interventions in terms of capital, technology, and facilitation of market linkages will be crucial in the current context.

> We are interested in engaging in cage culture fishing, but don't have the capital for it. (Focus group discussion, Batticaloa)

Culture fisheries suit the vulnerable groups in the post-war period, such as female-headed households, with the initial capital costs financed either through micro credit systems or livelihood assistance projects. The amount of heavy manual labor required for culture fishing is relatively low, which makes it more favorable to women and the elderly.

Although freshwater fish was not seen as a sustainable income generation activity due to its relatively low marketability compared to marine fish, at present there is a high and increasing local demand for inland/freshwater fish, and for brackish water species such as prawns and crab. This market can be capitalized upon.

New avenues for new generations?

The reluctance of the younger generations to engage in direct fish capture has become markedly more apparent in the recent past. Fishing traditions have been passed down through generations but the present younger generations seem to want to shift away from their 'inherited' livelihood in preference for livelihoods that represent 'better acceptance in the society'. They seem to go into other sectors (services or industry), fisheries-related support services or move to new forms of livelihoods such as tourism, using the fisheries-related skill-base.

As seen in a fishing village in Kalpitiya region in the North-Western coast of the country, some of the younger generation has left to seek employment in other sectors. Cultivating aloe vera is a secondary income source, while others have turned to tourism-related activities such as providing dolphin-watching and coral reef tours as an alternative source of income to fishing. On the whole, there is increased engagement in tourism-related jobs due to the end of the war and relaxing of security restrictions.

The switch to tourism-related activities such as providing tours is an alternative that is aided by the native knowledge of sea and the coastal areas and already having the basic equipment. However, they do not have financial resources to modify their boats according to the safety and comfort standards set by the tourist board, and expectations of the targeted tourist segment. In addition, poor infrastructure, inadequate lodging, regulations placed by the Department of Wildlife through permit issuance for visiting coral reefs, and security checks by the Navy pose hurdles to the tourism services they could provide. The community's lack of English further hinders their ability to communicate with foreign tourists. Despite these challenges, the fishers (especially the younger generation) see tourism as a good income source and employment generator for the future.

Way forward

Based on the current trends and the livelihood challenges presented above, this section looks at potential options to strengthen and improve the opportunities available to small-scale fishers in Sri Lanka.

Providing a platform for livelihood diversification

Diversifying income sources can prevent the risks related to seasonality and cushion adverse economic shocks. Certain vulnerable groups, such as families that are supported by beach seines, should be provided the platform to move into these diverse livelihood options. Introduction of high-value culture fishing can be one of the ways of diversification. There is a strong state interest towards culture fishing of high value fish species based on abundant brackish water resources that fetch higher incomes targeting the export market. In addition, small-scale, high-value aquaculture industries, for species such as sea cucumbers, lobsters and oysters, were present in an ad-hoc manner before the conflict in the North, and can be regularized and improved to provide better returns. The presence of favorable environmental conditions such as brackish water lagoons makes the area suitable for aquaculture. Using the zonal plans done for the East by the Ministry of Fisheries and Aquatic Resources in line with the ten-year plan for the fisheries sector, culture fishing can be developed beyond the model farms that have already been created in the East. The communities will have to be provided with the necessary skills training and initial capital to start this process.

Inland capture fisheries can be integrated with the agriculture system and promoted as a rural livelihood option, working through the already existing government structures of agriculture and irrigation. These factors and the presence of a large percentage of inland water bodies in the provinces, suggest that the inland fisheries sub-sector has a potential to grow in the North and East.

Introduction of cash crop cultivation or non-farm activities such as three-wheeler/motorbike repairing/servicing, information technology businesses such as 'communication centers' could be promoted, particularly to communities that have better access to main roads. An alternative/secondary livelihood option will provide fishing communities a reserve that they can fall back on during the fishing offseason, and move them away from the cycle of debt to the fish collector/trader.

Small and Medium Enterprises (SMEs) could be promoted, anchored around identifying more enterprising fisher communities/societies or individuals with initial capital. The more vulnerable fishers can then be linked to these SMEs which would give them a more stable income than the unproductive non-motorized day-boat fishing.

In the areas where the coastal fisheries are not profitable and the tourism industry is booming, communities can service the tourism sector. This provides an alternative source of income which is supported by native knowledge of the sea and the coastal areas. It assures them a continuous income, and one that seems to be preferred by the younger generations.

Creating linkages between coastal fisheries and deep-sea fishing through support services

Youth who are unwilling to engage in ocean-going fishing activities, and fishers in the less productive coastal sub-sector, should be encouraged to join the chain of support services related to fishing. Construction of boats, boat cleaning and maintenance, fish net manufacture, ice production, and supplying other services required by the industry such as transport, food provisions, and mechanical parts for the boat engines should be promoted with financial capital along with technical training from the relevant state authorities.

Promotion of support services locally will also boost the expanding fisheries sector in the conflict-affected areas, while also providing a larger range of jobs with wider benefits to the community. The shifting of fishers from the coastal sub-sector to the other services will also have added benefits of reducing over exploitation and resource depletion, thereby ensuring sustainability of the fishing industry.

Creating market linkages

The monopoly of the collector/wholesaler can be reduced by facilitating direct market linkages with wholesale markets inside and outside the district. The involvement of the private sector, such as hotels and restaurants, could be sought for other high-value culture fish and fish products. Awareness of the profitability of the culture fish industry should be made among the private sector, encouraging them to invest in these coastal regions. However, culture fishing should be introduced and managed with caution, with clear resource management mechanisms, in order to minimize possible damage to the environment.

Final thoughts

Over 30 years of war has taken its toll on the lives and livelihoods of the people of North and East Sri Lanka. There is now an opportunity to redevelop this area, and the fisheries sector has been identified as an important economic driver of this process. Resettled communities are engaging in small-scale fishing because of their previous experience, salvaged fishing gear and equipment, and the relatively low capital requirement. However, the revival of small-scale fisheries in the North and East faces several constraints: vicious cycle of debt, lack of capital to incorporate appropriate technology in culture and capture fishing, and the reluctance of the younger generation to engage in direct activities such as capturing fish. In addition, the government policy encouraging offshore fishing gives a hint of de-prioritization of the coastal/near-shore fisheries.

The discussion in this chapter reveals that there is a need to strengthen the small-scale fishing sector, recognizing its importance in terms of the number of people depending on it as a means to restart and re-establish their lives and livelihoods. In order to survive, small-scale fisheries in post-war Sri Lanka need to evolve from traditional primary production activities to more efficient, diverse and higher-value income-generating activities. Diversification is seen as a way of eliminating the shocks of seasonality by providing alternative income sources. Diversification can also help to captivate the younger generation and provide additional income which can address the problem of debt cycle in the long run. Providing support services to capture fishing can create alternative opportunities to those engaged in the low-value coastal sector. Coastal fisheries communities, especially the youth, also have the potential to gain from the growing tourism industry; an avenue that can capitalize on existing fisheries related skills, equipment and knowledge. However, in order to allow small-scale fishers to gain from these opportunities, providing initial capital and technical support as well as creating market linkages are crucial and necessary at this transition stage of rehabilitation and recovery.

References

Anonymous (2007) Sri Lanka poverty assessment engendering growth with equity: opportunities and challenges. Colombo: The World Bank Colombo Office.

Munas, M., Lokuge, G. and Thalyasingam, P. (2009) Agriculture and fisheries livelihoods in the North and East, Colombo. Colombo: Centre for Poverty Analysis.

Panayotou, T., ed. (1985) *Small Scale Fisheries in Asia: Socio Economic Analysis and Policy*. Ottawa: International Development Research Centre.

CHAPTER 3

Overcapitalization in a small-scale trawler fishery: a study of Palk Bay, India

Jyothis Sathyapalan[1], Jeena T. Srinivasan and Joeri Scholtens

This essay focuses on the causes of overcapitalization in the small-scale trawler fishery of Palk Bay, India. The study observed that the enforcement of incentive-blocking measures is insufficient and inefficient. Violations or non-compliance of various rules and regulations are widely observed. Many fishers are operating banned gears or adopting fishing practices like pair trawling. Misuse and misappropriation of subsidized diesel was also noted. In such a situation, incentive-adjusting measures are desirable, and there is a need for fisheries regulation to move in this direction. This is the real challenge of small-scale trawler fisheries like Palk Bay.

Introduction[2]

Regulation is a highly debated topic in fisheries management. Due to instances of intensive fishing and slow collapse of fish stocks, there have been widespread calls for global reduction in fishing fleet sizes and fishing efforts (Mullon *et al.*, 2005). It is reported that many world fisheries are biologically as well as economically unsustainable (Garcia and Newton, 1997). In India too, mechanization of vessels brought about drastic changes in the sector. More and more indigenous crafts were motorized, and the use of different types of gears was taken up to meet growing demand. Trawling gained importance for exploiting demersal fisheries in India, with almost 50 per cent

---

1 Corresponding author: Centre for Economic and Social Studies, Hyderabad India, sjyothis@cess.ac.in

2 This essay is based on the study "Fishing Fleet Reduction and Its Livelihood Implications: A Case Study of Palk Bay Resource Users in the East Coast of Tamil Nadu, India" supported by UNTRS and FAO of the United Nations in 2007-08 with advise from SIFFS Trivandrum, CESS Hyderabad and MARE Amsterdam. The usual disclaimers apply.

of the total catch coming from this method (Devaraj *et al.*, 1997; 1999). Moreover, the Indian government encouraged mechanization via various subsidy programmes and loans to fishers and their co-operative organizations (Srivastava *et al.*, 1991). Presently, there are too many fishing vessels, generating excess fishing effort in various areas, especially where valuable species occur (Somvanshi, 2001). The Marine Fisheries Regulation Acts of Indian maritime states provide various rules and regulations, some of them are meant for blocking fishers' incentives to invest excess capital in the fisheries. However, there have been enforcement problems in terms of widespread violations of rules and regulations threatening the sustainability of fisheries. 'Incentive blocking' measures attempt to restrict the level of fishers' activities in some way, whereas incentive adjusting measures attempt to address the property rights issue where an environment can be created in which fishers benefit within harvesting constraints, and allow the market to assist in reducing overcapacity (Ward *et al*, 2004). The important incentive blocking instruments are limited entry, buyback programmes, gear and vessel restrictions, aggregate quotas, non transferable vessel catch limits and individual effort quota (IEQs). Group/community fishing rights, Territorial use rights, individual transferable quotas and taxes and royalties constitute important incentive adjusting instruments. Our study focuses on the Palk Bay fisheries on the east coast of Tamil Nadu, India, and explores the factors leading to the failure of incentive-blocking measures in the trawler sector. We argue that current measures are insufficiently and unsuccessfully enforced, and should be replaced by 'incentive-adjusting measures.' Moving in this direction, however, is a challenge faced in the small-scale trawler fisheries in India, as shown in this essay.

The Palk Bay fishery

Palk Bay is situated between northwest Sri Lanka and southeast India, and connects the Bay of Bengal to the northeast with the Gulf of Mannar to the south (Figure 1). Studies report that it is a very shallow flat basin with an average depth of 9m, and maximum depth of 15m. It spreads across four districts of Tamil Nadu state–Ramanadhapuram, Puthukkottai, Tanjore and Nagapatinam– covering coastal area of 270 km, or approximately 27% of the total coastal area of the state.[3] This area was a troubled location since the start of civil war in Sri Lanka in 1983. The civil war has had a deep impact on the

3 The present study excludes Naga patinam, leaving a total area of 107 km.

fishing operations of this area (Vivekanandan, 2004, p. 3). Although fishing operations were restored in Sri Lankan waters in 2002, altering the situation in Palk Bay, the operations of Indian fleets in Sri Lankan waters pose a grave threat to the livelihood of Sri Lankan fishers. In some instances, there have been clashes reported at sea between fishers of the two sides (Vivekanandan, 2004, p. 4). Nevertheless, there have been organized efforts by both sides to engage fishers from Tamil Nadu and Sri Lanka in a dialogue on the issue of coexistence, and also to work out solutions for the conflict between the two groups in terms of sharing peace and resources. The total number of fishing villages in Palk Bay is 169, with a total population of about 140,000 as per the fisheries census of Tamil Nadu in 2000. Both traditional and non-traditional communities are involved in fishing. Traditional communities include Ambalakarar, Paravar, Pattinavar, Kadaien, Paruvatha Rajakulam and Padiachy, while non-traditional communities include Muslims, Theevar, Nadar, Konar etc. It has been reported that about half of the fishing population are literate.

Crafts, gears and fishing practices

The fishing practices of Palk Bay can be broadly categorized into mechanized and non-mechanized. Mechanized crafts consist entirely of trawlers and are locally known as launches or simply boats that account more than 2000 in the study area. The length of the trawlers varies between 28 and 50 feet. All boats have a portable icebox with a capacity of 200 to 500 kg. In the mechanized sector, both single and pair trawlers operate and they both use bottom and surface nets. They use an inboard engine with a capacity ranging from 68 to 120 horsepower. Non-mechanized boats can be broadly divided into motorized and non-motorized boats. Motorized boats are equipped with outboard motors – these include the wooden craft (*Vathais*) and the fiberglass crafts and inboard motors called *vallam*. The marine fisheries census categorized motorized vessels into dugout, catamarans, plank-built, ring seiner, fiberglass, ferro cement, and others. Non-motorized crafts are mainly dugout, catamarans, and plank built. The gear used by the non-mechanized fishers varies across seasons. Some of the main ones are 'Thallu Valai' (trawl net), 'Nandu Valai' (gill net for crab), 'Adi Valai' (drag net) and 'Thoondi' (hooks & lines). Destructive fishing practices prevailed in the non-mechanized sector with the use of *surukku madi* (purse seining) and dynamite, though both are legally banned. Generally, the use of purse seining is quite common during the month of March as the chances of being caught are very small.

Figure 1 Location and infrastructural facilities of Palk Bay Landing Centres

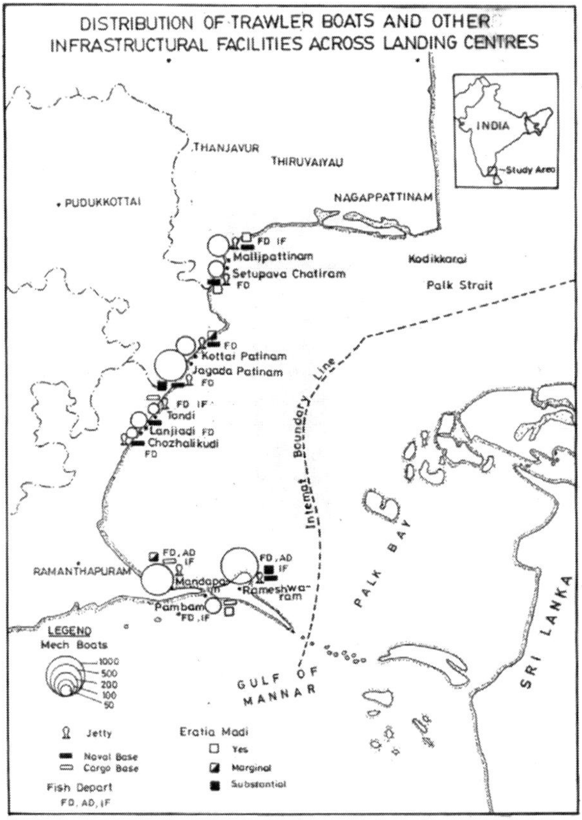

The physically contained and politically troubled geography of Palk Bay has some implications for the fishing grounds used by fishers. It has become clear that the large number of trawlers competing for limited resources have 'forced' many Indian fishers to Sri Lankan waters to obtain sufficient catch. This is done despite the fact that crossing the international boundary line can result in confrontations with the Indian or Sri Lankan navies, or with local fishers (Bavinck and Karunaharan, 2006). The *Rameswaram* trawlers head in an eastern direction where fishing grounds are rich in marine resources (behind the Sri Lankan boarder which is just one hour away from their berthing place) where the average fishing ground available per trawler is a mere 2 to 3 km^2.

Rules and regulations in Palk Bay fisheries

The rules and regulations of the Palk Bay fishery can be classified based on its temporal and spatial characteristics. While the rule pertaining to the spatial restriction is three nautical miles, that of a three to four day trip has a temporal dimension. In addition to these rules, they observe a fishing 'holiday' generally known as the 45-day ban (and also locally referred to as government strike). Fisheries regulations are provided in Table 1.

Table 1 Fishery regulations and their rationale in Palk bay (Sathyapalan et al. 2007)

Regulations	Rationale
– Three nautical mile rule – Three to four day rule – Banned gears	– Regulations stemming from conflict with artisanal fishers (mainly between drift net and trawlers)
– 45-day ban	– Regulations stemming from resource degradation
– Fishers are not allowed to cross the International Boundary Line – Fishers have to possess an identity card, issued by the Fisheries Department – If a trawler heads to sea it requires a fishing token, issued and signed by the Fisheries Department – A maximum of five crewmembers are allowed to man each trawler – One can only receive tax-free diesel if one possesses a fishing token	– These are regulations exclusively for Palk Bay area arising out of the proximity to Sri Lanka

One of the important clauses of the Tamil Nadu Marine Fisheries Regulation Act of 1983 relates to the three nautical mile rule that aims to restrict the operation of mechanized vessels, and demarcate fishing grounds for the operation of country crafts (vallam, Vathai, FRP crafts) at the exclusion of trawlers. This rule aims to limit or restrict the entry of crafts into Palk bay which in turn block fishers incentives to add more fishing vessels into Palk bay. But violation of these rules has been observed in many places throughout the area (Bavinck 2003, Scholtens 2006).

In order to solve the resource use conflicts between artisanal (non-mechanized) and trawler fishers targeting the same species within the same fishing territory, alternative night schedules were introduced for both groups in 1993 (Bavinck, 2003). This regulation aims to guarantee exclusive rights to trawlers on alternative days. In a way this rule adjusts fishers'incentives to add more vassals into Palk Bay. Generally, all trawlers obey this rule. Based on the tokens issued by the Fisheries Department in 2006, it is calculated that a mechanized trawler goes to the sea for an average 100 days a year (out of the 135 legally allowed days).

The limited number of days that fishers venture to sea is an indicator of overcapitalization as well. Many artisanal fishers head to sea for six days per week, although on trawler days they keep near the shore, despite the belief that the rule is actually meant to restrict trawlers, and not to limit their operational scope. In 2001 the Government of Tamil Nadu introduced an annual ban on mechanized boat fishing, lasting from April 15th to May 29th, which became known as the 45-day ban. During this period, mechanized boats are strictly forbidden to enter the sea. This restrictive measure was aimed at the conservation and regeneration of fish stocks and is well obeyed throughout Tamil Nadu.

Table 2 Reasons for opposing and favoring pair trawling in Palk Bay (Sathyapalan et al. 2007)

Opposing pair trawling	Favoring pair trawling
– It is an ecologically destructive fishing method – It reduces the catch available for other trawlers and small-scale fishers – It reduces fish prices, thereby affecting other fishers – It adds strain on the limited labor available – Pair trawling used to be operated from January to March, but now it is operated throughout the year	– The net is not destructive as it is used for catching migratory pelagic fishes – It targets only fish and not prawns which are targeted by the 'ordinary' fishers. Hence it has no impact on the catch of others or on the price – All the protest is just a matter of jealousy of smaller owners, as we have strong boats and big amounts of money are involved – In *Pamban*, pair trawling is also allowed for three months per year. In *Rameswaram,* we need the same. In return, we can promise not to enter restricted areas throughout the rest of the year – Small-scale fishers no longer oppose pair trawling – With regard to the criminal act: how can you punish innocent crewmembers. This is ridiculous – "If pair trawling is banned (in practice) we shall fight against it. Otherwise, it is possible to increase the horsepower of the motor and operate the net with one boat. This is already being done by boats in Tuticorin"

The government ban on pair trawling is an important incentive blocking mechanism in Palk bay. We found that many trawl fishers practice pair trawling in Palk Bay region, although the pair trawling net (Erattai madi) has been banned since 2000 by a government order. The fisher's reasons in favor and against the use of pair trawl nets are as shown in Table 2.

Disobedience and misuse

Several factors account for the failure of incentive-blocking measures in Palk Bay. First, there is the possibility of getting subsidized diesel without going to sea. The diesel can either be sold to outsiders or fellow boat owners or – in the case of an owner having more than one boat. Secondly, if one sells one's boat (for example to a boat breaker) but keeps his registration certificate book with him, there is a possibility of fishing tokens and subsidized diesel being issued on a continued basis. This diesel can be sold for a little margin either to fellow boat owners or outsiders. According to the Fisheries Department, this problem was solved in September 2006 when an extensive physical identification of boats was carried out and all vehicle registration certificates (RC books) that were not connected to an existing boat were cancelled. In Rameswaram landing centre, for example, 40 RC books were cancelled in 2006. Large sums of money were reportedly offered to the Fisheries Department by some powerful owners to continue their illegal activities, but accepting this allegation was difficult as inspectors from different districts carried out the physical identification. However, these misuses appear to occur especially in the larger landing centres where it is almost impossible for the Fisheries Department to monitor whether boats with fishing tokens really venture to sea. Evidently, these activities have to be carried out in secrecy because without substantial influence or facing local 'enemies', it is very difficult for boat owners to abuse this scheme, as their local 'enemies' would be eager to lodge a complaint to the Fisheries Department.

Institutional arrangements

The primary responsibility of fishery management rests with the Tamil Nadu Fisheries Department. The institutional duty of the Fisheries Department is both enormous and complicated. First of all, it has to implement a number of welfare schemes for the fishing community. The main concern here is the provision of 1,500 litres of tax-free diesel per month per mechanized boat.

Secondly, they are to maintain the records of the fishing operations. Thirdly, they have to implement a wide range of fisheries regulations, including those under the Tamil Nadu Marine Fisheries Regulation Act of 1983. Add to this, if conflicts arise, fishers can appeal to the Inspector of Fisheries to solve them. Finally, the Fisheries Department has to provide fishers with identity cards and daily fishing tokens with which they can identify themselves as genuine fishers. This extensive and complex procedure implies that the officials must simultaneously act as welfare providers, registration officers and police officers. In the concerned districts, the Fisheries Department has six offices; three located in the southern cluster and the other three in the remaining clusters. These offices are headed by an Inspector of Fisheries with 2-8 staff members, while the large landing places of Rameswaram and Mandapam are headed by an Assistant Director. The offices are generally poorly equipped and often face the problem of open vacancies.

Given the sensitivity of the area, the Coast Guard as well as the Indian Navy have a substantial presence in the Palk Bay. Across the districts under study, there are four naval bases, one in each cluster. The navy personnel make use of nine hired blue painted fishing boats (40 ft) on which they patrol daily or twice-daily along the coastal areas. The Coast Guard is located on the southern shore of Mandapam and has thus – strictly taken – no base in the Palk Bay. However, modern hovercrafts operating from Mandapam make regular checks in the southern areas of the Palk Bay. Additionally, the Coast Guard patrols with several ships permanently along the international borderline. As far as the fisheries in the Palk Bay are concerned, the responsibilities of the Coast Guard and the Navy do not differ significantly. More importantly, their principal duty is to safeguard the sovereignty of the national waters, and prevent smuggling activities, and this falls under the authority of the central government. This implies that they are not legally committed to the enforcement concerns of the fishing regulations designed by the Tamil Nadu government, and limits their cooperation with the Fisheries Department to security issues, such as checking the fishing tokens and preventing fishers from crossing the Sri Lankan border. Although fishers have nothing to fear from these security forces as long as they possess appropriate documents and stay on the Indian side of the border, they are generally afraid of them. Reasons include a general dislike of 'uniformed North Indians' and the fact that the Navy sometimes takes possession of some high value prawns. For landing centers where fishers often 'have to' head to Sri Lankan waters – like Rameswaram and Jagada Patinam – their fear is obvious as they need to circumvent their boats and ships in order to succeed. Recently, the Coast Guard has launched a 'community interaction program', with a view to improving relations with the fishers. At local meetings they express their ability and willingness to speak Tamil and also assist fishers in emergency situations. By doing so, they hope to establish a mutually beneficial relationship with these prime actors on the sea.

Another important but unofficial player in the institutional field are the boat owner associations. In each landing center, one or more of these associations is present depending on the heterogeneity of the owners' backgrounds. For example, in Jagada Patinam, where virtually all owners belong to the Pattinavar caste, there is one association with all owners as members. Rameswaram, on the contrary, has thirteen associations, divided along the lines of religion, caste and political affiliation. In total, there are about thirty associations spread across ten landing centers. The operational scope of the boat owners associations has basically two dimensions: dealing with internal and external matters. To start with the latter, lobbying for the collective interests of the boat owners at the district and state level is of vital importance. In the young history of the mechanized boat sector, this kind of lobbying has proven successful in many instances, with the allotment of tax-free diesel as the most recent example. Their second commitment relates to internal affairs, with a focus on the settlement of disputes between the trawler owners and the traditional boat owners. Boat associations are generally acknowledged by the Fisheries Department as vital bodies in representing the interests of the fishers; and the interaction between them is substantial. In those landing centers where boat owners are well organized (like Jagada Patinam, Malli patinam and Rameswaram), they even have the power to negotiate the scope of regulations with the Fisheries Department. Boat owner associations usually have a board headed by a president and a secretary, and tend to raise a small tax on sea-going boats (in the form of prawns or money) in order to secure their financial needs. The capacity of these associations tends to be a function of their fragmentation. In this respect, in landing centers with a high number of associations, competition for prestige tends to impede effective cooperation in collective lobbies against the outside world. As a result, their institutional strength differs significantly from one landing place to another.

Conclusion

The above discussion on the characteristic features of the Palk Bay fishery reveals various issues related to fishery management. As far as the location is concerned (geographically enclosed), it is very important to reduce the fishing capacity of the area. It is also important to note that the majority of boat owners are of the opinion that the overall catch per trip is considerably low. Some boat owners attribute this to destructive harvest practices and the uncontrolled increase in the number of trawler boats. Moreover, the fishing area per boat becomes considerably low, indicating increased competition for fishing space. This point towards a long-term problem of Palk bay fishery that is overcapacity and overcapitalization. The study also found that the present rules and regulation has very limited impact on controlling the fleet size of Palk Bay which was evident from the increase trawlers, practice of pair trawling and use of banned destructive gears in this area. Hence, further adding of vessels to the present fleet size might be counterproductive, and therefore, attention has to be given in reducing the cost of fishing rather than increasing the catch by adding vassals. In such a situation, incentive-adjusting measures are desirable, and there is a need for fisheries regulation to move in this direction. This is the real challenge of small-scale trawler fisheries like Palk Bay.

References

Bavinck, M. (2003) The spatially splintered state: myths and realities in the regulation of marine fisheries in Tamil Nadu, India. *Development and Change*, 34(4), 633-657.

Bavinck, M. and Karunaharan, K. (2006) Legal Pluralism in the Marine Fisheries of Ramnad District, Tamil Nadu, India. *IDPAD Working Paper No 2*. Amsterdam: Centre for Maritime Research.

Devaraj, M. and Vivekanandan, E. (1999) Marine capture fisheries of India: challenges and opportunities. *Current Science*, 76, 314-32.

Devaraj, M., Kurup, M. N., Pillai, N. G. K., Balan, K., Vivekanandan E. and Sathiadhas, R. (1997) Status, prospects and management of small pelagic fisheries in India. In Devaraj, M. and Martosubroto, P., eds., *Small pelagic resources and their fisheries in the Asia Pacific region*. Thailand: RAP publications, pp. 91-198.

Garcia, S. M. and Newton, C. (1997) Current Situation, trends and prospects in world capture fisheries. In Pikitch, E. K., Huppert, D. D. and Sissenwine, M. P., eds., *Global Trends: Fisheries Management*. Bethesda, Maryland: American Fisheries Society Symposium 20, pp. 2-27.

Government of India (2004) Marine Fisheries Policy. Department of Agriculture, Government of India: New Delhi.

Grafton, R. Q. and Nelson, H. W. (2004) The effects of buy-back programs in the British Columbia salmon fishery. *International Workshop on Fishing*

Vessel and Licence Buyback Programs, March 22-24. La Jolla: University of California.
Mullon, C., Fréon, P. and Cury, P. (2005) The dynamics of collapse in world fisheries. *Fish and Fisheries*, 6, 111-120.
Sathyapalan, J., Jeena, T. S., and Scholtens, J. (2007) Maintaining a viable small scale fishery: a case study of trawler sector in the Palk Bay. *Monograph No 5*. Hyderabad: CESS.
Scholtens, J. (2006) *Fishermen in a dire strait*. Bachelors program thesis. Amsterdam: University of Amsterdam.
Somvanshi, V S. (2001) Issues concerning marine fisheries and fisheries management in India. Fish code MCS\legal: report of the national workshop on fisheries management Goa, India, February 12-17, 2001. Rome: FAO, pp. 21-32.
Srinivasan, J. T. (2005) State regulation versus co-management: evidence from the Cochin Estuarine fisheries in India. *Environment and Development Economics* 10(1), 97-117.
Srivastava, U. K., Dholakia, B. H., Vathsala, S. and Chidambaram, K. (1991) Fisheries sector: review and analysis. In Srivastava, U. K., Dholakia, B. H., Vathsala, S. and Chidambaram K., eds, *Fisheries sector of India*. New Delhi: Oxford & IBH Publishing Co. Pvt. Ltd., pp. 24-59.
Vivekanandan, V. (2004) Fishing for a favour, netting a lesson: report of the goodwill mission of Indian fishermen to Sri Lanka. Thiruvananthapuram: South Indian Federation of Fishermen Societies.
Ward, J. M., Kirkley, J. E., Metzner, R., and Pascoe, S. (2004) Measuring and assessing capacity in fisheries: basic concepts and management options. FAO Fisheries Technical Paper No 433/1. Rome: FAO.

CHAPTER 4

Vulnerability and adaption of traditional fisheries to climate change

Abdellatif Khattabi and Guy Jobbins[1]

This essay discusses the implications of climate change for small-scale fisheries communities, drawing on the experiences of an adaptation research project in Morocco. It briefly introduces the physical impacts of climate change and conceptual approaches to understanding vulnerability, and then moves on to likely implications for fishing communities. The essay then explores how the resilience and adaptive capacity of fishing communities can be enhanced, particularly by improving infrastructure, strengthening social institutions, sharing information, and supporting fisheries communities through enabling policies, drawing on findings of the Moroccan cases study. The essay puts small-scale fisheries in the context of broader coastal zone issues and poverty reduction in the face of climate change.

Introduction

The scientific consensus is that the global climate is changing. It is unclear how long the global climate will continue to change, partly because the extent of future change depends on international actions to reduce emissions of greenhouse gases, such as carbon dioxide. However, even if all emissions were stopped today, the momentum of climatic changes already underway will continue for decades into the future.

Research indicates that the effects of a changing climate will be felt differently in different regions of the world. Mostly, the world will be hotter and wetter, although some places will experience more droughts, and others will have colder winters (IPCC, 2007a). Although climate change is most often thought of as occurring in the atmosphere, the oceans are also changing

1 Corresponding author: International Development Research Centre, Middle East and North Africa Regional Office, Egypt, GJobbins@idrc.org

due to the role they play in the global climate system. The consequences in marine and coastal areas will include more intense tropical cyclones, rising sea levels, an increase in storm surges, and changes in the intensity and timing of coastal upwelling.

Globally, numerous small-scale fishers live in poverty, and their livelihoods rely on natural resource systems which are already fragile or stressed. Climate change puts additional stress on the resilience of these communities, which are already facing considerable economic, social, and environmental pressures. This does not mean that coastal and fishing communities are helpless in the face of climate change, but it does mean that work is needed to help these communities to identify their vulnerabilities, to strengthen their resilience, and develop appropriate adaptation strategies.

The purpose of this essay is to briefly introduce and discuss climate change impacts, vulnerabilities and adaptation in a general way, and then to reflect on an adaptation strategy for fisheries prepared by the Adaptation to Climate Change in Morocco project (ACCMA)[2] in Morocco.

Climate change impacts and fisherires vulnerabilities

What is climate change?

The climate is the collection of patterns of temperatures, precipitation, humidity, winds, and seasons that exist in a place over a long period of time – usually measured in decades. It is distinct from weather, which is what happens on shorter timescales of days, weeks and months. Climate is shaped by latitude, the presence of nearby bodies of water and their currents, and landscape, and in turn shapes ecosystems and human economies that rely on them. Changes in climate can therefore have a complex set of consequences for human societies, potentially affecting food supply, availability of water, changing patterns of disease, and the frequency or intensity of extreme weather events, such as storms.

Moderate climatic variation over short timescales is a normal phenomenon. However, increasing research has demonstrated long-term trends showing that climate change is taking place, and that the magnitude of change may have profound consequences for human societies.

2 Adaptation aux Changements Climatiques au Maroc (http://www.accma-maroc.com/), supported by the Climate Change Adaptation in Africa Program, a joint initiative of Canada's International Development Research Centre and the United Kingdom's Department for International Development.

The Intergovernmental Panel on Climate Change (IPCC), a large-scale international scientific effort, released four successive assessment reports on climate change (1990, 1995, 2001 and 2007), with each report involving more scientists and more data, and reaching stronger conclusions about the evidence for, and implications of, global climate change[3] (e.g., IPCC, 2007a, 2007b). There is also increasing evidence that human emissions of greenhouse gases since the industrial revolution in the developed world are responsible for these changes.

The current indications are that, depending on actions taken to reduce greenhouse gas emissions and the rates of economic growth, average global temperature could increase between 1.1 and 6.4C, with a best estimate of 2 to 4C, by 2100, that sea levels will increase by up to 60cm, that there will be more frequent heat waves, heavy rainfall, and a strong possibility that there will be an increase in droughts, tropical cyclones and extremely high tides (IPCC, 2007a).

These findings are global, in the sense that they refer to the likely consequences to be most commonly experienced. However, the climate is a complex system, and increases in average temperatures may include colder winters as well as much hotter summers. As the global climate changes, it is likely to become less stable; meaning that patterns of winds and seasons may be disrupted, and that extreme weather events may become more intense and more common. The local effects of climate change, therefore, will be highly context dependent and cannot necessarily be deduced from projected global changes. What is clear is that the impacts of climate change will include long-term stresses (such as rising sea levels, increasingly warm summers, or changes in average annual precipitation) and also climate shocks (such as extreme storms, droughts, and floods). It also likely that gradual stresses could lead to sudden, irreversible changes, as thresholds are reached beyond which ecosystems cannot cope.

Physical impacts on marine and coastal environments

The marine environment has a central role in the regulation of global climate through the exchange of heat between the oceans and atmosphere. While most of the public discourse regarding global climate change has focused on the atmosphere, in terms of carbon dioxide levels and changes in atmospheric temperatures, the marine environment is also changing. Driven by absorption of heat and carbon dioxide from the atmosphere, the oceans are becoming hotter and more acidic, changes which will further

3 The IPCC's Fifth Assessment Report will be released in 2014.

stress marine ecosystems. Coastal areas, in particular, will increasingly feel the effects of rising sea levels, and changes in the intensity and frequency of weather events such as floods and storms.

These ongoing physical changes in the marine environment will have impacts on fisheries. Temperature is a major environmental factor determining the distribution of species in the ocean, and hence the assemblage of species found in a particular place. Observations since 1961 indicate that oceans have absorbed more than 80% of heat added to the climate system, and that increases in ocean temperatures have been observed to depths of 1500 metres (IPCC, 2007b). Scientific studies (e.g., Perry *et al.*, 2005) and numerous anecdotal accounts from fishers have observed species moving into new waters following temperature changes. This can affect ecosystem function through the introduction of new "invasive" species and through the disappearance of species with important roles in regulating productivity (e.g., Harley *et al.*, 2006).

Just as the exchange of heat between ocean and atmosphere drives the weather, it also drives marine currents. There is also evidence (e.g., Narayan *et al.*, 2010) supporting the Balukin hypothesis (1990) that climate change will affect the intensification of coastal upwelling systems. Any changes to the intensity, frequency, duration or timing of upwelling events could have significant implications for fisheries dependent upon them, particularly in Latin America and Africa.

The oceans also have an important role as a "sink" for atmospheric carbon dioxide, according to some studies absorbing about one-third of human CO2 emissions since 1800 (Sabine *et al.*, 2004). However, the increased concentration of carbon dioxide in the ocean is inducing chemical changes which result in greater acidification (Doney *et al.*, 2009). This is of concern because marine organisms, particularly at a larval stage, are sensitive to acidity, as it inhibits their ability to lay down skeletal material and develop normally. Coral reefs and marine plankton, the biological foundations of small-scale fisheries in developing countries, are particularly vulnerable (e.g., Donner *et al.*, 2005).

As the global climate warms, a long-term issue will be sea level rise. This is caused by the expansion of ocean waters due to heating and by increased melting of ice in glaciers and at the poles. The IPCC Fourth Assessment Report projected that sea levels would rise by between 18 and 59 cm by 2100[4]. Since the release of the report in 2007, further research has

4 Note that these projections do not include the contribution of rapid changes in ice sheets, as the authors could not agree on assessments.

indicated that these projections were conservative, and some have suggested that the range of sea level rise could be between 75 and 190 cm by 2100 (Vermeer and Rahmstorf, 2009). While less important than changes in temperature and acidification for the productivity of marine fisheries, sea level rise will affect coastal fishing communities. Coupled with expected changes in the frequency and intensity of weather events, sea level rise will threaten physical infrastructure, such as harbors, jetties, buildings and roads, and coastal ecosystems and agricultural areas, amongst other possible impacts.

Understanding vulnerability

There are two principal approaches to thinking about vulnerability to climate change (for a good review of concepts on climate vulnerability and adaptation, see Smit and Wandel, 2006). The first is drawn from the field of managing disaster risk, and views vulnerability as a function of exposure to a physical impact, the degree of sensitivity to an impact, and the capacity to adapt to the impact. For example, if considering the vulnerability of a community to coastal flooding, one could examine exposure in relation to the village's height above sea level, projections of sea level rise for the local area, local wave dynamics, and the shape and slope of the coastline. Sensitivity might relate to how many people live in the flood zone, whether crops and fields are located in the flood zone, how remote the village is, how much property and critical infrastructure might be destroyed or damaged, whether people in the village are wealthy enough to rebuild, and so on. Adaptive capacity might include factors such as the village's ability to, or influence the government to, construct flood protection infrastructure, develop an early warning system, build a walled harbor, relocate farmlands, change the designs of dwellings, and so on. This approach to understanding vulnerability requires recognizing the consequences of different kinds of impacts, as a community may not be particularly vulnerable to coastal flooding yet face devastation if fishing yields were to drop.

An alternate approach to vulnerability is to consider it as inherent to a community, rather than as resulting from a specific physical impact. This approach tries to systematically understand how social conditions, historical circumstances, and the political economy of individuals, organizations, and institutions prepare communities to cope and adapt to change (e.g., Adger, 1999). For example, a community without institutions such as a village council, a fishers' union or cooperative, or even a place of worship, is less able to decide and act collectively to address problems. One advantage of the *social vulnerability* approach, as this is known, is that it is independent of specific climate impacts, and asks broader questions about the ability of a community to cope with change. This is important because climate change is not the only stress communities are facing. In fact, the impacts of a changing climate will interact with changes in demographics, markets, technology, social pressures, and many other factors, in ways which cannot always be anticipated.

The two approaches are not necessarily exclusive, and many researchers have combined and refined these conceptual approaches. An example is assessments of 'place' vulnerability, which embed the understanding of social vulnerability in its geographical context, and usually include physical impacts (e.g., Cutter *et al.*, 2000; Wu *et al.*, 2002). One key lesson that has been learned from research projects on climate change is that adaptation, and vulnerability, to climate change is local and context-specific (e.g., Carter and Jobbins, 2009). Hybrid models can be appropriate tools for examining the vulnerabilities of a specific target community, which is an important first step to understanding how they can best strengthen resilience and develop adaptation strategies.

Implications of climate change for fishing communities

Although each community will have its own vulnerabilities, due to social and geographic contexts, it is useful to consider the types of vulnerabilities that small-scale fisheries may experience in general. There are multiple pathways by which climate change might affect coastal fishing communities. In addition to the impacts outlined above – changes in marine productivity and coastal flooding – others might include changes in supplies of potable water, disease burdens, or the availability and price of food crops. It is important to stress that not all communities will experience the same impacts, and that not all changes will necessarily be negative. Successful adaptation to climate change is about taking advantage of new opportunities, as much as it about responding to threats. Developing appropriate adaptation strategies requires understanding how changes in different factors – marine productivity, food availability, disease, inundation of lands, and so on – combine to create both opportunities and vulnerabilities.

Fisheries at high latitudes, and in ecosystems such as upwelling systems, coral reefs, rivers and lakes, appear most sensitive to climate-driven changes in productivity. We can expect that fisheries already stressed by pollution, overfishing and declining stocks, loss of nurseries, and other factors of environmental change will be sensitive to additional stresses resulting from climate change and ocean acidification. From the perspective of fishing communities, vulnerabilities relate to their ability to cope with changes in yields, catch species, and so on.

Fishing communities established in deltas, low-lying islands and coral atolls are particularly exposed to rising sea levels, with the risk of flooding, saline intrusion and coastal erosion. The vulnerability of such communities is highly context-specific. Some areas have been left more vulnerable by the loss of mangroves, coral reefs, and wetlands that act as natural sea defenses, controlling erosion and buffering wave energy. Some communities have already adapted to floods by constructing stilt houses or building on high ground. Indeed, the livelihood strategies of many agrarian communities include fishing during floods as a means of survival (e.g., Moench and Dixit, 2004).

A crucial point in understanding the vulnerability of fishers is that, in general, small-scale fishers are socially, economically, and politically marginalized. Poverty reduces the ability of people to weather crises, and means that they have fewer resources to adapt to new conditions. While this is true for individuals and communities of small-scale fishers, it is also generally true for the nations in which they live. According to a recent study by Allison *et al.* (2009), of the 33 nations most vulnerable to the effects of climate change on fisheries, 22 are Least Developed Countries, primarily in Africa. People in these most vulnerable nations were twice as dependent on fish protein in their diet as people elsewhere (27% versus 13%, respectively). Fisheries also made significant contributions to their economies through earnings from exports, employment in fishing, processing, marketing and retail sectors, as well as providing a social safety net and vital nutrition for millions of the poorest members of society.

In the future, climate change will mean that communities dependent on fisheries are likely to experience increased vulnerability due to greater volatility in catches and incomes, changes in quantities and/or quality of fish for consumption or sale, and risks to their safety if, for example, they are forced to fish in heavy weather or farther from their home base. Generally speaking, these impacts will lead to changes, both positive and negative, in the price of production and marketing, the sale prices of products from capture fisheries and aquaculture, and a heightened risk of damage to, or loss of, infrastructure, tools and dwellings.

It is important to reiterate that the effects of climate change will interact with other changes affecting fisheries, such as changes in technologies, global markets, populations, and the pressures of declining fish yields. Small-scale fishing communities already under pressure may not be directly challenged by the long-term changes implied by climate change for some time. Other stresses they experience in the next two to three decades are likely to be more pressing than the direct effects of temperatures on ecosystems, sea level rise, and ocean acidification. However, these pressures will continue to mount and, as the long-term stresses accumulate, communities will be exposed to the increasing effects of climate variability: more intense storms and floods, changes in upwelling systems, and so on. There is, therefore, both a long- and a short-term dimension to the implications of climate change, and in the responses needed to safeguard fishing communities.

Identifying solutions

Adaptation, adaptive capacity, and resilience

There are a number of different means for reducing vulnerability to the effects of climate change and increased climate variability. They might include national or regional strategies, or concrete measures taken by communities or individuals. Strategies might anticipate impacts, or be implemented remedially, after impacts occur. They might be planned as the result of political decisions, or they might occur spontaneously by individuals and groups, perhaps as a result of trial and error.

In comparison to remedial actions, anticipatory measures are more likely to be effective and avoid the full costs of suffering impacts in the first place. However, it can be difficult to build consensus for actions to address threats that lie in the future, especially when such actions are costly. Particularly for individuals, communities and nations already struggling with poverty, addressing immediate development challenges is, understandably, more likely to take priority over future problems, especially if there are remaining uncertainties about the local consequences of climate change.

Initiatives for adaptation to climate change should not waste resources. Perhaps more importantly, they should not exacerbate development challenges, such as poverty or lack of opportunities. One approach is to identify "no regret" solutions, particularly for planned anticipatory measures. A "no regret" strategy is one which has advantages regardless of future climate change, and often addresses key drivers of vulnerability affecting a community, rather than implementing a specific adaptation to a particular hazard.

Attempts to reduce vulnerability can be categorized into three broad, sometimes overlapping, groups. The first consists of actions to build resilience. In the second category are measures strengthening the community's capacity to adapt. The third category consists of adaptations; specific actions to address particular climate impacts.

Actions to strengthen resilience are intended to increase the capacity of communities to cope with adverse circumstances, and reduce their vulnerability in general. Resilience is often treated as an intrinsic part of adaptive capacity, as it reflects the extent to which a community can be unaffected by change, and it provides a buffer of time during which adaptation can take place. The ability of small-scale fisheries communities to cope with crises depends on their access to markets and credit, access to government services and transport services, physical security, sources and diversity of income, healthcare and nutrition, and many other factors that, in short, are good indicators of social and economic development. It follows that many actions to increase resilience to climate change and climate disasters are similar to actions to improve socioeconomic conditions in general, although the priorities would be identified through a climate vulnerability analysis. A key point in considering the resilience of fishing communities is their dependence on the fishery itself. Actions which strengthen ecological resilience, such as improving fisheries management and enforcement, can also improve the community's resilience.

Strengthening adaptive capacity relies on supporting key factors that shape how a community responds to change. Clearly, one aspect of this is having sufficient resources to invest in new technology, infrastructure and so on, but there are other factors that are equally important. Key amongst them are processes and institutions for creating, sharing and using information. This is to say that, in response to change, people are required to innovate. Building adaptive capacity might include actions such as developing ongoing collective experiments, implementing monitoring and evaluation processes, improving levels of education, improving access to and use of climate information, creating or strengthening institutions for decision-making, or building relationships with research institutions, government departments and other stakeholders (e.g., Folke *et al.*, 2002). Again, actions strengthening adaptive capacity are similar to the kinds of actions that might improve social and economic conditions more generally.

Specific adaptations, whether planned or spontaneous, are the outputs of adaptive capacity, and respond to specific changes. As sea levels increase, for example, some will adapt by constructing sea walls and dikes, others will adapt by retreating away from the shoreline. Changes in seasonality, yields, catch species, and so on will prompt fishers to adapt their gear and strategies in response. Successful adaptations can be defined as changes in technology or behavior which mitigate or avoid problems without further increasing vulnerability. By contrast, maladaptation is a response to change that increases vulnerability. An example of maladaptation might be, for example, responding to decreasing fisheries yields by using destructive techniques such as smaller mesh sizes, or dynamite fishing. When considering adaptations to sea level rise and flooding, the development of infrastructure and planning for coastal protection has to be carefully assessed so as not to increase vulnerability. Sea walls will, when breached, stop waters from draining out again. Engineers and planning laws typically specify the level of risk that such protection must be built to withstand. This is usually expressed as an event that might happen once every century, or once every thousand years. However, in a changing climate, with increasingly variable weather and extreme events, what was a one in one-hundred year event may become a one in fifty, or one in ten, year event. This is an important reason for strengthening adaptive capacity rather than focusing solely on specific adaptations. Adaptation to climate change is best understood as a process requiring continuous innovation rather than a series of final, completed adaptations.

Adaptation to climate change in Northeast Morocco

Having discussed adaptation, adaptive capacity, and resilience in general, we now present elements from an adaptation strategy defined for a fisheries community in Northeastern Morocco, as part of the ACCMA project. This project, conducted in the coastal zones of Nador and Berkane provinces between 2007 and 2010, looked at coastal adaptation in general, not just for fisheries. However, fishing is an important activity in the region, and Morocco's fisheries are considered vulnerable to climate change (Allison *et al.*, 2009).

The coastline of Nador-Berkane has some 45 artisanal fisheries sites, and includes the Lagoon of Nador which is a major area of fishing activity. More than 900 boats were registered in the region in 2007, only 20% of which were located in the area's three fishing ports. The remaining 80% of vessels were distributed amongst other sites on the coast, marked by a complete lack of basic infrastructure and trade. Boats are very homogenous, ranging from 3m to 6.7 m in length and an average age of 16 years. Most boats have small outboard engines between 8 and 15 horse power, although larger vessels can have larger engines, while the poorest fishers may rely only on oars. Three main types of gear are in use: trammel nets, the pallanza (a form of set net), and the beach seine. It is a mixed fishery, with eels, anchovies, seabream, mullet, cuttlefish, and octopus all important catch species.

Unlike some other areas of Morocco, fishing in this region is an entirely male activity with women playing no role in the supply chain. The fishing population is relatively poor and less likely to be literate than the regional average, and it is aging, with approximately 70% of young fishers leaving the profession in the last ten years. The importance of fishing to livelihood strategies varies locally. In some areas, particularly the lagoon of Nador, fishing is generally an exclusive profession. In others, especially along the less hilly eastern part of the coast, fishers combine fishing with subsistence agriculture or small commercial activities.

In a survey of 224 fishers conducted at the beginning of the research, 74% expressed concern about climate change, noting changes to regular weather patterns in recent years. In particular they reported increased frequency of floods, changes in seasonal patterns, such as crop harvests and fish spawning, and wind patterns. The adaptation strategy was developed with the participation of fishers, government officials, academics, and non-governmental organizations, and fishers' cooperatives and associations. In a series of workshops, participants were provided with downscaled climate change projections, and then invited to identify potential impacts on the community, discuss adaptation options, and identify resources available in the community and those needed from outside. Key criteria for agreeing on the elements of the strategy were that they be socially acceptable, and economically and technically feasible. The resulting plan systematically addressed livelihoods, including socioeconomic development, ecosystem protection, infrastructure development, and reduction of vulnerabilities to climate change.

Construction of landing points

Along the coast only three ports cater to artisanal fishers. Fishers identified the construction of designated landing points as key to helping them reduce journey times, increase safety, and support marketing of quality fish. Such landing sites should be strategically located to allow as many fishers as possible to use the facilities. These would include slipways and maintenance areas, supporting infrastructure such as access roads, drinking water systems and telephones, rooms for storing supplies, and administrative offices for the authorities to monitor catches and provide training. The facilities would also include important infrastructure relating to marketing, such as cold rooms, ice production and storage, and a fish market. The fishers also identified the importance of having a social space, such as a cafeteria or prayer room, to create a sense of belonging and identity for the users. The provincial fisheries authorities agreed to construct two such landing points, financed by national endowments for development.

Development of retail outlets

Improving income from fishing was consistently identified as a key need and, in addition to the landing points discussed above, stakeholders identified the development of retail outlets as important. Due to their distribution along the coast, and the cost of transportation, most fishers currently sell at reduced prices to wholesalers who come to their landing sites or in the port of Beni Nsar. The development of retail cooperatives could improve prices by selling directly to consumers in towns, while reducing transport costs. Establishing retail outlets properly equipped with storage facilities would also allow better compliance with a 2005 governmental decree on the sanitary inspection of seafood, ensure high quality products, and improve consumer confidence.

Supporting infrastructure

Nador-Berkane is prone to heavy flooding during the winter season, which can leave communities in hilly areas isolated for up to weeks at a time. Floods have increased in frequency and intensity in recent years, with severe impacts on the affected communities. Stakeholders identified access to markets and healthcare as key priorities during such times, attributing several childbirth deaths to the inability to access medical clinics or midwives. Flood protection for road networks, including basic tarring and surfacing of vulnerable rural roads, was identified as a priority.

Port redevelopment

Of the three ports in the area developed for artisanal fishing, two were highlighted as needing improvements to properly protect vessels from the effects of increasing storms. In the port of Sidi Hsaine additional breakwaters were needed; while at Sidi Ali port, modeled projections of increased wind and wave intensity in the future meant that work would be needed to raise the existing breakwaters.

Access to subsidies for equipment

Fishers identified a number of areas in which their working conditions could be improved by new equipment. This included equipment for the refrigeration, storage and transport of catches. It also included the modernization of the fleet by replacing traditional wood-constructed boats with modern fiberglass designs which are both lighter and more stable at sea, improving safety and reducing fuel costs. A key concern was improving access to credit and subsidies for upgrading equipment.

Improved fisheries management

Stakeholders stressed the importance of better managing fisheries to increase their resilience to the impacts of climate change. A number of issues and recommendations were identified, including improved monitoring and enforcement to reduce illegal fishing and trawling in spawning grounds, the restoration of key habitats including wetlands, and establishing a network of small marine protected areas or "no-catch" zones. The potential for artificial reefs and no-catch zones to improve fish stocks was highlighted as an option for experimental trial.

Another question was whether changing temperatures and other environmental conditions was changing spawning times, therefore requiring a change in the regulations for the rest period of the fishery. Fishers, scientists and officials agreed that this was a possibility, but that too little was currently known and that conditions could change in the future. They concluded that a process of adaptive management linked to continued research would be required.

Protection of Nador Lagoon

The Nador Lagoon is an important fishing site, fish breeding ground and nursery, and an area of natural heritage. The lagoon also provides a natural sea defense for the city of Nador. It is, however, subject to pollution by sewage, industrial waste, and agricultural discharges. The disappearance of *Posidonia oceanica*, a seagrass, from the lagoon concerned many fishers because of its importance in primary production.

Other stakeholder groups consulted during the project also raised the issue of the Nador Lagoon, and a plan was developed to reduce pollution in the lagoon. This will involve the construction of a water treatment plant, a drainage system and diversion channels, shoreline cleaning, and a program with farmers to reduce agricultural pollutants including pesticides. The plan is now being developed as a major project with support from an international donor.

Diversification of sources of income for fishers

Reducing the vulnerability of fishers to income volatility was another priority. In addition to fluctuations in catches and market prices, fishers were vulnerable to losing their vessels and gear in bad weather. Poor fishers were particularly vulnerable, as they were under more pressure to risk fishing in bad weather, generally had slower and less seaworthy boats, and could not afford to replace lost equipment. These issues were more pressing for fishers around the lagoon and the western, hilly part of the coast. Two possible options were identified for diversification of their income: ecotourism and agriculture.

The potential for ecotourism was considered strongest in three key areas of outstanding natural beauty which are listed in tourism development plans for the region. The proximity to Europe and the support of the Moroccan government mean that tourism growth in the region is expected to be strong. Fishers could earn extra income from working as fishing guides and possibly renting accommodation to hikers in the remote Cap des Trois Fourches. However, it was recognized that training would be necessary, and that the Ministry of Tourism's cooperation must be enlisted to ensure that benefits to fishing communities were included in tourism planning.

Historically, the hilly terrain, limited fresh water supplies, lack of access to markets, required labor, and land rights issues have been constraints on fishers adopting agriculture in the region. However, advances in water management and soil conservation, and improved transport linkages have reduced some of these obstacles. Some saw the growing and future demand for niche products such as organic agriculture as an opportunity that could be exploited by gaining support from local agricultural officials and forming a cooperative with local farmers.

Strengthening the organizational capacity of fishing communities

Stakeholders recognized that fishers' cooperatives had multiple benefits, including marketing, access to microloans and credit, and providing a channel for communicating with the authorities. Fishers in the region should be encouraged to create new cooperatives, and those cooperatives which already exist should be further strengthened. In addition to increasing access to credit, fishers also prioritized safety training and improved fishing techniques, strengthening relationships with fisheries and other state authorities, and the development of codes of conduct for artisanal fishers. The potential role of cooperatives in promoting general education and literacy, access to scientific information and research institutions on fisheries and climate change, and in helping enforce regulations was also highlighted. The promotion of literacy and general education was heavily prioritized by the young, who were less committed to fishing as a life-long occupation.

Accompanying measures

The strategy also identified a range of supporting policies and actions at local and national levels to reduce the vulnerability of fisheries communities. These included the implementation of integrated coastal zone management, greater consultation of fisheries organizations by authorities responsible for coastal development and planning, more research and engagement from research institutions, and greater availability of climate information.

Reflections

The strategy developed by the ACCMA project reflects local conditions, in which climate change is expected to place a gradual stress on the fishery, rather than lead to abrupt change. Most of the options identified by the strategy are not specific to addressing climate vulnerabilities, but reflect development constraints already impacting fishers. Addressing these priorities will have immediate benefits to the community and their capacity to cope with and adapt to future changes.

Of the climate-specific priorities, the majority concerned protection from storms and floods, issues which already affect the communities, and which are expected to worsen in the future. Beginning to address them now is a no-regrets strategy, as it will have both short- and long-term benefits. Likewise, conducting research into spawning times and developing adaptive management is a potential gateway to involving fishers in research more broadly, and building consensus for fisheries management.

The strategy contains a mix of specific adaptations (flood defenses, port redevelopment), options to increase resilience (income improvement and diversification, improving safety, improved fisheries management and enforcement), and measures for building adaptive capacity (strengthening fisheries organizations, increasing research). The strategy is also anticipatory. In addition to considering current challenges faced by fisheries, introducing the question of future changes in climate acted as a catalyst for thinking through other changes likely to happen in the area, such as increased urbanization, aging of the fishing community, and changes in markets as the regional economy grows. The strategy also recognizes that while fishers and fishing communities have agency and the ability to strengthen their own organizations and institutions, they are also reliant on others. The future of fisheries in the region will depend on choices regarding coastal planning, pollution control, the protection of coastal environments, and a host of other decisions by a range of government authorities and other stakeholders.

It is not yet clear what the outcomes of the ACCMA fisheries strategy will be for Nador and Berkane, or whether the fishing communities there will be resilient and adapt to climate change. However, the elements of this strategy give them the best options that can be determined at the moment, and these can be reviewed and updated in the future, as information and predictions improve. As the strategy identifies actions that also address current needs, it has been easier to build consensus and develop momentum for adaptation.

In the coming decades, climate change will further stress many fishing communities already under pressure. It is probable that communities already stressed by resource degradation, poverty, and social and political marginalization will be the most affected by climate change impacts. In the short-term, climate change impacts may be less significant than more immediate pressures, and the incentives for adaptation may appear marginal. However, we suggest that any effort to improve the sustainability, livelihoods, natural resource management, and resilience of fisheries communities should also consider climate vulnerabilities and how to reduce them, to ensure that communities are best prepared to meet the uncertainties of the future.

References

Adger, W. N. (1999) Social vulnerability to climate change and extremes in coastal Vietnam. World Development, 27 (2), 249-269.

Allison, E. H., Perry, A. L., Badjeck, M. C., Adger W. N., Brown, K., Conway, D., Halls, A. S., Pilling, G. M., Reynolds, J. D., Andrew, N. L. and Dulvy, N. K. (2009) Vulnerability of national economies to the impacts of climate change on fisheries. *Fish and Fisheries*, 10(2), 173-196.

Carter, S. and Jobbins, G. (2009) Le programme ACCA: renforcer la capacité d'adaptation grâce à la recherche-action participative. Québec: Institut du développement durable et des relations internationales (IDDRI) et Institut de l'énergie et de l'environnement de la francophonie (IEPF).

Cutter, S., Mitchell, J. T. and Scott, M. S. (2000) Revealing the vulnerability of people and places: a case study of Georgetown County, South Carolina. *Annals of the Association of American Geographers*, 90(4), 713-737.

Doney, S. C., Fabry, V. J., Feely, R. A. and Kleypas, J. A. (2009) Ocean acidification: the other CO_2 problem. *Annual Review of Marine Science*, 1, 169-192.

Donner, S. D., Skirving, W. J., Little, C. M., Oppenheimer, M. and Hoegh-guldberg, O. (2005) Global assessment of coral bleaching and required rates of adaptation under climate change. *Global Change Biology*, 11, 2251-2265.

Folke C., Colding, J. and Berkes, F. (2002) Building resilience for adaptive capacity in social-ecological systems. In Berkes F., Colding, J. and Folke,

C., eds., *Navigating Social-Ecological Systems: Building Resilience for Complexity and Change*. Cambridge: Cambridge University Press.

Harley C., Hughes A., Hultgren K., Miner B., Sorte C., Thornber C., Rodriguez L., Tomanek L. and Williams S. (2006) The impacts of climate change in coastal marine systems. *Ecology Letters*, 9, 228-241.

IPCC (Intergovernmental Panel on Climate Change) (2007a) Climate Change 2007: Impacts, Adaptation and Vulnerability, Contribution of Working Group II to the Fourth Assessment Report of the Intergovernmental Panel on Climate Change. Parry, M. L., Canziani, O. F., Palutikof, J. P., van der Linden P. J. and Hanson, C. E., eds. Cambridge: Cambridge University Press.

IPCC (Intergovernmental Panel on Climate Change) (2007b) The Physical Science Basis, Contribution of Working Group I to the Fourth Assessment Report of the Intergovernmental Panel on Climate Change. Solomon, S., Qin, D., Manning, M., Chen, Z., Marquis, M., Averyt, K., Tignor, M., and Miller, H. L., eds. Cambridge: Cambridge University Press.

Moench, M. and Dixit, A., eds. (2004) Adaptive capacity and livelihood resilience: adaptive strategies for responding to floods and droughts in South Asia. Boulder, Colorado: Institute for Social and Environmental Transition (ISET) UEA/Nepal.

Narayan, N., Paul, A., Mulitza, S. and Schulz, M. (2010) Trends in coastal upwelling intensity during the late 20th century. *Ocean Science*, 6, 815-823.

Perry, A. L., Low, P. J., Ellis, J. R. and Reynolds, J. D. (2005) Climate change and distribution shifts in marine fishes. *Science*, 308, 1912-1915.

Sabine, C. L., Feely, R. A., Gruber, N., Key, R. M., Lee, K., Bullister, J. L., Wanninkhof, R., Wong, C. L., Wallace, D. W. R., Tilbrook, B., Millero, F. J., Peng, T. H., Kozyr, A., Ono, T. and Rios, A. F. (2004) The oceanic sink for anthropogenic CO2. *Science*, 305, 367-371.

Smit, B. and Wandel, J. (2006) Adaptation, adaptive capacity and vulnerability. Global Environmental Change, 16(3), 282-292.

Wu, S. Y., Yarnal, B. and Fisher, A. (2002) Vulnerability of coastal communities to sea-level rise: a case study of Cape May County, New Jersey, USA. *Climate Research*, 22(4), 255-270.

Vermeer, M. and Rahmstorf, S. (2009) Global sea level linked to global temperature. *Proceedings of the National Academy of Science*. 106(51), 21527-21532.

CHAPTER 5

What restructuring? Whose rationalization? Newfoundland and Labrador's Memorandum of Understanding on its fishing industry

Deatra Walsh[1]

In July 2009, Newfoundland and Labrador's Department of Fisheries and Aquaculture, the Association of Seafood Producers, and the Fish Food and Allied Workers' Union signed a Memorandum of Understanding (MOU) aimed at designing plans for the restructuring and rationalization of an industry argued to be over capacitated and economically unsustainable. For nearly 18 months, working groups met to discuss and decipher industry proposals, financial analyses, and marketing possibilities. In February 2011, the final MOU report was released and promptly rejected by the Provincial Government on the grounds that it did not adequately address restructuring, focused too much on rationalization, and did not contain a long-term vision for the industry. This essay traces the history of events that led to the MOU process, describes the responses of different parties to the result, and reflects upon the meaning and reasoning behind this result. I argue that to make sense of the dispute that followed the report's release it is important to address the lack of clarity on what restructuring meant as a concept throughout the process. While rationalization is a type of restructuring, the brand of rationalization proposed in the MOU does not necessarily represent the vision of all of those participating in and affected by the industry.

Introduction

Located adjacent to the Atlantic Ocean, Canada's easternmost province of Newfoundland and Labrador has relied on the fishery as its economic core for

1 Corresponding author: York University, Canada, deatraw@yahoo.com

hundreds of years. While oil and gas development have recently taken center stage in the provincial economy, the fishery remains an important source of employment for tens of thousands of people working directly as harvesters or as fish plant workers, and for thousands of others employed in management, as union officials or within the shipping and distribution sectors. Like other fisheries throughout the world, Newfoundland and Labrador's industry is not without its problems. It is challenged by a dual management structure whereby the Federal Department of Fisheries and Oceans (DFO) conducts fisheries science, manages stocks, sets harvesting quotas and establishes policies, while the provincial Department of Fisheries and Aquaculture (DFA) sets policy and practice in the processing sector. More importantly perhaps, it is challenged by ongoing concerns regarding resource availability and the structure of the industry itself. In many ways, the Newfoundland and Labrador fishery is nearly always in the midst of crisis (see Ommer, 1994).

In July 2009, following a "shrimp crisis" due to price disputes, the economic recession, and the 2006 federal-provincial Fisheries Industry Renewal Strategy, the DFA signed a Memorandum of Understanding (MOU) with the Association of Seafood Producers (ASP) and the Fish, Food and Allied Workers' Union (FFAW). Led by representatives from each group and an independent chair from Memorial University's Department of Business Administration, the MOU steering committee oversaw four working groups tasked with soliciting proposals from industry stakeholders on ways to rationalize and restructure the inshore and nearshore fisheries. These activities focused on generating and deciphering financial information about the harvesting and processing sectors, and exploring seafood sales and marketing options. It was "designed to provide the level of analysis required to inform the debate on the rationalization and restructuring initiatives necessary to ensure the long-term stability of the province's fishing industry" (Report of the Independent Chair, 2011, p. i). DFA provided the funding to support this initiative, including tendering two consulting firms to conduct financial analyses in the processing and harvesting sectors. The independent chair was responsible for writing the final report. While representatives of DFO did sit on the various working groups, the department was not an official signatory on the MOU.

Overdue by nearly six months, the final report was released in February 2011. To the surprise of the ASP and the FFAW who were under in the impression that they had done what was requested of them, the report was formally rejected by the provincial Minister of Fisheries Clyde Jackman and Premier Kathy Dunderdale on the grounds that it focused too much on rationalization, did not adequately address restructuring, and contained no vision for the long-term future of the fishery in the province (Hunt, March 8, 2011). With MOU members unable to agree on how to move forward, the MOU remained at an impasse as of March 2011 (Fitzpatrick, March 22, 2011).

In this essay, I discuss the history of events that led to the MOU process, describe the responses of different parties to the result, and reflect upon the meaning and reasoning behind this result. I pay close attention to the discourse of rationalization and restructuring as it relates to this report and the events preceding it. Using materials from media coverage and press releases before and after the MOU release, the document itself, and post-release reactions, I argue that to make sense of the dispute that followed the report we should examine the lack of clarity on what restructuring meant as a concept. While rationalization is a type of restructuring, the brand of rationalization proposed in the MOU does not necessarily represent the vision of those participating in and affected by the industry. Instead, the proposed rationalization appears to target small-scale fishing enterprises in rural areas, thus supporting larger scale operations argued to be more equipped to compete in the global marketplace. I ask if this form of rationalization is the only solution to Newfoundland and Labrador's fishing industry woes, or if another restructuring vision is possible.

The essay is divided into five sections. I begin with a brief discussion of research on industrial restructuring; then move to a contextual introduction on change in the Newfoundland and Labrador fishery since 1949. In the next section, I introduce the MOU, and the events surrounding its instigation. This is followed by an examination of its demise. In both sections, I trace the use of the language of renewal, restructuring, and rationalization. I have chosen language as one means with which to enter this discussion because it is eventually language, and thus meaning, which becomes the bone of contention for the MOU's rejection. In conclusion, I ponder the language-based restructuring-rationalization debate that has pervaded media coverage of industry and government relations regarding what to do next with the MOU, and speculate on other political circumstances that may have led to this outcome.

Getting a handle on restructuring

Restructuring, understood generally as reorganization related to various sectors of the economy, has been a common process throughout the 20th and 21st centuries – but for different reasons and with different circumstances related to capital, production, and labor. We can look most readily to

Marx ([1887]; 1995) and more recently to Polanyi (1944) for perhaps the most succinct macro-theoretical accounts of the shift from small-scale production methods, based on kin groups and local geographies, to wider scale industrial capitalistic models based on mass production throughout the 18 and 1900s. The rise of Fordism in the 1940s and 50s, with its "huge factory buildings, heavy machinery and massive labour forces" (Bauman, 2000, p. 57), marked a further shift in this direction, propelling industrial expansion throughout the world and in western countries of the global north in particular. Post-war Fordist-style development represented the heyday of employment possibilities and opportunities, and was fuelled by massive resource extraction in mining, forestry, oil and gas, and fisheries throughout the world to feed a growing manufacturing sector, rapidly expanding urban populations and, thus, increased consumer demand. Fordism evolved hand-in-hand with the development of monopoly capitalism as well as the rise of the Keynesian welfare state (Jessop, 1992; 1994).

For employers, Fordist-type restructuring ensured wealth accumulation. For employees, the industrial shift led to more stable and better wages, and the onset of organized labor that protected and represented their interests. It marked the beginning of what is now known as the standard employment relationship, where workers have one employer, statutory benefits and entitlements, and work year-round on the employer's premises with the expectation that this will continue indefinitely (Vosko, *et al.*, 1997; Fudge and Vosko, 2001). Bauman (2000) has described this as a time of "heavy capitalism", such that infrastructure, individuals, and labor were predominantly immobile and perceived as both solid and everlasting in their ability to provide the right conditions for a better economy and working life.

Throughout the latter part of the 20th century, industries began to shift their activities and focus from mass production and mass labor to leaner, tighter and "lighter" (Bauman, 2000) operations that were more flexible, from both a capital and human resource perspective. This shift was associated with technological changes, the erosion and transformation of the welfare state, and industrial and policy changes designed to increase the mobility of global capital and, with that, employment through reduced trade barriers and larger processes associated with what is now known as globalization (MacDonald *et al.*, 2008).

The conceptualization of Fordism and post-Fordism, as separable eras in our recent economic history, has afforded us a language for examining industrial and related employment change in the post World War II era (McDowell, 1991). As an academic concept, restructuring emerges from left-wing writing and thinking focused on making sense of this change as it relates predominantly to class and gender relations. Massey and Meegan (1982) identified three forms of restructuring or production reorganization associated with post-Fordism in particular. Intensification includes minor technological improvements or the reorganization of how materials are processed, but does not feature significant investment. Investment/technological change includes substantial capital inputs into new or existing structures in an effort to make them more efficient. The final form of post-Fordist restructuring is rationalization, an industry-wide reduction in production capacity, usually through plant and other closures. Drawing on Massey and Meegan's work to examine plant closures and the use of rationalization as a competitive market response, Maclachlan writes that it "is undertaken to eliminate excess capacity and increase the rate of profit by permitting the redeployment of proceeds from disinvestment to more profitable activities" (1992, p. 130). Massey and Meegan's work reminds us that rationalization is only one form of restructuring. Alternative strategies include combinations of intensification, investment, and rationalization in different segments of regional and now global production systems.

A quick overview of restructuring processes in the Newfoundland and Labrador fishing industry leading up to the MOU process shows that, in recent years, the dominant discourse has focused on rationalization and the argument that it can lay the groundwork for the development of a stable, profitable industry less prone to recurrent crises, less reliant on government support, and thus more likely to attract private investment and generate stable livelihoods for workers and fish harvesters. There are also counter discourses from some researchers, community groups and, to some degree, from union representatives which argue for a sustainable livelihoods approach focused on the coping and adjustment strategies of households within the context of industry change; a bottom-up localized knowledge approach to management; and the role that fisheries play in rural and coastal communities (Allison and Horemans, 2006). This is akin to other writings on resilience and restructuring in coastal areas (see Ommer and Sinclair, 2006; Ommer, 2007) critical of the economistic stance emphasizing pure rationalization. In these counterarguments against neoliberal policies or trickle-down economics, emphasis is on the risks to rural communities of privatized fisheries and the risks to the resource sustainability.

The Newfoundland and Labrador fishery

Since 1949, the Newfoundland and Labrador fishery has undergone rapid change. In fact, the Fordist to post-Fordist shifts which occurred over the span of 100 or more years elsewhere in the world actually occurred in half that time here. In 1949, Newfoundland and Labrador officially joined Canada as its 10th province. Among other modernizing developments at that time, including communication and transportation infrastructure, a widening public service, and the pursuit of new ventures in economic diversification, the province's fishery underwent industrialization (Cadigan, 2003; 2009). Wright (1998) argues that events preceding 1949 also contributed significantly to this industrialization shift. These included a war-time demand for fish, the technological advancement of refrigeration, and a growing number of urban consumers who demanded more pre-packaged, processed, and often frozen food (see also Kurlansky, 1998). Investment in industrialization and expansion was made possible, however, by the federal and provincial governments. Both the processing and harvesting sectors benefitted from increased capital (Sinclair, 1987), with the state's belief that these investments would modernize an otherwise "backward" industry (Wright, 1998) that, prior to this time, depended on fisher-merchant relations, and independent commodity production (Sinclair, 1983; Clement, 1986; Ommer, 1994). New gear and faster boats increased the capacity of inshore harvesters (Neis, 1991) and the building of an offshore trawler fleet, along with the foreign fishing fleets already harvesting outside of the Canadian 12-mile zone, meant that offshore harvesting also increased. With the introduction of the 200-mile limit in 1977, expansion in the industry occurred after a period of decline (Shrank, 2005).

Despite expansion throughout the 1950s and 60s, and again in the late 1970s, restructuring, increased regulation, and reduced capacity were recommended to address what seemed to be constant crises in the fishery. Throughout the 1970s and 80s, stock stability was an increasing concern for those in the industry. After inshore stocks fell in 1982 and offshore stocks showed decline in 1986, DFO began to question its stock population estimates (Hutchings *et al.*, 1997; see also Shrank, 2005). Shrank (2005) identifies three reports commissioned between 1969 and 1982 which called for reductions. Following the closure of the Northern Cod fishery in 1992 after the collapse of the resource, the Cashin report recommended a 40-50% reduction in harvesting and processing capacity (Shrank, 2005). Adjustment programs leading up to and following the moratorium were introduced by the Canadian federal government to assist fishers to get out of the industry, and to support displaced workers affected by cuts in the industry. These included the 1989 Atlantic Fisheries Adjustment Program, the Northern Cod Adjustment and Recovery Program, and The Atlantic Groundfish Strategy, also known as TAGS. The total cost of these programs was about three billion dollars (Shrank, 2005), and even though they did reduce some capacity in the industry, the post-1992 fishery saw increased capacity in two sectors – crab and shrimp. With harvesters no longer able to fish cod, many turned their attention to these species of shellfish which were increasing in abundance. Once a "modest" industry (Shrank, 2005), the shellfishery quickly became a lucrative harvesting replacement for its groundfish predecessor and now comprises most of the landed catch value in the industry.

Overcapacity: moving from the language of renewal to restructuring and rationalization

Pundits argue that overcapacity remains one of the most pressing problems in the Newfoundland and Labrador fishery, in both the harvesting and processing sectors. As noted above, this has been the conclusion of several government-funded reports, as well as economists working on the subject (Schrank *et al.*, 1995; Shrank, 2005), and has been specifically identified as a persistent challenge in the fisheries renewal process (Gov of NL; Gov of Canada, 2006). Shrank (2005, p. 411) states that the story of "too many fish plants, too many fish plant workers and too many fishermen for a commercially viable industry" is an old one. Restructuring to reduce this overcapacity has therefore been deemed the answer, although Shrank concludes that the industry has consistently missed the opportunity to do so and therefore the possibility to become a commercially viable industry that is "truly sustainable" (2005, p. 418). In more recent years, rationalization – the distinct third form of Massey and Meegan's production reorganization – has been emphasized as the main means of achieving the desired restructuring needed for commercial viability and sustainability.

The 2009 Memorandum of Understanding was not initiated because of the pressing need to restructure the industry. Rather, it was the result of a dispute between harvesters and processors over fish prices – deemed yet another crisis that had beset the industry (DFA, July 14, 2009). In June of that year, following the global economic recession, a rising Canadian dollar, and uncertainty in the global seafood market, shrimp processors in Newfoundland and Labrador announced that they could no longer afford to buy shrimp at the then price of 46 cents per pound – previously set by a government appointed pricing panel (CBC News Online, June 2, 2009). Eleven companies with processing plants located across the province shut their doors (ASP, June 2, 2009). No shrimp was to be processed until a better – and lower – price agreement could be reached between these processors, represented by ASP and the harvesters, represented by the FFAW. Plant workers remained without employment; leaders and activists in communities across the province decried the negative impacts for rural areas (CBC News Online, June 8, 2009).

Concerned about the stalemate and its community impacts, the provincial government intervened. On 17 June, 15 days after the plant closures, then provincial Fisheries and Aquaculture Minister Tom Hedderson announced that the provincial government would waive processing licensing fees for the 2009 season if the ASP and FFAW could reach a price agreement. In addition, he said that the government would make provisions for those who had trouble qualifying for Employment Insurance (the federal employment benefits system); maintain the provincial government's commitment to contribute up to 30% of industry rationalization costs over the long-term (cost-shared with the federal government); and undertake a financial analysis of the industry as well as review proposals on seafood marketing (DFA, June 17, 2009). These last two elements – financial analyses and marketing strategies – were to become the basis of the MOU between all parties. Signed only a month later, in July 2009, the MOU represented a working agreement between the provincial government, the ASP and the FFAW. The processors and harvesters did agree on a price and, as a further compensatory act, the government agreed to provide rebates for processing licenses in both 2008 and 2009. The MOU began with a plan to establish a steering committee and four working groups to address the key areas of harvesting, processing and marketing: the Financial Analysis Working Group, Harvesting Sector Rationalization Working Group, the Processing Sector Restructuring Working Group, which also had a Worker Adjustment Committee Sub-Group, and the Seafood Marketing Working Group.

As stipulated in the signed MOU, the Terms of Reference and Request for Proposals on the financial analysis were to be established by the MOU steering committee in conjunction with the Financial Analysis Working Group. This group was to prepare a plan to sample harvesting costs and earnings, including the identification of the required fleet, species and geographic breakdowns on a statistically significant basis. For the processing sector, processor costs and earnings were to be sampled also on a statistically significant basis (MOU, p. 2). Contracts for these analyses were awarded in November of 2009 to Deloitte (harvesting) and Grant Thornton (processing).

Deloitte developed a survey for distribution to the harvesting sector. The data gathered from this were to be augmented by DFO data on landed volume and value. In the end, Deloitte was unable to get an adequate sample size from completed harvester surveys and, in addition to the data provided by DFO, purchased financial data on harvesters from an accounting firm (unidentified) to complete the analysis. Grant Thornton's analysis of the processing sector was underpresented by smaller processing firms, and had limited data on the groundfish and pelagic processing plants, both small and large (Report of the Independent Chair, 2011).

Fisheries industry renewal

Although initiated as a result of the shrimp pricing crisis, the MOU was quickly branded by DFA as an effort to "address the long-term structural problems" in the industry and a means by which to navigate "long-term development" in the fishery (DFA, July 14, 2009, n.p). The MOU was also deemed a continuation of the Fishing Industry Renewal Strategy, a consultation and review process that began in 2006 with a series of meetings with industry stakeholders and the federal government. At that time, marketing was identified as a key issue requiring attention. Danny Williams, premier of the province in 2006 and at the time of the establishment of the MOU, stated that other "structural" issues in the fishery must be addressed. This included dealing with the effects on profitability of a weak American dollar, international competition and, again, overcapacity in the harvesting and processing sectors (Exec Council and DFA, May 31, 2006). A federal-provincial working group aimed to meet with stakeholders in a series of consultations to focus on what Williams referred to as industry renewal; i.e., change needed to ensure that the industry is economically sustainable (Gov of NL; Gov of Canada, 2006). Key dimensions of this renewal were harvesting and processing restructuring, addressing technological issues in the industry, and exploring new opportunities in the fishery including marketing (Exec Council and DFA, May 31, 2006). In October, the provincial government released a renewal discussion document for stakeholder review. It was to be used as the basis for a series of public consultations on the renewal across the island. The purpose of the process was stated in the opening letter:

> This Discussion Paper provides options that would redefine how we harvest, process and market our fish. It offers ideas for a made-in-Newfoundland and Labrador solution that includes a vibrant small boat fishery and addresses the needs of plant workers. We believe a renewed fishing industry will be competitive and prosperous in the

> international marketplace. A renewed industry will help fishers and plant workers get higher and more stable incomes, and communities to have a more secure future. However, we must work together to manage the pace of change and help those impacted by renewal of the province's fishing industry. It will take courage and foresight, and both governments are committed to working with you and each other to make it happen (Gov. of NL; Gov. of Canada, 2006, n.p.).

In the introduction to the document, restructuring – via rationalization – was identified as the means to overcome the structural problem of overcapacity. Furthermore, it was stated that the desire was to maintain an "economically viable, internationally competitive and ecologically sustainable industry" over the long-term, particularly in rural areas where "industry renewal and restructuring represents the best opportunity for the fishery to remain the primary economic driver" (Gov. of NL; Gov. of Canada, 2006, p.2). This document and the process upon which it was built, was couched within a language of renewal, and indicated commitment to rural communities, the small-boat fishery, and to plant workers through consultations with individuals representing all of these groups.

Nearly a year later, in April 2007, the Government of Newfoundland and Labrador announced that both it and the Government of Canada were committed to 16 initiatives associated with the renewal process. These included national policy changes on fleet rationalization through license combining strategies, additional funding for fisheries science at both the provincial and federal levels, provincial investment in industry research and development, as well as in marketing strategies and options (DFA, April 12, 2007).[2]

It should be noted that the language of renewal, restructuring and rationalization moved in tandem during this renewal process, with these terms often appearing together in the same sentences. Restructuring and rationalization, however, were given a somewhat positive tone because of the linkages to renewal. Furthermore, restructuring and rationalization were presented as having distinct and distinguishing properties. Rationalization was associated with capacity reduction, whereas restructuring was presented as broader and farther reaching in terms of organization.

2 Additional details regarding these initiatives and investments can be found at the Fishing Industry Renewal Web site (http://www.fishaq.gov.nl.ca/industry_renewal/index.html).

From renewal to rationalization

When the provincial government first announced that it had reached the MOU agreement with the ASP and FFAW, the release stated that the work was building upon the fishing industry renewal process. The release did not include reference to rationalization. However, the signed MOU document specifically stated that one of its purposes was to establish working groups to develop proposals on *rationalization* models for both the processing and harvesting sectors (MOU, 2009, p. 1, emphasis added). None of these groups were formed around the language of renewal *per se*, nor was the language of renewal introduced as a focal point in the documentation (media releases and the MOU itself). The exception was that the MOU was recognized as a continuation of the renewal process.

Subsequent releases on the status and progress of the MOU predominantly referred to the process as one addressing restructuring (see DFA, July 15, 2010). When the final MOU report was released in February 2011, rationalization and restructuring were seemingly lumped together as one process. The introduction of the report stated that: "the primary objective of the MOU is to develop proposals to rationalize and restructure the industry in a manner that would ensure the viability of the remaining fishing and processing enterprises" (Report of the Independent Chair, 2011, p. 1). Even so, proposed changes to the harvesting and processing sectors, as indicated by the document's table of contents, were both referred to as rationalization. Only the harvesting working group, one of four established for the MOU process, was named the rationalization working group. The processing sector working group was focused on processing *restructuring*, not rationalization, even though it was also analyzing "several options for industry rationalization" (Report of the Independent Chair, 2011, p. iii).

This examination of the language used in and following the MOU prompts one to question whether it was addressing industry rationalization, or was it indeed focused on restructuring, or both? Furthermore, one wonders when and how the language of renewal was lost in the MOU, despite the fact that it was deemed to have grown from the commitment to renewal. The interchange between rationalization and restructuring in the MOU final report, the documentation leading up to its release, and among the individuals involved, suggests a lack of clarity on their meaning and intended usage. In fact, there was no discussion on what these terms mean as core concepts within this process; and there never has been.

The demise of the MOU

In the week following the release of the MOU, Minister Jackman appeared on the local Newfoundland and Labrador radio program, the *Fisheries Broadcast*. During his interview with host Jim Furlong, the program host, Jackman explained that restructuring had to be part of the process. He said he was willing to look at rationalization, despite the fact that it was occurring naturally without investment from the government. However, a discussion on what the future fishery would look like was still needed. This, he argued, was not addressed in the MOU (CBC Fisheries Broadcast, Feb. 28, 2011).[3] On the program the next day, new Premier Kathy Dunderdale similarly argued that the MOU was a report on rationalization and not restructuring. When the host questioned her on what the latter would look like, she responded that a desired plan on restructuring has to address a broad section of the industry, not just the shareholders involved, which she said was the problem with the MOU. Dunderdale went on to say that the MOU report addressed plant owners and license holders, not plant workers, crewmembers, or the communities in which they live (CBC Fisheries Broadcast, Mar. 2, 2011).[4] This rhetoric of omission continued in other published articles and responses to the MOU.

Community groups supporting sustainable community-based fisheries argued that communities were ignored throughout the MOU process [5] and several people applauded the government's rejection of the MOU because of the recognition that rural communities would be decimated if the mass reductions to fleets (between 50 and 80% in some rural areas of the province) went ahead as recommended (see Boyd, March 22, 2011). With this rejection, the other two parties involved in the process say they are at a loss of what is left to address, and argue that they have done what they were asked to do (see Fitzpatrick, March 8, 2011; Fitzpatrick, March 22, 2011).

Conclusions

I began this essay with an introduction to industrial restructuring. Drawing upon the work of Massey and Meegan (1982), I noted that rationalization is considered a form of restructuring specific to addressing overcapacity

3 http://www.cbc.ca/podcasting/pastpodcasts.html?60#ref60.
4 Ibid.
5 see http://www.communitylinkages.ca/MOU.html

and reduction. It is not, however, synonymous with restructuring. In fact, restructuring over the course of Fordist and post-Fordist economies, and the shift between them, has taken on many different forms that include expansion, technological change and investment, as well as rationalization.

In the case of the Newfoundland and Labrador fishery, restructuring – after the 1970s and several crises in the industry related to stock decline – has taken on a distinctive negative tone and has been closely associated with rationalization because the contention is that overcapacity is at the root of the province`s problems with the fishery. This overcapacity is thus argued to make profitability difficult among stakeholders in the industry. A lesser discussed point is that it continues to tax the resource itself. In other words, the problems of overcapacity continue to be argued from the point of view of the economic sustainability of the industry, rather than the sustainability of the resource itself. Global competition, problematic currencies (i.e., a weak American dollar) and marketing obstacles were cited as reasons for the pricing issues which preceded the MOU. While dealing with these issues was also a part of the MOU process, via the Seafood Marketing Working Group, the restructuring and rationalization debate did emerge as the problematic point.

It appears that underlying the MOU was indeed a focus on rationalization as capacity reduction, particularly in inshore fleets located in more rural locations of the province. The report was therefore about restructuring; but a particular form of restructuring that has emerged through neoliberalist ideologies and has been supported by the rhetoric of globalization, whereby the only means through which viability can be achieved is through efficient, lean, vertically integrated operations. Akin to its agricultural counterpart, fisheries have become subsumed by the mantra of "aquabusiness". In the agricultural sector, it is evident that forms of agribusiness have decimated small producers, and eliminated the potential of the family farm to provide local resources and local business opportunities. Within the rationalization and restructuring discourse that has subsumed the MOU and many of the reports before it, aquabusiness *is* the way of business. Restructuring, despite contentions to the contrary in the post-MOU discussions, has no other form; thus restructuring and rationalization are effectively the same process. Based on comments by both Minister Jackman and Premier Dunderdale, it is not clear whether the government actually agrees with this or whether they fear that a public admission to this effect would prove disastrous in a general election in October 2011. Furthermore, both Jackson and Dunderdale entered the MOU process as it was nearing completion. The MOU began under the leadership of then Premier Danny Williams and Minister Tom Hedderson.

For the ASP and FFAW, rationalization is the only obvious restructuring choice to deal with the immediate problems in the industry. There are other voices around the restructuring table, however, who have indicated that rationalization need not be the means to ensure a viable and ecologically sustainable industry. These are the supporters of small-scale sustainable fisheries with a social, community conscience. It is within this discourse that the true distinctions between restructuring and rationalization have been drawn; the problem is that these voices were absent from the MOU process.

Acknowledgments

I would like to acknowledge the financial support of the Community-University Research for Recovery Alliance (CURRA) at Memorial University. The CURRA is funded by the Social Science and Humanities Research Council of Canada (SSHRC) and Memorial University, with additional financial and in-kind support from numerous community partners and groups. (www.curra.ca). I would also like to thank Drs Barbara Neis and Peter Sinclair for their earlier comments on this essay.

References

Allison, E. H. and Horemans, B. (2006) Putting the principles of the sustainable livelihoods approach into fisheries development policy and practice. *Marine Policy*, 30, 757-766.

Association of Seafood Producers (June 2, 2009) Shrimp producers in Canada remain shut in face of economic circumstances. St. John's, NL. http://seafoodproducers.org/files/shrimp-plants-closed.pdf (accessed March 2011).

Bauman, Z. (2000) *Liquid modernity*. Cambridge: Polity Press.

Boyd, D. (March 22, 2011) Good riddance: letter to the editor. The Telegram Online Edition. http://www.thetelegram.com/Opinion/Letters-to-the-editor/2011-03-22/article-2352963/Good-riddance/1 (accessed March 2011).

Cadigan, S. (2003) The moral economy of retrenchment and regeneration in the history of rural Newfoundland. In Byron, R., ed., *Retrenchment and regeneration in rural Newfoundland*. Toronto: University of Toronto Press, pp. 14-42.

Cadigan, S. T. (2009) *Newfoundland and Labrador: A History*. Toronto: University of Toronto Press.

CBC News Online (June 2, 2009) N.L. processors shut down shrimp operations. http://www.cbc.ca/news/canada/newfoundland-labrador/story/2009/06/02/shrimp-closed.html (accessed March 2011).

CBC News Online (June 8, 2009) Shrimp season may be lost, N.L. fisheries minister warns: Gloom hanging over N.L. shrimping towns, mayor says. http://www.cbc.ca/news/canada/newfoundland-labrador/story/2009/06/08/shrimp-town-dispute-608.html (accessed March 2011).

Clement, W. (1986) *The struggle to organize: Resistance in Canada's fishery*. Toronto: McClelland and Stewart.

Department of Fisheries and Aquaculture (April 12, 2007) Renewing the Newfoundland and Labrador Fishing Industry. St. John's, NL: Government of Newfoundland and Labrador. http://www.releases.gov.nl.ca/releases/2007/fishaq/0412n03.htm (accessed March 2011).

Department of Fisheries and Aquaculture (2009, June 17) Province willing to eliminate licensing fees if industry reaches agreement. St. John's, NL: Government of Newfoundland and Labrador. http://www.releases.gov.nl.ca/releases/2009/fishaq/0617n10.htm (accessed February 2011).

Department of Fisheries and Aquaculture (2009, July 14) Agreement reached on long-term development of fishing industry. St. John's, NL: Government of Newfoundland and Labrador. http://www.releases.gov.nl.ca/releases/2009/fishaq/0714n06.htm (accessed February 2011).

Dolan, H. A., Taylor, M., Neis, B., Ommer, R., Eyles, J., Schneider, D. and Montevecchi, B. (2005) Restructuring and health in Canadian coastal communities. *EcoHealth*, 2, 195-208. doi:10.1007/s10393-005-6333-7.

Executive Council, and Department of Fisheries and Aquaculture (May 31, 2006) Premier reports progress on issues since meeting of fishing industry stakeholders. St. John's, NL: Government of Newfoundland and Labrador. http://www.releases.gov.nl.ca/releases/2010/fishaq/0305n04.htm (accessed February 2011).

Fitzpatrick, A. (March 22, 2011) Letter delivered: MOU at impasse. The Telegram Online Edition. http://www.thetelegram.com/News/Local/2011-03-22/article-2353037/Letter-delivered%3B-MOU-at-impasse/1 (accessed March 2011).

Fitzpatrick, A. (March 8, 2011) Jackman to spell out direction for fishery MOU. The Telegram Online Edition. http://www.thetelegram.com/Business/2011-03-08/article-2311157/Jackman-to-spell-out-direction-for-fishery-MOU/1 (accessed March 2011).

Fudge, J. and Vosko, L. (2001) Gender, segmentation and the standard employment relationship in Canadian labour law, legislation and policy. *Economic and Industrial Democracy*, 22, 271-310. doi:10.1177/0143831X01222005

Government of Newfoundland and Labrador, Association of Seafood Producers and Fish Food and Allied Workers' Union. (2009) Memorandum of understanding. http://www.releases.gov.nl.ca/releases/2009/fishaq/MOU%20Fishing%20Industry.pdf (accessed March 2011).

Government of Newfoundland and Labrador and Government of Canada. (2006) Canada-Newfoundland and Labrador fishing industry renewal: a discussion paper. http://www.fishaq.gov.nl.ca/industry_renewal/fs114_5_2006e.pdf (accessed March 2011).

Hayter, R. (2004) "The war in the woods": Post-Fordist restructuring, globalization, and the contested remapping of British Columbia's forest economy. *Annals of the Association of American Geographers*, 93(3), 706-729.

Hunt, C. (March 8, 2011). Jackman rejects MOU report. The Coaster. http://www.thecoaster.ca/News/2011-03-08/article-2311621/Jackman-rejects-MOU-Report/1 (accessed March 2011).

Hutchings, J. A., Walters, C. and Haedrich, R. L. (1997) Is scientific inquiry incompatible with government information control? *Canadian Journal of Fisheries and Aquatic Sciences*, 54, 1198-1210.

Jessop, B. (1992) Fordism and post fordism: critique and reformulation. In Scott, A. J. and Storper, M. J., eds., *Pathways to regionalism and industrial development*. London: Routledge, pp. 43-65.

Jessop, B. (1994) Post-Fordism and the state. In Amin, A. ed., *Post-Fordism: A reader*. London: Blackwell Publishers, pp. 251-279

Kurlansky, M. (1998) *Cod: A biography of the fish that changed the world*. New York: Penguin Books.

MacDonald, M., Neis, B. and Murray, G. (2008) State policy, livelihood protection and gender on Canada's east coast. *International Journal of Canadian Studies*, 38, 149-180.

MacLachlan, I. (1992) Plant closure and market dynamics: competitive strategy and rationalization. *Economic Geography*, 68(2), 128-145.

Marx, K. ([1887]1995) *Capital: An abridged edition*. Oxford: Oxford University Press.

Massey, D. and Meegan, R. (1982) *The anatomy of job loss: The how, why and where of employment decline*. London and New York: Methuen.

McDowell, L. (1991) Life without father and Ford: The new gender order of post-Fordism. *Transactions of the Institute of British Geographers, New Series*, 16, 400-419.

Neis, B. (1991) Flexible specialization: what's that got to do with the price of fish? *Studies in Political Economy*, 36, 145-175.

Ommer, R. (1994) One hundred years of fishery crises in Newfoundland. *Acadiensis*, 23(2), 5-20.

Ommer, R. (2007) *Coasts under stress: Restructuring and social-ecological health.* Montreal: McGill-Queen's University Press.

Polanyi, K. (1944). *The great transformation: The political and economic transformations of our time.* Boston: Beacon Hill Press.

Power, N. G. (2005) *What do they call a fisherman? Men, gender, and restructuring in the Newfoundland fishery.* St. John's, NL: ISER Books.

Report of the Independent Chair: MOU Steering Committee (2011) Newfoundland and Labrador Fishing Industry Rationalization and Restructuring. http://www.gov.nl.ca/fishaq/publications/mou.pdf (accessed March 2011).

Schrank, W. E., Skoda, B., Parsons, P. and Roy, N. (1995) The cost to government of maintaining a commercially unviable fishery: the case of Newfoundland 1981/82 to 1990/91. *Ocean Development and International Law*, 4(357), 390. doi:10.1080/00908329509546067

Schrank, W. E. (2005) The Newfoundland fishery: ten years after the moratorium. *Maritime Policy*, 29, 407-420.

Sinclair, P. R. (1983) Fishers divided: The impact of limited entry licensing in northwest Newfoundland. *Human Organization*, 42(4), 307-313.

Sinclair, P. R. (1987) *State intervention and the Newfoundland fisheries.* Aldeshot: Avebury.

Sinclair, P. R., and Ommer, R. E., eds. (2006) *Power and restructuring: Canada's coastal society and environment.* St. John's, NL: ISER Books.

Vosko, L. F., Zukewich, N. and Cranford, C. (1997) Precarious jobs: a new typology of employment. *Perspectives on Labour and Income Online Edition*, 4(10).

Wright, M. (1998) *A fishery for modern times: The state and the industrialization of the Newfoundland fishery, 1934-1968.* Oxford: Oxford University Press.

Section 3:

Livelihoods and communities

CHAPTER 6

Women's contribution in small-scale fisheries in the European Union

Katia Frangoudes[1]

Women's contribution in European small-scale fisheries is high but still invisible in the public space in many Member States. This essay attempts to assess the contribution of women within this male-dominated industry in the absence of gender-based statistics. It first presents formal employment and unrecognized work of women in fisheries. A second section reviews the many cases where women initiate or contribute to the diversification of activities by fisher families. The essay also looks at the role of national or European legislation in enabling recognition of women's work and in facilitating or preventing income diversification in family-based fishing enterprises.

Introduction

Since the late 1990s, the General Directorate of the European Commission in charge of Fisheries (currently Maritime Affaires and Fisheries, DG MARE) organized conferences and ordered studies concerning women's role in fisheries. In spite of ten years of action by the European Union (EU), women's contribution in fisheries is still unknown to the majority of fisheries managers at national and European levels. The initiatives developed by the European Union towards women in fisheries are the response to the Union's objective to implement the principle of equal opportunity in all EU policies, including the Common Fisheries Policy (CFP). DG MARE introduced this dimension in some fundamental fisheries sector regulations such as the European Fisheries Fund (EFF) regulation of 2007. Article 11 of this regulation calls

1 Corresponding author: Université de Brest, UEB, UMR AMURE, Katia.Frangoudes@univ-brest.fr

for equality between women and men, and asks Member States to promote gender perspectives in all stages of its implementation at the national level.

The institutional recognition of fisher women's economic contribution in the industry brought fisher women into the EU public space. But this doesn't mean recognition of women's contribution by everybody. And if today you ask European or national civil servants about women's contribution in fisheries the response is often "there are no women in fisheries". For them, only people working on fishing boats are contributing to fisheries, and women don't have the physical strength to work on fishing boats. Following the same logic, women are considered to be staying at home to care for the children, the family and the community. Nobody seems to see the work they perform from the household for the fisheries enterprises and, for some of them, on board boats or on the shore.

This essay discusses the economic contribution of women in small-scale fisheries enterprises in Europe. The first section presents two sides of the coin; the recognized work and the unrecognized work. It discusses how recent legal evolution has helped to move the borderline between the two. The second section links the recognition of spouses and life partners' contribution to fisheries enterprises and women's initiatives in diversification of family activities.

Women's economic contribution in small-scale fisheries

Women's economic contribution in fisheries is divided into two categories: the recognized work in fishing, aquaculture or processing through formal employment, and the unrecognized work by fishers' wives, mothers and sisters who are in charge of different tasks within the family enterprise. The latter tasks are linked to the land business of the boats. It is for these reasons that some researchers call these women "on-shore crew" (Symes and Frangoudes, 2001; Van Ginkel, 2009). Research conducted in Brittany (France) on women's contribution in family fisheries enterprises showed that fisher's wives in small-scale businesses make a higher contribution than those in enterprises with large boats (Frangoudes and Keromnes, 2008).

Recognized work of women in fisheries

Searching for a gender-based statistical picture of employment in fisheries and related activities in the EU is not an easy task. Ideally, we should have for each member state the number of men and women working as employee or employer in production (fishing and aquaculture), distribution and processing for all aquatic products, marine and freshwater. We should also have employment data in relation to the management of the industry (organizations, administration, and research) and ancillary activities (supply of goods and services). Part- and full-time should be differentiated with indications of full-time equivalent employment. This is basically the structure applied to the information stored by EUROSTAT (European Union statistic office). In practice, there is much heterogeneity across countries in the definition of statistical categories, and many countries do not provide information. In most cases, gender data are not available. When they exist, they often differ from one source to another. This has been stressed by all studies conducted at the European level, such as the reports produced by McAllister (2002) or Salz *et al.* (2006) financed by the European Commission. These are most recent reviews on the subject based on EUROSTAT, cross-checked with national administrations or industry sources.

According to the available data there are about 97,000 women representing 25% of the total declared employment in the fishing industry (390,000). They are more numerous than men in the processing industry (57%); an industrialized component of the seafood industry providing more than 120,000 jobs in the EU. Women are relatively important in aquaculture where they represent 32%, many of them being self-employed. This category includes women gathering shellfishes on the shore, like the Galician or Portuguese mariscadores.

In capture fisheries, women represent only 5% of statistically recorded employment. The number of women working on all types of vessels was 8,696 in 2003. It is impossible to know the distribution by category of fleet – coastal or deep sea – and by status – owner or crew. From qualitative data gathered by the AKTEA network[2] in meetings and fieldwork, it appears that women are employed mainly on board coastal vessels. These vessels do daily trips and are usually operated by couples. Women often become crew on family boats because of lack of male crew or lack of other work prospects for women in rural areas. Working on the family fishing boat gives the opportunity for flexible working hours, which may be combined with childcare. And they are not all officially recorded for this work. The example of fisherwomen in Portugal illustrates how their work has become more and more formally recognized. Fisheries in the estuaries of Tejo and Sado have always involved women.

2 AKTEA is a European Network of Women's Organizations in Fisheries and Aquaculture (www.fishwomen.org)

In the Tejo estuary, women work with their husbands in small boats practicing estuary fisheries, and they are absent from their home for several months of the year even if they are not fishing far away from the community. Fisheries in this area are a traditional activity within the different communities around Tejo and, as women say, "their mothers were also fishing with their husbands". Some of the women we met in a meeting organized in Villa Franca de Xires told us that they were born on their parents fishing boat and that some among them also gave birth on the boat. Until 1998, women working on boats were not recognized as professional fishers even if they were working full-time on family fishing boats. With the help of the fisher's national union, Tejo women took a training course and passed the necessary examination to be recognized as professional fishers. On passing the examination they were awarded a diploma, which entitled them to be crew on fishing boats.

In Sado estuary, wives of fishers turned to fisheries activity after the decline of rice farming. Likewise, women of Carrasqueira community worked with their husbands on boats because of lack of other type of employment. Family life is organized around fisheries activity. Children are left with grandparents to permit their mothers and fathers to leave early in the morning. Children go to school with the help of grandparents or neighbors. Women claim the creation of kindergarten within the community or other form of childcare because they don't want to be dependent on the family.

Despite this high contribution of women in fisheries harvesting, Portuguese fisheries authorities ignore the presence of women within the fisheries sector. Unlike Tejo women, 200 women from Azores islands practicing fisheries as a job are not registered in the national statistics.

Invisible work of women in fisheries / without remuneration

Many European fishers' wives play an important role in onshore activities. They undertake the administrative work of the fishing enterprise: they go to the bank, oversee harbor administrative duties, and take charge of fish auctions, book-keeping and cleaning boats. They also sell fish or mend fishing nets. Women's contributions to the fisheries enterprise increases when the industry or the family business is faced with a crisis. Women view their contribution as a way to save money, since their input means that the husband does not need to employ another person. Women do not perceive this contribution as a real job because they can do it at home while they look after the children.

Today, three options are available for women working in the fishing enterprise. They can work without a salary and social benefits; they can register as self-employed and achieve some social protection; or they can be employed by the fishing enterprise and be entitled to a salary with all the social benefits.

In some Member States (MS), wives who actively contribute to the family business have a recognized status, for the role they perform, which allows them to benefit from social welfare insurance, in their own right. By doing this, the MS give recognition to the status of the partner as someone who contribute to fisheries enterprises and also to communities. The legal framework for the status attributed to fishers' wives finds its legal basis in the European Council Directive 86/613 on "the application of the principle of equal treatment for men and women working in a self-employed capacity, including agricultural workers". First and foremost, this directive is aimed at equal treatment for both men and women who work in a self-employed capacity in agriculture and, secondly, it is about recognition for work carried out by a spouse. Through this directive, the Council requests MS to take into consideration the role played by the spouse in a self-employed enterprise and, furthermore, to study the conditions necessary for her legal recognition. This legal recognition ensures her access to social security, health care, old age pension, maternity benefits, replacement services, disability and incapacity benefits. Some texts, mainly EU parliament resolutions, suggest that MS should be **obliged** to register assisting spouses, and called on the Commission to revise and strengthen Article 6 of the Directive.

Finally, in July 2010 a new directive was adopted by the European Parliament, and the Council directive (2010/41/EU) "on the application of the principle of equal treatment of men and women engaged in an activity in a self-employed capacity" replaced directive 86/613/EEC. The main innovation of the new directive is that it introduces the notion of life partners who can claim collaborative spouse status when their life partnership is recognized by a state. The second novelty in the directive is that MS "shall communicate all available information concerning the application of this directive to the Commission by 5 of August 2015" (article 15). It will be interesting to examine in a few years which MS apply this directive, as it focuses more on the category of self-employees in general than specific groups like agriculture or fisheries.

Diversification of fisheries household activities, a new objective of the European Union

Diversification in fisheries is concerned with extending the earning potential of fisheries, beyond and aside from capture. When the fisher's family income is under threat, it is often women who instigate diversification activities to secure sufficient family earnings. Their principal objective is to stabilize and increase, where possible, the income derived from the family fishing business. The decision to undertake ancillary activities is usually made by the family, as these activities often require a reorganization of family habits. The development of new activities contributes to the sustainable development of artisanal fishing communities, often by creating employment opportunities for women.

Diversification strategies have, historically, been common in many coastal areas; fishers have combined fishing with other activities such as agriculture, sailing or maritime commerce. Nowadays, diversification of activities can be divided into two categories: traditional sale and processing and those related to tourism. The examples discussed here are gathered from written or oral sources from workshops and fieldwork.

Sale and processing of fishing products

The sale of fish directly from the fisher to the customer is a win-win situation. By cutting out the middleperson, a fisher can obtain a higher price for his catch and the customer often pays a lower price for fresh produce. Direct sales thus enable fishers to improve their income. Small-scale fishing boats or small aquaculture units can often sell all or large part of their production by directs sales due to the small quantities which they catch. Recent research in Brittany shows that women who operate direct sales are the wives of small-scale fishers or shellfish farmers (Frangoudes and Keromnes, 2008). Selling has become a real profession by which they earn their own wages and pay social security contributions. But developing direct sales is not necessarily an easy venture. The direct sale of fish products requires equipment that meets the hygiene standards required by the EU. This equipment (refrigerated vans, tiled workshops, etc.) is costly and requires substantial investment. Direct sales as practiced in some MS can be contrary to legislation, such as the obligation to go through auctions for statistical and/or taxation purposes. In some countries, like Greece, the legislation protects small retailers against any form of unfair competition. So, fishers or other members of their family are not permitted to engage in direct sale. Nevertheless, these sales may be practiced because of derogations or tolerance from the local authorities.

There are many different ways to develop the sale of fisheries products. Itinerant sales on foot or by refrigerated van, fish shops and collective initiatives are illustrated here with examples from various parts of Europe. In Portugal, women who sell fish on foot are confronted with particular problems. The women involved in this activity are extremely poor and often have no other opportunities for income. These saleswomen go to wholesale markets every morning, where they buy a small quantity of fish on credit. They then transport the fish in a simple basket balanced on their heads. This activity is forbidden in numerous Portuguese towns for hygiene reasons. The women are not permitted to use public transport to reach their markets. To circumvent this prohibition, women from Matosinhos go to the fish market before five o'clock in the morning and then take the bus, illegally. There is a certain tolerance towards this "basket selling" because the women do not use the buses during rush hour. Nonetheless, the activity is considered illegal by Portuguese authorities and these women risk being stopped and fined. The police can confiscate and destroy their fish and make them pay a fine on the spot. Despite the legislation against this activity, the poorest families continue to practice this form of itinerant sales because they need the income for the survival of their family. Daily, each woman sells between 25-30 kg of fish. They consider themselves professionals and they would like to be permitted to contribute to social insurance so that they could become eligible for pension.

In certain areas, women carry out street sales of sea products with refrigerated vans. They seek business by occupying fixed spots along a route or by touring the neighboring towns and villages. In Corsica, for example, women go by van into the mountain villages to sell their products. During the summer months there are plenty of tourists along the coast and the sale of fish is fairly easy. For the rest of the year, however, local customers on the coast cannot absorb the production and the women travel around the island and up into the mountains to sell. Portuguese women also practice street selling from vehicles. To do this they must obtain a special permit issued by the municipality. However, certain municipalities in Portugal do not grant this licence because they do not want to have this type of selling in their area. In Brittany (France), local legislation allows vehicles to park at the side of the road and sell products. However, people with refrigerated vans sell their fish at fixed points of sale throughout the towns, usually in marketplaces. The municipality grants them a licence, which permits them to set up an iced stall on market day.

An obstacle to itinerant selling is the price of the equipment. The purchase of a refrigerated vehicle, in compliance with European standards, is indispensable but represents an enormous investment for artisan fishers. Sometimes the cost of a refrigerated vehicle (around US$ 65,000 in France) can be higher than the value of a small-scale fishing vessel and/or licence. Fisher families seldom benefit from financial assistance from the public authorities for the purchase of such equipment.

Opening fish shops, fishmongers and market stalls are other common ways in which fish are sold directly to the public. Typically, family-run fish shops sell only the catch of the family boats. While this activity may be lucrative when fishing is good, it presents difficulties when the catch is insufficient or when potential customers prefer a wider variety of fish. In France, for example, the sale of products from boats other than one's own is prohibited unless the fish shop is specifically declared a fishmonger. The investment required for the establishment of fishmongers is high, as the shop must be fitted to meet European standards and must include tiled floors, refrigeration units and ice-making facilities. Nevertheless, fishmongers are sometimes partially financed by public authorities through the fisheries structural fund. In Scotland, the wife of an oyster farmer diversified from selling oysters in bulk to selling to small retailers and direct to the public at game and seafood fairs, and recently established a website and hopes to develop online sales.

The development of direct sale of local production is not only a matter of private initiative. Throughout Europe there are numerous collective initiatives by local authorities or fisher's organizations. In Southern France, fish markets were established by local authorities with the objective of attracting people from neighboring urban centres and tourists, particularly during the summer months. Fish markets attract people because customers can more easily assess quality fish and the prices are sometimes lower than in fish shops. An example from Corsica illustrates the difficulties faced when embarking to establish a fish market, but also highlights how this type of project can develop female employment where opportunities are limited.

The need to modernize the marketplace of Ajaccio was the impetus that led the local fishing community to come up with a new way of selling their fish. Together with the Corsican Regional Fisheries Committee, the local fisher's organization presented a modernization project, in compliance with European health standards. The project to renovate the market, as supported by the local fisher's organization, had twenty selling pitches to be distributed between the members interested in direct selling. Since there were insufficient pitches for each of the fishers, their organization suggested that three different boats support each pitch. Fishers' women (wives, mothers or daughters) were invited to become a part of this project and to operate the sale pitches. Each sale pitch is occupied by one, two or three women who sell their family catch. These women have gained customer loyalty because of the freshness and price of their fish. The quality of their produce is their strong selling point.

Fish processing is another activity often undertaken by wives with the objective of adding value to the catch. Women from different MS have explored ways of upgrading catches and improving family income. Many of them have set up small-scale family processing units where they produce higher value foods which are greatly appreciated by consumers: some of the product is of restaurant quality. Successful initiatives to establish processing units or workshops have a common characteristic: they begin as small-scale endeavors, often in the family kitchen.

A woman from Brittany (France), fisher woman and seaweed farmer, has created a processing unit for the fish she catches as well as for the seaweed she grows. She began tentatively but today has a modern processing unit and produces a wide variety of delicatessen foods. Another example of a fisher woman from Finland who is processing herring and makes marinated filets shows that women start usually in small workshops, and only when they control their job well they expand their activity. She started in her kitchen with only the herring caught by the family boat, and then she established a big workshop, and now processes the catches of all the community boats and employs five people during the peak season.

The products of these small processing units often benefit from an image of quality and, as a result, frequently fetch higher prices than industrialized products. These enterprises aim for placement in the delicatessen section of the food industry.

Diversifying fisheries activities with links to tourism

New activities linked to the presence of tourists have appeared within the fisheries industry in the last few years. Tourist fishing excursions is one of these activities; others are restaurants set up by fisher's wives; organized visits of shellfish farms; or the establishment of tourist accommodation such as bed and breakfasts.

Some fishing families open restaurants as a way of adding value to their catch. Eating in a fish restaurant run by a fishing family usually means quality fish for tourists (Frangoudes, 2004). This activity brings in considerable income, which often exceeds that of the first activity of the household, fishing. The woman can be the sole person in charge of the restaurant or she can share this responsibility with other family members or employees. Restaurants are usually linked to the family boat, which provides the necessary fish for their restaurant. If a single boat cannot provide sufficient quantities to the restaurant, the women buy from other local boats. The quality of fish attracts customers ready to pay a high price for fresh fish. Tourist guides often highlight and recommend the restaurants belonging to fishing families (Pascual and De la Cruz Modino, 2004).

Al fresco takeaways are another alternative found in many fishing communities. In France, women cook fish or shellfish at the side of the road or in "night markets" established during the summer where they rent a special space to install chairs, tables and provide meals for tourists at night. They may cook fish, and shellfish, which have come directly from the boats and/or offer marinated and smoked fish. In the Danube Delta (Romania), women cook fish caught by their husbands and sell to tourists on the beach or close to the boat. This activity doesn't require a lot of investment and the necessary skills are simply an extension of their primary role. In many cases, women truly enjoy this activity because it expands their social network and makes them feel less isolated. The income earned is indispensable to the household.

In areas where there is high tourism potential, as in many coastal areas, it is possible to develop activities for tourists. In comparison to land farmers, fishers do not have under-used buildings to upgrade, but their boat and the fishing activity can present complementary business opportunities. Such usage has implications, not least of which are legal.

Throughout Europe, fishing families have added to their incomes by renting out rooms to visitors. A Finnish fisher family restored the family house and rented the rooms to tourists (Frangoudes and Escallier, 2004). Similar examples are found throughout Europe, especially in coastal areas along the Mediterranean and the Atlantic. In Romania, on the edge of the Black Sea between Constanza and the Bulgarian border there are few hotels, so fisher families supplement their income by renting rooms to tourists, who are attracted to the area because of its outstanding natural beauty.

In Portugal, some fishing families rent out their own houses during the summer and go to live in the cabin where they stock their fishing material. In other places, fishing families rent out rooms in their homes. In Greek islands, fishing families, like other families, offer rooms to tourists. Women welcome the tourists, provide information about local sites, prepare breakfast, and clean the rooms. Renting rooms to tourists requires an investment, and not all fishing families can undertake this activity. Even those families who do undertake such activities are not overwhelmed because the tourists are usually present for only two months of the year.

Tourists today often want to learn about the natural and cultural heritage of the places they visit. Many show a great interest in the coastal area and in professions related to the sea. In some places there is a strong demand for fishing excursions and organized visits to fish farms. Small-scale fishing enterprises can sometimes be adapted easily to accommodate the new demands of "eco-tourism". At times there are legal implications that make this type of diversification difficult. The following examples illustrate some of the advantages and constraints of diversifying into eco-tourism activities.

Fishing excursions, particularly during the summer, can represent new sources of income for fishers. In places where this activity is already in place, fishers are financially satisfied because tourists are generally willing to pay high prices for the opportunity to fish from a professional boat, in the company of a "true fisher". To carry out this activity, fishers must ensure that their vessel complies with the safety standards required for the transport of passengers. It may be necessary to purchase life-vests and other safety equipment to guarantee the security of passengers. It is also an obligation to pay extra insurance.

The development of these activities more often than not needs to overcome cultural and legal obstacles. First, fishers do not always like taking tourists on board a working vessel and, secondly, national fisheries legislation often forbids this kind of activity. Italy is alone in the EU in that it has introduced "pescaturismo" in its national fisheries legislation, and this permits some categories of working fishing vessels to carry tourists. Following the Italian experience, fishers from other Mediterranean countries – Spain, Greece and France – decided to explore "pescaturismo" in their countries. They quickly discovered that it was not possible for them to do the same. Fisheries legislation in these MS does not allow fishers to develop "pescaturismo" for a variety of reasons.

The following example from Sardinia demonstrates how a fisher family implements the law in practice and involves women in this new activity. Women have no on-board role but are in charge of all the work on shore. They are "under the umbrella" (Mondardini, 2005) which means that they have a stand in harbors where they promote the trips organized by the family boat. Wives give all practical information concerning the trip and in some cases they play the role of tour guide. They inform tourists about the maritime environment or the area, the fishing gears used by the boat, etc. Women perform all the tasks except steering the boat. Prior to undertaking this activity the women receive training, where they learn how to receive the tourist, how to inform people about the marine ecosystem, etc.

Experiences around Europe show that visits to shellfish farms or other aquaculture units may be viable ways to diversify income. A good example is the work of the association of shellfish farmer spouses in *Marennes Oléron* (France) where a project of seaside tourism has been successful. The aim was to upgrade the image of oyster farming for tourists who visit their region during the summer. Accordingly, the association organized a circuit that takes visitors to the oyster farms by boat and then offers them oyster-tasting.

While visits to shellfish breeding beds are organized by the oyster farmers and their families in France, this was not the case in Galicia, Spain, where tours were provided by enterprises outside of the fishing community. Women mussel farmers in Galicia regret not having had the foresight to envisage this kind of diversification. Recently, however, a group of women shellfish gatherers, following the example of the French oyster farmers, have formed their own tourist enterprise.

Women shellfish gatherers of Campados (Galicia, Spain) founded a cultural association in 2004. The objective of the association is to develop the tourism dimension of their professional activity. To get qualifications, women took training courses in techniques for public speaking, history, architecture and urbanism in the area, as well as biology, management, English and traditional cuisine. But their professional experience as shellfish gatherers remains the most important skill.

The association is composed of 17 women shellfish gatherers and two women net-menders, and is open to all women doing work related to the sea. Members of the association act as local guides and show the tourists the shellfish banks, the historical area of the village, and explain their heritage, the fish and shellfish auction system, the functioning of the shellfish treatment plant, and the handling of sea-related products. They also explain the fishing techniques used in the *Rías* (estuaries in the coast of Galicia) and the most appropriate traditional cooking techniques for each fish or shellfish. At the end of the tour, the tourist receives a small net-bag made by the net-menders (also women) with different typical shells. The visit lasts about 90 minutes, and the price is 3 euros per person. Their aim is to improve the image of their products by making people aware of the difficult nature of their work.

Discussion

The above presentation on women's involvement in fisheries, fisheries enterprises and in diversification of activities has raised many issues of interest for women in fisheries. The first issue is the EU principle of equal opportunities between women and men in the fishing industry. Women play an important role in the viability of fisheries enterprises but many of the services they provide are not legally recognized, on the grounds of deeply rooted beliefs that these are a natural extension of their domestic role, and that women are born to remain in the shadow of men. This lack of recognition prevents them from access to many social benefits. The reform of directive 86/613 will probably modify the situation, as MS will be obliged to implement it in their legislation. The lobby capacity of fisher women organizations at national and European levels will be very important in this regard.

Women's participation in diversification of activities is another issue. Since women control the household budget, they are usually the first to notice a deficiency in the household income. Accordingly the impetus to diversity usually emanates from the woman, as was the case in the majority of examples presented in this essay. Initial efforts at diversification are usually an extension of the woman's core role, i.e., related to food, shelter and education. Tourism provides an alternative to traditional ways of diversification based on the development of direct sales or value-added products. Successful expansion requires seed funding and business support including training opportunities. State agencies most often target the owner of the fishing boat, when means of supporting diversification are considered.

This essay demonstrates that targeting the women associated with fishing enterprises may well result in greater benefits, not only to the immediate family but also to the local community. Legal constraints and cultural reluctance may have to be overcome to further develop diversification. However, women in fisher families are largely unrecognized resources who have shown considerable initiative, skill and enterprise when challenged by dwindling incomes. Networking and exchanges of experience are other elements to aiding the emergence of diverse and successful initiatives. Networking is also important to strengthen women's position in the public space. Fisher women organizations emerged by the end of the 1990s in different MS and in 2006, a European network of Women's Organizations in Fisheries and Aquaculture, AKTEA, was established. All these organizations aim to improve and strengthen women's position within the industry. They claimed participation in the decision-making process concerning resource management and they obtained seats in Fisheries Regional Advisory Councils. The AKTEA network, through scientific presentations and exchanges of experiences, promotes the improvement of women's self-confidence. The objective is to train the leaders of women's organizations to be ready to lead national organizations and the European network. In a recent public hearing organized by the fisheries commission of the European Parliament, the Commissionaire in charge of fisheries indicated that the Commission is willing to support a European network. Women must be ready to respond to this invitation.

References

European Council (1986) Council Directive 86/613/EEC of 11 December 1986 on the application of the principle of equal treatment between men and women engaged in an activity, including agriculture, in a self-employed capacity, and on the protection of self-employed women during pregnancy and motherhood. OJ L 359, 19.12.1986, p. 56-58.

European Parliament (2003) Women in rural areas of the European Union in the light of mid-term review of the common agriculture policy. Report A5-0230/2003.

European Parliament (2007) European Parliament Resolution of 27 of September 2007 on equality between women and men in the European Union. 2007/2065 (INI).

European Parliament and European Council (2010) Directive the 2010/41/EU, on the application of the principle of equal treatment between men and women engaged in an activity in a self-employed capacity. O.J. of European Union, L 180/1, 15/7/2010.

Frangoudes, K. and Escallier, C. (2004) Women in public space. Brest workshop February 20-21, 2003 Programme FEMMES, Q5TN-2002-01560. http://www.cedem.eu/actes-proceedings.php

Frangoudes, K. (2004) Tourisme et la question de genre dans le secteur de la pêche en Corse. Revista Interdisciplinar dos Centros e Nucleos da Unicamp, Brésil: Multiciência.

Frangoudes, K. and Pascual, J., eds, (2005) *Women in Fisheries and Aquaculture: Lessons from the past, current actions and ambitions for the future.* AKTEA International Conference, November 10-13, 2004. CD-ROM. http://www.cedem.eu/actes-proceedings.php

Frangoudes, K. and Keromnes, E. (2008) Women in artisanal fisheries in Brittany, France. *Development*, 51, 265-270.
Frangoudes, K. (2008) The role of women in the sustainable development of European fisheries areas. European Parliament, Policy Department B, study PE 389.586.
Pascual-Fernández, J. and De la Cruz Modino, R. (2005) Mujeres, reservas marinas y estrategias de diversification en las problaciones litorales: el caso de los restaurantes de pescado. In Frangoudes, K. and Pascual, J., eds., *Women in Fisheries and Aquaculture: Lessons from the past, current actions and ambitions for the future.* International Conference AKTEA, November 10-13 2004. Santiago de Compostela, Galice, Espagne: CD-ROM. http://www.cedem.eu/actes-proceedings.php
Richiero, D. (2004) Economic diversification of shellfish enterprises: blue tourism. In Frangoudes, K. and Escallier, C. Eds., *Women in public space.* Brest workshop February 20-21, 2003. Program FEMMES, Q5TN-2002-01560. http://www.cedem.eu/actes-proceedings.php
Salz, P., Buisseman, E., Smit, J. and Vos (de), B. (2006) Employment in the fisheries sector: current situation, European Commission. DG Fisheries, FISH/2004/4.
Symes, D. and Frangoudes, K. (2001) The social organisation and reproduction of inshore fisheries. In Symes, D. and Phillipson, J., eds., *Inshore Fisheries Management.* Dordrecht, The Netherlands: Kluwer Academic Publishers, 159-175.
Van Ginkel, R. (2009) *Braving Troubled Waters: Sea Change in a Dutch Fishing Community.* Amsterdam: Amsterdam University Press.

CHAPTER 7

Fish and fisheries in the evolution of Newfoundland foodways

Kristen Lowitt[1]

Food is central to the formation of individual and collective identities. The term 'foodways' is used to refer to socioculturally informed patterns of food use. Newfoundland's cultural identity is linked to the sea and seafaring traditions, reflective of an economic history tied to the fishery which produced fish for subsistence and for sale. Cod was a staple food in the diet. Settlers developed a system of occupational pluralism, providing for themselves those necessities they could produce, such as food. This essay uses foodways as a framework for looking at the links between everyday practices, such as procuring and eating food, with broader social and cultural identities. Through a review of various Newfoundland writings, I piece together a picture of traditional Newfoundland foodways. Given the historical importance of fish and small-scale fisheries, I place these at the centre of my discussion. I argue that traditional Newfoundland foodways express a distinctive identity tied to the sea as well as the land. I follow a chronological approach, beginning with the foodways of the early fish planters and continuing into the twenty-first century. Here, I consider contemporary changes in traditional foodways, including the rise of 'culinary tourism' and the present-day importance of small-scale fisheries to the food security of Newfoundland's coastal communities.

Introduction

The study of food has received considerable attention from scholars of many academic traditions including anthropology, sociology, history, folklore, and geography, among others. Food has been used as a lens through which

1 Corresponding author: Memorial University of Newfoundland, Canada, klowitt@mun.ca

to explore changing patterns in economies, social relations, and cultural development. More recently, the term 'foodways' has come into popular use among social scientists to refer to socioculturally informed patterns of food use. Foodways refer to overall patterns of food use, including ways of procuring, preparing, presenting, and eating food (Smith, 2007). The term refers to tangible (material) and intangible (attitudes, rituals, customs, traditions) aspects of food (Smith, 2007; Thursby, 2008).

At the core of the study of foodways is the specific relationship between people and their food practices. Cultures, clans, families, and the identity of individuals are shaped by the foods they select and the ways food is prepared and served; further, the foods selected are often determined by the region where people live (Thurbsy, 2008). Traditionally, it was anthropologists and folklorists who studied foodways as part of the broader cultural system; however, as the recent interdisciplinary fields of food studies and food history have emerged, foodways have been deemed worthy of focused study in and of themselves (Smith, 2007).

In this essay, I explore Newfoundland foodways. Newfoundland and Labrador is Canada's most easterly province and is comprised of the island of Newfoundland and mainland Labrador, located northwest of the island. It is also Canada's newest province, joining Confederation in 1949. It has a population of just over five hundred thousand people, the majority of which live on the Avalon Peninsula where the provincial capital of St. John's is located. In this essay, I focus my discussion to foodways on the island of Newfoundland, recognizing that foodways in Labrador are shaped by a different set of influences, including a larger Aboriginal population.

As in other Maritime regions, Newfoundland's social and cultural identity is intimately linked to the sea and seafaring traditions, reflective of an economic history tied to the small-scale fishery which produced fish for home consumption and for sale. Kin-based units have worked together in small-scale fishing enterprises for generations, and cod was a staple food in the diet from the time of the first fish planters. However, immigrants arriving to Newfoundland soon discovered that an economy based on the fishery alone could not support them. So, they turned to the land, working out a "delicate balance", using farming to supplement low and uncertain incomes in the fishery (Cadigan, 2002, p. 242).

I use foodways as a framework for looking at the links between everyday practices, such as producing, procuring, eating and preparing food, with broader social and cultural identities. The essay is based on a review of various texts and collections from Newfoundland including ethnographies, food folklore writings, food history and anthropological writings, and cookbooks. From these sources, I piece together a picture of the traditional foodways of Newfoundland. Given the importance of fish and fishing to Newfoundland, both for sustenance and livelihood, I place it at the centre of my discussion. Ultimately, I argue that traditional Newfoundland foodways are important sites for the articulation and expression of a distinctive identity tied to the sea as well as the land. The essay follows a chronological approach, starting with the first fish planters and settlers of the seventeenth and eighteenth centuries. I then describe the seasonal cycle of small-scale fishing and food production characteristic of Newfoundland foodways from the late eighteenth to mid-twentieth centuries. Lastly, I look at changing patterns in the fishery and foodways across the later twentieth and early twenty-first centuries, including the rise of culinary tourism and the present-day importance of small-scale fisheries to the food security of Newfoundland communities.

Tracing the origins of traditional Newfoundland foodways

Newfoundland foodways have their origins in the food traditions of the early migratory fishers and settlers who came to Newfoundland to work in the cod fishery. The original English, French, Scottish and Irish settlers, along with the migratory fishers before them, brought their own foodways which they adapted to the Newfoundland environment. From the beginning, cod was a staple in the diet. Historian Peter Pope (2004) explains that the practice of catching and processing cod into a salted dried food is much older than the Newfoundland fishery. It is not known who developed the salt cure method for fish, but it was used by Breton fishers to process hake in late-medieval times. The dry salt cure worked particularly well in the temperate climate of the Atlantic and produced a stable product suitable for export, with the first recorded cargo of salt cod leaving Newfoundland in 1502 on the Gabriel of Bristol (Pope, 2004).

Prior to the early seventeenth century, the Newfoundland fishery was largely seasonal, with few Europeans overwintering. By 1510, Bretons, Normans, Basques and the English had established seasonal fishing settlements in Newfoundland. In the early seventeenth century, the English began establishing a resident fishery. The most obvious social distinction in Newfoundland society at this time was between masters and servants. As Pope describes, Newfoundland planters may best be understood as resident boatkeepers who were the masters, or in some cases mistresses, of household production units; the other broad class of residents was the servants, usually male, employed by the boatkeepers to work in the fishery. The planters managed competing household production units within the salt fish industry. While English families were the most established in the seventeenth century, data suggest a similar social structure existed among the small population of French households later in the seventeenth century, consisting of 'habitants' and 'engages' (Pope, 2004).

Sheilah Robert's (2003) book *For Maids Who Brew and Bake* is a collection of seventeenth century Newfoundland recipes, gathered from old English manuscripts and recipe books. The collection lends insights into what families of the time were eating. The fish and seafood recipes indicate that early settlers most likely ate salt cod with mustard, butter, or vinegar. While salt cod was a staple, other types of fish including capelin, herring, mackerel, eel, halibut, salmon and trout were also eaten and usually prepared stewed, stuffed or baked. This fish was often served on "sops", slices of bread with melted butter poured on top, as the potato had not yet come into general use. Many seventeenth century families were also fond of shellfish and ate mussels, periwinkles, lobsters, crabs, shrimps, and prawns. A favorite method of cooking shrimp was to "seeth them in equal parts water and ale, salt and savoury" (p. 29). Another seventeenth century recipe details how to prepare barnacles. Entitled 'barnacle with short broath,' the recipe reads "Dreste and lard it, then seeth it with water, and season it well, when it is half sod, put to it a quarte of white wine, and seeth it well, then serve it with parsley over it" (p. 29).

Aside from readily available fish and seafood, censuses from this time indicate that most families maintained gardens and small livestock (Omohundro, 1994). Immigrants to Newfoundland brought their Old World horticultural knowledge with them (Thirsk as cited in Omohundro, 1994). Women coming to Newfoundland maintained their traditional contributions to household production in baking, brewing, dairying, and caring for animals. Data from the period indicate that the agricultural effort was much greater in households in which there was a female present (Pope, 2004). However, women in seventeenth century Newfoundland were more than housewives. As Pope points out, they also had a strong identity as economic participants in the fishery, with some of the largest plantations operated by women. Other staple foods were imported to the island. In 1677, the most commonly imported food provisions were bread, flour, salt beef, peas, oil, sugar, molasses, rum, and salt (Gray, 1977). According to Pope, the diet suggested by these imports was "unexceptional," typical of foods from the Old Country (2004, p. 365). Nonetheless, these items were sufficiently precious that meals alone may have been wages enough for fishing servants who stayed on the island over winter and were referred to as "dieters" (Handcock as cited in Omohundro, 1994, p. 95).

As both resident and migrant fishers in the late seventeenth century faced the threat of war with France, planters further intensified their reliance on local resources, an important part of which included a greater diversification in diets to incorporate more locally-available sources of food, including seal meat. In reference to this new dietary addition, Commodore John Graydon is cited as saying "which they and none but they could eat," also adding "such people such stomachs" (Pope, 2004, p. 427). In the early eighteenth century, Irish immigration intensified along with the extension of settlements in Placentia Bay and Bonavista and Notre Dame Bays to the north. Thus, as Pope says, "Newfoundland's traditional culture had emerged, in a form still remembered, strongly shaped by these developments in subsistence and migrations" (p. 427).

Seasons of traditional Newfoundland foodways

I refer now to 'traditional' Newfoundland foodways in the sense that the pattern of food provisioning across many Newfoundland communities comes into focus around the end of the eighteenth century, with the establishment of the planter's household unit, and remained until the 1950s with the influx of more modern goods, services, and cash in the post-Confederation days (Porter, 1995). The Newfoundland fishing enterprise remained small-scale, located within the household, with the typical family unit becoming smaller and evolving to exclude servants (Pope, 2004). As migrants continued to come from England and Ireland throughout the eighteenth century, they had to be not only fishers, but also hunters, trappers, and gardeners in order to survive (Omohundro, 1994). Settlers developed a system of occupational pluralism in which they provided for themselves those necessities they could produce, such as fish and food, purchasing as little as possible on credit against fish in the merchant store (Ommer *et al.*, 2006). The merchant played a central role in traditional foodways, serving as the go-between to the outside world and the shopkeeper (Omohundro, 1994).

Newfoundland foodways thus developed as part of an adaptive survival strategy, producing as much food for home consumption as possible. Some ethnographic studies from the past several decades lend insights into traditional patterns in food provisioning. This includes John Omohundro's (1994) book *Rough Food* about home production on the Northern Peninsula; and James Faris' (1972) book *Cat Harbour* about this settlement on the northeast coast. I draw closely upon these books, as well as some other research and recipe books, to paint a picture of what Omohundro has called the 'seasons of subsistence' in traditional Newfoundland foodways.

Across Newfoundland, the summer season from June to August was the most active time for household food production and commercial fishing. Seedlings were started earlier in the spring, usually indoors by the women, and sacks of potatoes saved for seed were brought inside from the root cellar and left to sprout (Omohundro, 1994). Once the ground was thawed, digging and planting began in late May and early June. Potatoes, turnips, and cabbages were among the staple crops grown (Faris, 1972; Omohundro, 1994). The annual capelin run in June was well-timed to provide ample fertilizer for the potato beds (Omohundro, 1994). Most of the responsibility for gardening fell to the women and children (Faris, 1972; Murray, 1979; Omohundro, 1994; Cadigan, 2002). For many, keeping animals was another important aspect of a diversified household strategy (Faris, 1972; Omohundro, 1994).

For many communities across the island, the summer was organized around the annual inshore cod migration, and it was common for families to migrate to summer fishing stations (Porter, 1995). Both men and women were busy with the summer fishery, men in the catching of fish and women in the home processing. Until the late nineteenth century, the most common method of fishing cod was handlines, a fishing line with a hook attached to one end. Starting in the late 1800s, cod traps became more common (Higgins, 2007). The cod trap was made up of a net 'leader' designed to direct fish into a net box, or 'trap,' which was moored in a specific spot called a 'berth' (Faris, 1972). Once anchored in the water, the traps fished while unattended (Higgins, 2007). Communities developed customary ways of regulating the fishery. These systems ranged from being implicitly understood to explicit community laws, and helped ensure that the greatest number of households possible had the potential to earn a living from fishing (Matthews, 1993). For example, in Cat Harbour the selection of cod trap berths took place early in the spring each year by a draw which assigned a berth to each man or crew for the entire summer fishery (Faris, 1972). Throughout the summer, women worked on shore making fish harvested by their male relatives into salted products for the commercial and household economies (Ferguson, 1996). Like the men who had well defined fishing crew structures (Faris, 1972), women's work was governed by a number of patterns with, for example, married women working together, often directed by a more experienced or senior woman. Keeping an orderly flake and taking good care of the fish were matters that affected personal identity, status, and esteem of women in a community (Ferguson, 1996).

While cod was a mainstay of the household diet, the proper washing and drying of fish was critical to making a living in the salt fishery. Women worked hard to produce good-looking, quality fish and their work substantially increased the value of the fish product. In his study of salt-cod processing in Newfoundland, Mark Ferguson (1996) describes the various classes of salted fish that were made. Historically, the two most basic cures were the light-salted and the heavy-salted fish. The heavy salted cure was used mainly on offshore fishing ships, as its higher concentration of salt allowed for shorter drying times. Lightly-salted fish was produced by households in the inshore fishery. It required the longest and most complex drying phase, and was preferred in most markets because of its higher quality. As Ferguson explains, the lightly-salted fish reverted closer to its original form when soaked in water, since it could take up more water than the heavier salted fish. However, the making of light-salted fish varied between regions, based on the modes and amounts of salting. The two distinct inshore styles were fish pickled in brine and fish dry-salted in bulk, with the latter method being more common. In the making of dry-salted fish, further variation existed. For example, external influences such as the weather could affect the amount of salt and drying time required. Ferguson thus suggests it "...might be possible to make a case for the existence of regionally distinct salt fish – whether or not conscious [salting] techniques played any role" (1996, p. 197).

While cod was the center of the summer fishery, communities and households diversified to various extents into other fisheries (Faris, 1972; Omohundro, 1994). A salmon fishery developed near the end of the eighteenth century, partly in response to the drop in the value of cod at that time (Butt, 1998), and in the late nineteenth century a string of lobster fisheries developed along the island's North Shore (Higgins, 2009). While lobster is one of the most expensive food items available on the market today, it was not historically held in such high regard (Mariana, 1999; Johnson, 2007). It was one of the most commonly found crustaceans making it "fit for the tables of the poor" (Mariana, 1999, p. 186).

Another important summer activity was berry picking, which started in July with the first wild strawberries and continued into the fall (Omohundro, 1994). From the marshlands, marsh berries, bakeapples, currants, and cranberries were harvested, while raspberries, blueberries, blackberries, and sarsaparilla came from the barrens (Faris, 1972). Berries were a very important fruit source, and would be "put up" in jars to last until the next season (Faris, p. 33). They also offered some protection against vitamin C deficiency.

The fall was the season for harvesting crops and getting in the winter's diet or "rough food," understood among Newfoundlanders to be "your staples, your winter's diet" (Omohundro, 1994, p. xiii). Before roads started coming through to many communities around the second half of the twentieth century, rough food was set aside for six to eight months (Omohundro, 1994). Accordingly, the fall fishery placed more emphasis on fishing for home consumption, and was often less complicated and formally arranged than the summer fishery. Also integral to the winter's diet was the men's hunting and trapping rabbit, caribou, moose and other wild game which would continue during the winter months (Faris, 1972; Omohundro, 1994). Because food was relatively bountiful, the fall was the season for 'scoffs'. A scoff is a cooked meal either on sea or ashore, usually done in the evening often as part of an impromptu party. They were often lavish spreads, featuring home produce and wild foods (Dictionary of Newfoundland English, 1990; Omohundro, 1994).

Winter was a slower season in home food production, with more time for home cooking and some sea-bird hunting and trapping (Omohundro, 1994). Families who lived on exposed headlands often moved into the more sheltered head of the bay for the winter (Porter, 1995). Ice fishing began in March, and sealing in April-May (Faris, 1972; Omohundro, 1994). Sealing has been an important component in adaption for northern outport communities over the past two centuries, undertaken for subsistence and commercially to pay for materials for the summer fishery (Omohundro, 1994).

Meal plans, based on this seasonal round of fishing and home food production, were typical across the Island. Omohundro (1994) describes a traditional meal plan common to the Great Northern Peninsula, the Strait of Belle Isle, the South Coast, and the Avalon Peninsula. This plan specified that on Tuesdays, Thursdays, and Sundays the main meal, served at midday, was a 'boiled dinner' comprised of salt beef, carrots, cabbage, potatoes and bread pudding or 'duff'. On Sunday, boiled salt beef was supplemented with moose, caribou, rabbit or duck. On Wednesday and Friday cod fish was served with potatoes. Sunday breakfast was fish and brewis (salt cod cooked with hard bread, originally developed for sailors at sea) and supper featured potato salad and hard-boiled eggs. On Monday, leftovers were eaten from Sunday. Saturday's dinner was either a pea soup made with salt beef, or pork and baked beans. While some imported foods are apparent, the prominence of wild and home-grown foods, including cod, rabbit, moose, duck, potatoes, cabbage, and carrots is readily evident, with the meal routine varying according to the availability of seasonal foods (Omohundro, 1994).

These traditional meals are likewise reflected in many collections of old Newfoundland recipes. *Fish and Brewis, Toutens and Tales* by Len Margaret (1980) is a set of traditional recipes and recollections from the community of St Leonard's on the Burin Peninsula. An entire section is devoted to "fish and other accompaniments, " with recipes for fish and brewis, basic salt fish and potatoes, as well as more specialty dishes such as cod tongues and fried cods heads. Many parts of the cod were traditionally eaten, including the heads, britches (roe), tongues, sounds (swim bladders), and puttocks (intestines) (Power, 2000). Some parts, such as britches, sounds, and puttocks were eaten more in the fall when they are harder, as compared to the summer months when they are softer in texture (Power, 2000). It should be noted that, in Newfoundland, the word 'fish' refers to 'cod,' while other kinds of fish are specified by their name (Jesperson 1974; Gray 1977). For example, a recipe calling for "salt fish and potatoes" would be understood locally as "salt cod and potatoes." Along similar lines, Faris (1972) found in Cat Harbour that a man who catches only lobster or salmon is not a fisher, but a "lobster catcher" or "salmon catcher" (p. 27).

One of the best known Newfoundland recipe books is *Fat-Back and Molasses*, a collection of old favorite recipes from across Newfoundland, edited by clergyman Ivan Jesperson (1974). The recipes in this collection reflect Newfoundland's rich sea and land activities, such as 'fishermen's fresh fish stew,' 'trapper's bread,' and 'sailor's duff.' Faris (1972) likewise found that expressions from the sea infiltrated the realm of food in Cat Harbour. For example, since the sea is salt as opposed to fresh water, it followed that any food that was 'fresh', such as fresh meat, was understood to mean free from salt (Faris, 1972).

Despite the importance of an adaptive strategy based on fishing and subsistence farming, these activities have not always had an easy relationship in Newfoundland foodways. By engaging in the fishery as well as subsistence production on land, Newfoundlanders were combining two distinct spheres of work (Nemec as cited in Omohundro, 1994). As Omohundro says, "Newfoundland's farming had to compete with many other demands for the family's labour during the mild months" (p. 146). Similarly, Faris (1972) describes Cat Harbour as organized around the hunting and processing of fish for sale and subsistence, with gardening only achieving priority in labor scheduling at limited times in May and September. Fishing was the main reason for settlement in many communities and made possible the purchase of nearly all essential inputs such as molasses, cotton, tea, iron, and flour (Omohundro, 1994). During early settlement, farming was actively discouraged because it would detract from the prosecution of the fishery (Murray, 2002).

Changing fisheries and foodways

While change and adaption has always been a constant in Newfoundland foodways (Hanrahan, 2001), more rapid changes to the small-scale fishery and to traditional patterns of food provisioning and consumption came in the 1950s with the influx of more modern goods, services, and cash in the post-Confederation days. There began a shift in foodways from the harvesting and consumption of local food to the commodification of food resources in a global economy (Turner *et al.*, 2007). An especially significant change in the small-scale fishery and Newfoundland foodways in the post-World War II era was the shift of women's home processing of dry salted fish to their employment in fresh fish processing plants. Historically, fishing effort was limited by the fairly small landings from traditional fishing methods and the limited labor and space available for making salt fish. The fish processing plants overcame many of these limitations, encouraging more fishing effort and in some communities replacing the seasonal, small-scale fishery with a year-round trawler-based fishery (Power, 2000).

There is some evidence that these changes in the fishery affected the personal foodways of Newfoundland families in terms of the quality and amount of fish they were eating. Sociologist Nicole Power (2000) interviewed fish processing workers on the Bonavista Peninsula, most of whom were women. The older women Power spoke with who had made salt fish in the 1950s and 1960s reported feeding fish to their family every day during the fishing season, and salting fish for the winter. They also thought that in the past relative to more recent years, cod fish had met good standards for quality in terms of size and texture. When women started working in plants, cooking fish for meals became more difficult because they spent less time at home (Power, 2000). Medical anthropologist Maura Hanrahan (2001) similarly noted that cod declined in the Newfoundland diet after the 1950s. Hanrahan goes on to suggest that "fish, salted or otherwise, will never again be at the core of the Newfoundland diet" (p. iv). Also in the 1950s, efforts in gardening declined further because of constraints on time related to women's work in fish plants, as well as a greater cash income to buy food (Omohundro, 1994).

In 1992, a moratorium was declared on Atlantic cod and other groundfish stocks. Communities across Newfoundland have since undergone substantial social and economic restructuring related to changes in the fisheries (for an overview of restructuring in Canada's coastal communities, see *Power and Restructuring* by Sinclair and Ommer, 2008). In 2000, Power noted that the women she spoke with said that their access to fish had become more constrained with the collapse of the cod fishery, since they couldn't as easily get high-quality fish from local fishers and many would not buy fish from the stores because of the high price. Following the depletion of groundfish stocks, there was a shift to more shellfish harvesting and processing. The consequences of these changes for food security, and the broader foodways of coastal communities, have not yet been explored in-depth. However, *Resetting the Kitchen Table* (Parrish et al., 2008) provides a good introduction to some of these issues.

In the context of rapid social and economic change, folklorist Andrea O'Brien (1999) makes a compelling argument that practicing traditional foodways is an important reenactment of identity. O'Brien looked at boil-ups, a traditional type of outdoor cooking practiced on the sea while fishing, or on the land, as a form of community and cultural validation in her home community of Cape Broyle. She writes, "In the face of rapid economic changes...boil-ups have become important as symbolic expressions and vital enactments of Newfoundland's distinctive culture and society for current residents and expatriates alike" (p. 67).

While fish, as Hanrahan (2001) suggests, may never again be at the "core" of the Newfoundland diet, I believe there remains an important role for small-scale fishing to feed people in Newfoundland and Labrador today. As a province, Newfoundland and Labrador faces challenges related to health and food security, including among the highest rates of food bank use and obesity in Canada. Being an island province also poses challenges in terms of ensuring a reliable supply of food because so much food has to be flown, ferried, and trucked in. However, a nutritious source of protein which we still have in relative abundance in various parts of the province, along with the knowledge and skills about to harvest and prepare it, is seafood. In 2009, Newfoundland and Labrador was Canada's third leading province in the export of fish and seafood, with the most valuable species being crab, shrimp, and mackerel (Fisheries and Oceans Canada, 2010). However, small-scale fishers, and others dependent on the small-boat fishery, are struggling to survive in the current export-focused industry. Long food supply chains are known to decrease the proportion of value in food production captured by primary food producers (Marsden *et al.*, 2000). Providing more direct local markets for the domestic consumption of seafood may open up new economic and livelihood opportunities for small-scale fish harvesters and processors, while enhancing the accessibility and availability of seafood in the personal foodways of Newfoundland families.

In the summer of 2009 I had the opportunity to spend time on the west coast of Newfoundland around Bonne Bay in Gros Morne National Park. I spoke with residents, food producers, food store owners, and restaurant operators about their ideas of the local food system. An important shift in eating patterns that many identified was the road put through to Deer Lake in the 1960s. As one resident from the town of Rocky Harbour said, "People came out of the gardens and went to the stores" (Cliff Butt, personal communication, July 2009). However, many people I spoke with also attested to the importance of fish in their diets, whether they attained it by fishing recreationally, keeping a share of the catch for their family if they were fishing commercially, buying it from the local fish plant, or enjoying a restaurant meal of 'fish and chips'.

As some aspects of traditional foodways decline and modes of food provisioning and consumption continue to undergo change, they are being taken up by the tourism industry. Culinary tourism is concerned particularly with the nature of food and eating (Long, 2004). In Newfoundland, traditional foodways are starting to be used by tourism operators as a way of promoting Newfoundland as a destination (Everett, 2009). As in other Maritime regions, seafood is an important aspect of traditional cuisine to which restaurants in Newfoundland are trying to cater. For example, Jacqui Hunter is the owner and operator of Java Jacks Restaurant and Café in Rocky Harbour, a town located in the world-class destination of Gros Morne National Park. When asked what types of food tourists are looking for, Jacqui says "people are looking for local seafood, fresh" (personal communication, July 2009). The well-known Bacalao Restaurant in St. John's describes itself as celebrating the cuisine and culture of Newfoundland and Labrador, offering "a unique and innovative presentation of the foods, flavours, produce and spirits of our province" (Bacalao, 2007). Bacalao sources local sustainably-caught potted cod from the Fogo Island Fish Cooperative, which is supplied by inshore vessels. The restaurant's innovative take on Newfoundland cuisine has received attention in popular tourist publications including Canadian Geographic Travel. There may be an emerging role for culinary tourism ventures in supporting small-scale fisheries. At the same time, we need to celebrate the food culture and heritage that lives on in common practice.

In conclusion, traditional Newfoundland foodways express a unique cultural identity and way of life tied to the sea and the land. The main focus of my essay has been on the integral place of the small-scale fishery in traditional Newfoundland foodways, having provided food and livelihood for many generations of Newfoundlanders. However, the land also provided important sustenance. Traditional foodways embodied a resource-based cycle of seasonally available foods from the land and sea and were an important part of an adaptive household strategy. While perhaps not as elaborately as in the past, elements of traditional Newfoundland foodways persist (Omohundro, 1994). Whether we're eating a dinner of fish and brewis, or harvesting wild blueberries on the barrens, O'Brien (1999) reminds us that practicing traditional foodways is an important reenactment of identity.

References

Bacalao (2007) http://bacalaocuisine.ca/ (accessed March 21, 2011)

Butt, J. (1998) The decline of the migratory fishery. Newfoundland and Labrador Heritage website. http://www.heritage.nf.ca/home.html (accessed March 16, 2011)

Cadigan, S. (2002) The role of agriculture in outport self-sufficiency. In Ommer, R.E. ed., *The resilient outport*. St. John's, NL: ISER Books, pp.241-262.

Crouch, M. and O'Neill, G. (2000) Sustaining identities? Prolegomena for inquiry into contemporary foodways. *Social Science Information*, 39(1), 181-192.

Everett, H. (2009) Vernacular health moralities and culinary tourism in Newfoundland and Labrador. *Journal of American Folklore*, 122(483), 28-52.

Faris, J. (1972) *Cat Harbour: A Newfoundland fishing settlement*. St. John's, NL: ISER Books.

Ferguson, M., (1996) Making fish: salt-cod processing on the east coast of Newfoundland, a study in historic occupational folklife. M.A. Thesis. St. John's: Memorial University of Newfoundland.

Fermor, B., Footer, A. and Simpson, A. (2010) Vegetarianism and identity in Santa Cruz. In Boyle, J., Colls, R., Evans, B. and Plater, A., eds., *Research Report Series No. 5*. University of Liverpool, pp. 39-44. http://www.liv.ac.uk/geography/undergraduate/fieldclasses/santacruz/SCRS2005/index_page.htm

Fisheries and Oceans Canada. (2010) Provincial and territorial breakdown of Canada's fish and seafood exports in 2009. http://www.dfo-mpo.gc.ca/media/back-fiche/2010/hq-ac10a-eng.htm (accessed March 29, 2011)

Godbout, G. (2008) Breton bread ovens of the Petit Nord: The archaeological landscape of foodways in the French fishing stations of Newfoundland. (M.A. Thesis). Memorial University of Newfoundland, St. John's.

Gray, P. (1977) Traditional Newfoundland foodways: origin, adaptation and change. M.A. Thesis. St. John's: Memorial University of Newfoundland.

Harris-Shapiro, C. (2006) Bloody shankbones and braided bread: the food voice and the fashioning of American Jewish identities. *Food and Foodways*, 14(2), 67-90.

Hashimoto, A. and Telfer, D. (2006) Selling Canadian culinary tourism: branding the global and the regional product. *Tourism Geographie*, 8(1), 31-55.

Higgins, J. (2007) 19th century cod fisheries. Newfoundland and Labrador Heritage website. http://www.heritage.nf.ca/society/19th_cod.html (accessed March 16, 2011)

Higgins, J. (2009) French involvement in the Newfoundland fishery. Newfoundland and Labrador Heritage website. http://www.heritage.nf.ca/society/french_fishery.html (accessed March 16, 2011)

Jesperson, I., ed. (1974) *Fat-back and molasses: A collection of favourite old recipes from Newfoundland and Labrador.* St. John's, NL: Ivan Jesperson.

Johnson, P. (2007) *Fish forever: the definitive guide to understanding, selecting, and preparing healthy, delicious, and environmentally sustainable seafood.* Hoboken, NJ: John Wiley and Sons Inc.

Long, L., ed. (2004) *Culinary tourism.* Lexington, KY: The University of Kentucky Press.

Margaret, L. (1980) *Fish and brewis, toutens and tales.* St. John's, NL: Breakwater Books.

Mariana, J. (1999) *Encyclopedia of American food and drink.* New York, NY: Lebhar-Freidman.

Marsden, T., Banks, J. and Bristow, G. (2000) Food supply chain approaches: exploring their role in rural development. *Sociologia Ruralis*, 40(4), 424-438.

Matthews, D. (1993) *Controlling common property: Regulating Canada's east coast fishery* Toronto, ON: University of Toronto Press Inc.

O'Brien, A. (1999) There's nothing like a cup of tea in the woods: continuity, community, and cultural validation in Newfoundland boil-ups. *Ethnologies*, 21(1), 65-83.

Ommer, R., Turner, N., MacDonald, M. and Sinclair, P. (2006) Food security and the informal economy. In Parrish, C., Turner, N. and Solberg, S., eds., *Resetting the kitchen table.* New York, NY: Nova Science Publishers Inc, pp.115-128.

Omohundro, J. (1994) *Rough food: The seasons of subsistence in Northern Newfoundland.* St.John's, NL: ISER Books.

Parrish, C., Turner, N. and Solberg, S., Eds. (2008) *Resetting the kitchen table.* New York, NY: Nova Science Publishers Inc.

Pope, P. (2004) *Fish into wine: The Newfoundland plantation in the seventeenth century.* Chapel Hill, NC: University of North Carolina Press.

Porter, M. (1995) 'She was skipper of the Shore Crew:' Notes on the history of the sexual division of labour in Newfoundland. In McGrath, C., Neis, B. and Porter, M., eds., *Their Lives and Times: Women in Newfoundland and Labrador.* St. John's, NL: Killick Press, pp. 33-47.

Power, N. (2000) Women processing workers as knowledgeable resource users: Connecting gender, local knowledge, and development in the Newfoundland fishery. In Neis, B. and Felt, L., eds., *Finding our sea legs.* St. John's, NL: ISER Books, pp.189-205.

Roberts, S. (2003) *For maids who brew and bake.* St. John's, NL: Flanker Press.

Scholliers, P. (2007) Twenty-five years of studying 'Un phenomene social total;' food history writing on Europe in the nineteenth and twentieth centuries. *Food, Culture and Society,* 10(3), 449-71.

Scoff (1990) In Story, G., Kirwin, W. and Widdowson, J., eds., *Dictionary of Newfoundland English.* Toronto, ON: University of Toronto Press, p. 438. Sinclair, P. and Ommer, R., eds. (2008) *Power and restructuring: Canada's coastal society and environment.* St John's, NL: ISER.

Smith, A., ed. (2007) *The Oxford companion to American food and drink.* New York, NY: Oxford University Press Inc.

Thursby, J. (2008) *Foodways and folklore.* Santa Barbara, CA: Greenwood Press.

Turner, N., Parrish, C. and Solberg, S. (2007) Introduction: food security in coastal communities in the context of restructuring: coasts under stress. In Parrish, C., Turner, N. and Solberg, S., eds., *Resetting the kitchen table.* New York, NY: Nova Science Publishers Inc., pp.1-12.

Tye, D. (2008) 'A poor man's meal': Molasses in Atlantic Canada. *Food, Culture and Society,* 11(3), 335-353.

CHAPTER 8

Preserving and managing Aboriginal small-scale fisheries: the experience of the Labrador Inuit

Lawrence Felt[1] and David Natcher

Spatially limited, smaller scale fisheries around the world involve aboriginal people. Such fisheries range from subsistence food fisheries that involve sale and barter to more commercial forms utilizing modern boats and technology of fairly modest size. Aboriginal fisheries have increasingly become formalized, and attendant rights legally entrenched in Canada and elsewhere by constitutional protection through treaty rights or, particularly with the Inuit of northern circumpolar regions, formal resolution of land claims agreements, and subsequent creation of regional aboriginal governments such as Nunavut, Nunavik, Inuvialuit and Nunatsiavut in Canada, the Greenland Home Rule government, and similar examples in Norway and elsewhere. The acquisition of newly acquired power and agency poses numerous challenges and opportunities in fisheries management. Focusing on Atlantic salmon (*Salmo salar L.*) and Arctic char (*Salivelinus alpinus*) we explore three challenges: (1) administrative 'stretch' or gap; (2) integration of traditional knowledge and science; and (3) effects of rapid environmental change. While each aboriginal experience is in some sense unique, our discussion based on Nunatsiavut, and to a much less extent the Nunavut, experience is a useful perspective to examine a number of shared challenges and opportunities facing many, if not most, aboriginal people.

Introduction

Aboriginal peoples constitute a relatively small percentage of Western populations; e.g., Europe (<1%), United States (2%), Canada (4%), Australia (2%) (Australian Bureau of Statistics, 2006; U.S. Census Bureau, 2006;

1 Corresponding author: Memorial University of Newfoundland, Canada, lfelt@mun.ca

Government of Canada, 2008). Even the Western-settled country of New Zealand has only 15% of its population self-identifying as Maori (Statistics New Zealand, 2006). These generally low populations in Western countries, however, belie thousands of aboriginal peoples residing outside urban areas who pursue fishing as an important source of food, nutrition, barter or income. In Alaska, Western and Northern Canada, Greenland, Northern Europe and Siberia, indigenous peoples are either a large and important plurality or an outright majority depending upon geographic location. As a consequence, any discussion of small-scale world fisheries would seem to necessitate their inclusion. The case for fishing is easily made. To illustrate, we offer examples from our work with the recently established Inuit regional government of Nunatsiavut on Labrador's north coast in Eastern, subarctic Canada focusing on two cornerstone species for the region, Arctic char (*Salvelinus alpinus)* and Atlantic salmon (*Salmo salar L.*). We explore three challenges and potential opportunities facing sustainable and effective future management of these fisheries. These are: (1) administrative 'stretch' or gap; (2) integration of traditional knowledge and science; and (3) effects of rapid environmental change. To do so, we examine new structures of resource co-management linking aboriginal peoples, their extensive knowledge and experience, as well as the ontological and epistemological foundation upon which they are based, with Western state models of fisheries management in emergent co-management models.

Aboriginal fisheries have increasingly become formalized, and attendant rights legally entrenched in Canada and elsewhere through constitutional protection in treaty rights or, particularly with the Inuit of northern circumpolar regions, formal resolution of land claims agreements, and subsequent creation of regional aboriginal governments such as Nunavut, Nunavik, Inuvialuit and Nunatsiavut in Canada, the Greenland Home Rule government, and similar examples in Norway and elsewhere. The acquisition of newly acquired power and agency poses numerous challenges and opportunities in fisheries management. While each aboriginal experience is unique in some ways, our discussion based on Nunatsiavut is a useful perspective to examine a number of shared challenges and opportunities confronting many aboriginal people who have secured or are pursuing greater authority in natural resource management.

Arctic char and Atlantic salmon: a Nunatsiavut illustration

Arctic char are the most important and widely distributed species in circumpolar regions. Above 75 degrees North, they are the only freshwater fish (DFO, 2007). For Inuit they have served important nutritional, economic and cultural purposes for thousands of years. Closely related to other cold water species, such as salmon and trout, two recognized subspecies are found as far north as Lake Hazen in Canada's Ellesmere Island in the Canadian high arctic (DFO, 2007). *S. alpinus erythrinus* is the most widely distributed and common species of char, while a smaller, dwarf subspecies *S. alpinus taranets* is frequently found in deep, coldwater arctic lakes. Known as Ikaluk in Labrador Inuttitut language, char, along with beluga whale and hood seal, are critical species profoundly connected to northern aboriginal people in a multitude of ways.

Highly adaptive, Arctic char readily adjust to a variety of freshwater ecosystems encompassing diverse hydrological and climate micro-environments. Where unimpeded access to saltwater is available most char become anadromous, displaying a lifecycle split between fresh and saltwater environments. The typical lifecycle begins with hatching in well oxygenated freshwater gravel, followed by a juvenile period in the surrounding area, and then migration to nearby saltwater estuaries for adult stages. Arctic char return to spawn in late summer and fall and overwinter in freshwater before repeating the cycle the following spring with migration back to the estuary. Prized for its rich, red flesh it is fished for income as well as subsistence provisioning, barter and exchange. In the fall, lines are strung with drying and curing char filets in permanent communities as well as seasonal summer fishing stations.

Atlantic salmon, *Salmo salar L.*, is arguably the second most important fish species to aboriginal people in the eastern Arctic. An anadromous species as well, the life cycle of salmon includes a much longer saltwater migration than char, in some instances thousands of kms, longer tenure in a saltwater environment, and more clearly demarcated lifecycle stages. Atlantic salmon begin life as eggs deposited in the gravel of clear flowing and well-oxygenated freshwater streams. Juvenile salmon spend between three and six or even seven years as adolescent parr in freshwater before migrating as 'smolt' to saltwater where, eschewing local estuaries, they travel several thousand kms where they spend anywhere from seven months to two years before returning to natal rivers as adults to renew their lifecycle (Taylor, 1985; Dempson *et al.*, 2001; Felt, 2008).

Arctic char and Atlantic salmon overlap to some extent in their distribution, though the latter prefers slightly warmer water, whether fresh or salt. In Labrador, char predominate above 56° north though they can be found at much lower abundance south to 51°. Atlantic salmon occupy a wider, slightly more temperate range from mid-Atlantic to subartic. Historically, Atlantic salmon were found from the Hudson River in New York State at 40° all the way to 56° north, though much of this range has been lost as a result of habitat destruction (Taylor, 1985; Felt, 2008). Today, Atlantic salmon in the Northwest Atlantic are limited to between 44° and 56° north except for an isolated northerly component in Ungava Bay, Northern Quebec. The most productive remaining range of Atlantic salmon lies within the lower two-thirds of Labrador between the most southerly Inuit community of Rigolet (54°10') to just north of its new capital, Hopedale (55° 28'). Since the retreat of the last ice age approximately 12,000 years ago, a succession of aboriginal people on Labrador's coast, including present Inuit and their ancestors, utilized the rich salmon rivers as an important nutritional, social and economic resource, and continue to do so.

The last fifty years have brought profound political, socio-economic, demographic and cultural changes to the Inuit and Kablunangajuit (settlers living on the Labrador coast, many of whom have intermarried with their Inuit neighbors; see Ben-dor, 1966; Kennedy, 1977, 2007). While still geographically remote in the sense that no roads link communities to larger Canadian populations, bigger and more powerful ships bring supplies for nearly six months of the year, new airports link communities to each other and more southerly areas on a daily basis, though fares are very high. Forms of modern state presence are now more visible and palpable to people as well. Government offices, Inuit and White alike, administering a wide array of programs, are evident in each community, and state sponsored initiatives to bring modern technology through television and internet reach virtually every household on the coast. Modern industrial job opportunities have arisen through resource developments such as the Voiseys Bay nickel mine just south of Nain. The buzz and scurry of resource exploration is widespread, symbolized by the pulsating whirl of helicopter blades heard in every community. Unemployment remains crushingly high, particularly during winter, but wider opportunities for wage labor have grown to now comprise the largest source of household income (Government of Newfoundland and Labrador, 2006).

Yet many of the old ways remain evident, particularly harvesting and curing for personal consumption or barter. Household harvested char and salmon are both frozen in modern freezers and hung outside to dry by strong summer and fall breezes to be preserved for long winters. While the commercial fishery for Atlantic salmon ended in 1998, household consumption and barter remains very strong, depending on availability and community. Upwards of 76% of households continue to harvest and over 90% benefit through sharing or bartering. In more northerly Nunatsiavut communities where arctic char dominate, similar harvesting and sharing levels prevail alongside a commercial net-based fishery (Felt *et al.*, 2011). Research dating back 25 or more years in Labrador (Usher, 1982; Brice-Bennett, 1977, 1986; Mackay and Orr, 1988; Williamson, 1997) and comparison with a recent Natcher *et al.* (2011) study, reveals stable to moderately increased harvesting of salmon and char respectively. Research from other northern jurisdictions in Canada, the United States and Northern Europe indicates a strong continuity in fishing for subsistence, exchange, barter and more limited commercial fisheries for the two species as well (Magdanz *et al.*, 2002; NWMB, 2004; Naves, 2009; Alaska Fish and Game, 2010)

Despite undeniable economic, technological and social changes, fishing continues to be a foundational activity for a wide range of aboriginal peoples in North America and Northern Europe. Evidence from South America, Africa, Southeast Asia, Australia and New Zealand support this general view as well. While the literature is too voluminous to adequately review here, the New Zealand experience is particularly well studied and important given the large percentage of Maori people there. As a result of original treaty rights supplemented by extensive lobbying, Maori have been successful in reasserting traditional subsistence and barter rights, have gained rights to co-management, and expanded their presence in selected commercial fisheries though numerous tensions remain, particularly where ITQs (Individual Transferable Quotas) and a larger Quota Based Management System prevail, built on Western conceptions of property and competition (Hersoug, 2002; Day, 2004).

Where Western peoples have colonized areas previously occupied by aboriginal peoples, fishing and other harvesting rights have typically been lost or significantly diminished. The struggle to reclaim, entrench and sometimes expand them has involved political battles and acrimonious litigation. Where older treaties existed, effort was required to gain contemporary statutory relevance; where treaty rights never existed, alternative strategies for recognition and legal entrenchment needed to be pursued. Where traditional homelands were beyond settled lands associated with colonization and usurpation, the strategy was largely oriented towards land claims negotiations through which control and governance of traditional aboriginal homelands was pursued, including rights to primary harvesting of renewable resources such as fish. While not every jurisdiction mirrored the Canadian experience, parallels are sufficient that a discussion of that process is useful to understanding the larger issue of aboriginal fisheries.

The entrenchment of aboriginal fishing rights

In Canada, legal recognition of aboriginal rights respecting subsistence and bartering of renewable natural resources such as fish are less than thirty years old. In Calder v. British Columbia, the Canadian Supreme Court in 1973 confirmed that aboriginal people's historical occupation of specified lands provided legal use rights to those lands despite subsequent European settlement. Nine years later, in1982, Canada repatriated and modified its Constitution, maintained in Great Britain by virtue of Canada's Dominion status. Section 35 of the updated Constitution formally entrenches those rights for aboriginal people, including First nations, Metis and Inuit. While entrenching such rights, Section 35 was unspecific in defining what such a right explicitly was, its extent and possible limitations on it. This clarification was to several subsequent decisions, both, incidentally, related to traditional fishing rights. Two cases are particularly noteworthy, R. v Sparrow in 1990 (Allain, 1996) and R. v. Marshall in 1999 (Caldwell, 1999). In R. v Sparrow, the Supreme Court of Canada held that inherent aboriginal rights, including the right to fish, that pre-existed repatriation of the Constitution in 1982, were not extinguished and, in fact, were protected in Section 35(1). Moreover, rights were explicitly defined as activities, practices, and traditions that are integral to the distinctive culture of aboriginal peoples in Canada.

Neither Section 35 nor Sparrow exhaustively defined "aboriginal rights" nor established limits associated with them. These issues were addressed nine years later in R. v Marshall (1999). In this case, the Canadian Supreme Court extended treaty rights to the modern era by including the right to sell and barter fish harvested under aboriginal title. In fact, the court ruled that such rights even survived any exclusive rights to fish that might have been rescinded by the same or a subsequent government (Caldwell, 1999). According to Caldwell:

> The court pointed out that the treaty only gave a right to trade for "necessaries". In a modern day context, this translates into a right to earn a "moderate livelihood". . . a moderate livelihood "includes such basics as 'food, clothing and housing, supplemented by a few amenities', but not the accumulation of wealth (Caldwell, 1999, p. 1).

To ensure fisheries remained sustainable the court, drawing on earlier rulings supporting the right of the Crown to set reasonable limitations to address conservation issues, concluded that aboriginal treaty rights, while inalienable and permanent, were not absolute and were limited by any necessary conservation requirements.

While the Sparrow and Marshall decisions were inclusive of all abortiginal peoples, Inuit people had few or any treaties with Europeans due largely to their more northerly, remote location. For them, legal recognition under Section 35 as well as the treaty-based court rulings based on First nation experiences provided added impetus for a different way of articulating Inuit rights generally, and fishing rights in particular. Given the remoteness of their traditional land, articulation of such rights was pursued through movements to acquire ownership and political control over large tracts of traditional land(s) and through such control Inuit management of fishing and other traditional activities.

These legal precedents added pressure on Canadian and provincial governments to conclude ongoing negotiations between Inuit leaders and the Federal Government (and provincial governments in the case of Labrador and Quebec) regarding traditional land claims and the creation of Inuit regional governments to govern these lands. The result has been four largely Inuit northern regional governments in Canada comprising approximately 25% of Canada's land mass, including arctic islands, and 50,000 people, about 85% Inuit. The first was the Western Arctic Settlement Act of 1984 establishing the Inuvialuit Regional Corporation (IRC) in the Western Arctic. This was followed by the Nunavut Land Claims Agreement (NLCA) completed in 1993 (NLCA, 1993; Berger, 2006) resulting in the creation of the largest Inuit regional government of Nunavut. Since then, regional Inuit governments of Nunatsiavut in Labrador (2005) and Nunavik in Quebec (finalized in 2010) have also been established.

While all four regional governments represent traditional Inuit homeland(s), only Nunatsiavut is ethnically Inuit in the sense that membership is limited to those defined as Inuit by the agreement. Nunavut, IAC and Nunavik are essentially Inuit regional governments by demography rather than legislation. All residents of each designated settlement area are deemed to be equal citizens and have equal rights to political participation, but not traditional terrestrial and marine harvesting rights. Since approximately 85% of Nunavut's 32,000 residents trace their roots to Inuit ancestry, as do more than 90% of Nunavik's 12,000, they are *de facto* rather than *de jure* Inuit governments as is IAC. Nunatsiavut, in contrast, defines citizenship and associated harvesting rights to those defined as Inuit beneficiaries under the Land Claims Agreement (LILCA, 2005; Procter *et al.*, 2011).

Regardless of how rights of access and use might be defined, an equally important issue is the institutional configuration through which management is implemented. In land claims' situations, management of renewable resources such as fish is implemented through some type of bi- or tri-partite, arms-length co-management board depending upon whether a species is covered under provincial or federal jurisdiction. Establishment of such boards is commonly included in the text of land claims agreements. Government representatives are typically scientists and managers drawn from relevant wildlife ministries. Aboriginal representation includes a certain number of respected elders, ministerial representatives from the land claims governments, and other aboriginals considered to possess particular knowledge about species under the board's mandate. Aboriginal members constitute at least half of the board's membership. The intention is to ensure an aboriginal 'voice' (s) is clearly embedded in the board's approach with a resulting aboriginal agency or capacity to influence resource management. To meet their mandate, boards undertake their own research as well as fund other work necessary for management plan development such as wild food harvest studies (NWMB, 2004). Resulting information is shared with Inuit regional governments as well for their own separate recommendations to provincial and national governments. Under such arrangements, there is a clearer mandate as well as considerable political pressure to approve board recommendations in the absence of extremely convincing contrary evidence.

Two relevant examples in Canada are the Nunavut Wildlife Management Board (NWMB) and the Torngat Wildlife, Plants and Fisheries Secretariat created as a result of the Labrador Inuit land claims settlement (LILCA, 2003, ch. 12-13). The NWMB, by terms of the Nunavut Land Claims Agreement, is mandated to "help ensure the protection and wise use of wildlife and wildlife habitat for the long term benefit of Inuit and the rest of the public of Nunavut and Canada" (NWMB, 2004). It consists of nine members, five of which are Inuit and/or reside in the region. Others include a representative, usually a scientist, appointed by the Canadian minister responsible for fish and marine animals, scientist and/or manager from the Canadian Wildlife Service, a public appointee and a neutral chair appointed upon the recommendation of the board itself.

The Torngat Wildlife, Plants and Fisheries Secretariat has a slightly different configuration. It consists of two co-management boards and a support secretariat comprising an executive director and several resource and support positions. Its mandate, similar to NWMB, is the conservation and sustainable utilization of fish, wildlife, plants and habitat within the Labrador Inuit Settlement Area, including catch controls, allocations, licensing, harbor management and aquaculture. Each joint board has six members including one from the Canadian government, one from the provincial and three appointed by the Inuit regional government. A chair is appointed based on the recommendation of each board.

Both the NWMB and Torngat boards resemble smaller versions of contemporary government wildlife departments and agencies and are explicitly science-driven. Unlike their government counterparts, however, they are mandated to collect and utilize other types of knowledge, most importantly traditional knowledge and experience held by aboriginal residents, in assessment and management plan development. The boards ensure that all relevant knowledge, information and perspectives, regardless of its epistemological and ontological basis, contributes to comprehensive management plans that reflect Inuit voices alongside positivist science.

The recognition of aboriginal rights, the creation of institutions to reflect those rights, and the construction of new co-management initiatives to ensure sustainable harvest consistent with traditional aboriginal practices represent a dramatic shift in government thinking and a significant increase in aboriginal agency. It is not surprising, therefore, that these changes pose enormous challenges and opportunities for newly formed aboriginal governments, their people, and the boards created to implement these changes. While these challenges find their way into multiple spheres of aboriginal life, we have spent a considerable time describing these new management boards because it is within them that we see many of the more important challenges and potential opportunities. We now turn to three such challenges

Challenge 1: Administrative stretch or gap

The creation of aboriginal regional governments and more meaningful and extensive participation in co-management boards places a significant pressure on the qualified labor available in aboriginal communities. Exacerbating the situation is the expectation that aboriginal polities structure their governance institutions along western state lines to facilitate 'government to government' relations. This means that no matter how small and scattered the population or vast the territory, such aboriginal governments must institutionally mirror their provincial and national counterparts. Thus, Nunatsiavut must have a Ministry of Environment with one or more deputy ministers, assistant deputy ministers, directors, etc. to serve the Minister despite a total population, including children, of approximately 7,000.

A number of ramifications result. Four important ones include: the merging of ministries and subsequent diffusion of roles leading to inefficient use of a limited number of skilled people, the undermining of traditional models of aboriginal governance, the difficulty in filling positions with fully qualified personnel, and frequent turnover as those with sought-after credentials continuously move in response to better opportunities (Jorgenson, 2007; Felt and Natcher, 2008). Our research with the Nunatsiavut government and the Torngat Secretariat provides numerous examples of each.

With a population of approximately 7,000 beneficiaries and a territory of 72,000 square kms, the new government decided to combine ministries. As a result, separate ministries were created for education and economic development, health and social development, finance and human resources, as well as Nunatsiavut affairs. At the beginning of our work we spent much of a day with the Deputy Minister (DM) responsible for our area of research. In a typical day, the DM's cell phone rang forty to fifty times. Two calls illustrate the range of issues responded to over the phone. One call was from a confused and irate older beneficiary from the most northerly community frustrated with the new requirement for a permit to collect driftwood. The other was from a high level bureaucrat of the European Union who, upon unsuccessful in reaching a government minister, was transferred to this DM for assistance in planning a special presentation a Nunatsiavut minister was invited to make before the EU assembly the following spring. While the request came for a minister in a different department, the receptionist, who served several different ministries, and not knowing who to forward the call to since higher level administrative positions in the relevant ministry were then vacant, passed the call to this Deputy since his office was just down the corridor.

A second concern is the replication of a hierarchical government structure similar to modern states. Traditionally, Inuit governance was far less formal and considerably 'flatter'. Elders provided much of the authority and guidance when travelling and formalized Elder Councils were promoted as more permanent settlement evolved alongside non-Inuit in the coastal communities. By the late 1950s, nascent Inuit political organizations emerged in Labrador, most notably the Labrador Inuit Association (LIA), and agitation for land claims resolution started in the early to mid-1970s. While explicit institutional models for a homeland government were not formalized in these negotiations since the objective was to secure land, it was implicit that some form of government reflective of Inuit values and tradition would be critical in building an Inuit society consistent with Inuit values and culture. 'Government to government' structures probably do not preclude this objective but they make it less direct and potentially more difficult to achieve. The same is likely true regarding management of wildlife harvesting.

Difficulty in filling positions and frequent turnover has meant a large number of government employees are not designated beneficiaries, and it is common for those who are beneficiaries to lack educational credentials and job experience. In our work, we are now dealing with our fourth manager of renewable resources in just less than four years. Each incoming manager had little knowledge of previous work, records were often difficult to locate, and considerable time has been spent reinforming them. As a result, there is little organizational memory and activities take much longer to complete. This will change as more and more beneficiaries graduate from high school and secondary schools to acquire technical skills needed to run modern governments. In fact, there is already evidence in the increased numbers of Inuit beneficiaries in university and trade schools. Established programs in nursing and education, delivered in part through the internet, have existed for some time and have graduated a number of nurses and teachers; a new program in social work has also just begun. Some progress is also being made in training the many other professions required for effective government but they are further away. Challenges typically create opportunities to build longer term capacity and we are quite optimistic that this is happening. In the present, those who are beneficiaries and possess appropriate skills and credentials are highly mobile within the government and parallel organizations such as the Torngat Co-management Secretariat.

The overall result is that necessary research in support of resource management is limited and delayed, with increasing reliance on contractual research by outside parties. With respect to Atlantic salmon, where there is some concern regarding stock health, it has proven difficult to start needed research.

Challenge 2: Integrating traditional knowledge and western science

A major objective of established as well as newly created aboriginal governments is to build capacity in preserving, protecting and vitalizing the culture, language and understandings that define them as an aboriginal people. In resource management, this means finding ways to make traditional knowledge and experience gleaned from hundreds if not thousands of years on the land, sea and ice relevant to resource management. The issue is complicated in that this needs to be done within a resource management regime premised on a western science paradigm and its positivist ontology and epistemology (Neis and Felt, 2000). The Nunavut government and the Nunavut Wildlife Management Board has been committed in principle to equal consideration of both traditional Inuit knowledge and modern science, though there has been some frustration and disagreement over how this might be best achieved. To understand the issues, a brief review of Inuit Knowledge is useful.

Inuit knowledge or *Inuit Qaujimajatuqangit* is composed of traditional understanding, experience and values held by Inuit derived from occupying the land, sea and ice. Dynamic and holistic, it builds on historical understanding to inform contemporary Inuit knowledge. In so doing, it links past with present and future and in so doing creates a contemporary society informed by its past yet capable to adapt and change to present and future challenges without losing its sense of historical place. The term is derived from the verb root *qaujima* meaning to know, and has a more general translation as 'that which has been long known' by Inuit (Thorpe *et al.*, 2002; White, 2006; Nadasdy, 2007; Armitage, 2009; Qaujisarvingat Centre, n.d.). *Qaujimajatuqangit* is collectively produced and shared, though some individuals, particularly elders, may acquire more detailed and extensive forms. It is expressed in story and anecdote and difficult, for a variety of ontological as well as epistemological reasons (see in this regard Wenzel, 2004; Ellis, 2005; Nadasdy, 2007; Armitage, 2009), to translate easily into western paradigms of science.

To the extent that Inuit Qaujimajatuqangit becomes incorporated into resource management, a delicate balancing must occur that recognizes and respects its uniqueness and great divergence from western scientific understanding, while simultaneously finding a means of utilizing it alongside some form of scientific orientation.

But exactly how might this be done? Can western researchers, all too often minimally conversant in local language, history and culture necessary for contextualizing knowledge, proceed to extract 'testable propositions' for their own use without devaluing it? Is such a process of scientization (Ellis, 2005), in which understanding communicated in stories is stripped of descriptive and contextual content for analytical and systematic elements, possible or even ethically acceptable?

Academic literature is mixed on the utility and appropriateness of such efforts. Several social scientists (Samson, 2003; Ellis, 2005; White, 2006; Nadasdy, 2007) are skeptical, even condemning, of such efforts characterizing them as flawed and disrespectful. In contrast many natural scientists respect and appreciate traditional knowledge but find it difficult to utilize unless it provides either new research questions or directs the investigators towards previously unacknowledged relationships for exploration using more traditional scientific techniques (Fischer, 2000). In the middle are social scientists who accept the distortion and incompleteness of extracting insights from traditional knowledge for use in some variety of scientific paradigm, but condone it because alternative and potentially insightful understanding may result. In siding with this middle approach, our view is similar to that of Armitage (2009), Menzies (2006), Wenzel (1999) and others. In particular, Wenzel notes similar features in both epistemologies, despite profoundly different origins, that offer some possibility of transportability and relevance to researchers. This may be one of the reasons that many Inuit leaders and bureaucrats have frequently expressed a keen interest to acquire scientific research skills as a complement to more traditional ways of knowing, as they increasingly participate in scientifically dominated co-management committees on a wide range of topics affecting them.

In our experience, successful utilization of aboriginal knowledge in any form ultimately depends upon understanding, transparency, respect and trust between academic and aboriginal people, as well as the knowledge itself. Our experience suggests three critical conditions to successfully connecting the very real differences between scientific and traditional knowledge and utilizing each. When all are met, it is possible to construct partial yet respectful and heuristic bridges between academic researcher and Aboriginal knowledge holder, as well as balancing critical inquiry with respect for difference and distinctiveness.

The first condition involves the sharing and understanding of what critical inquiry does and does not mean with Aboriginal research partners. Critical inquiry means questioning and asking why and how, not for the sake of being negative or disbelieving but of understanding. It is constructive not dismissive. We have found that as both Inuit knowledge and western science are dynamic knowledge systems continuously evolving and changing, this focus provides a reasonable foundation. When this has been done, many discussions have ensued about different ways of knowing about many things and their interconnectedness in mutually respectful ways. The second condition is to involve everyone as much as possible in all phases of research and encourage questions drawn from either or both scientific and traditional perspectives. The third is to admit that we are, perhaps of necessity, distorting and stripping traditional knowledge from its larger context and meaning to satisfy a more limited understanding and use. In this acknowledgment we are in considerable sympathy with Armitage's (2009, p. 14) comment that in utilizing Aboriginal knowledge for land use occupancy research we are "distilling, compartmentalizing, decontextualizing Aboriginal knowledge from the integrated, holistic knowledge of Aboriginal land users". In a conversation with an Inuit elder, the first author used the word 'parsimonious' to describe how he searched for what the elder considered simple, superficial generalities of little use. The elder still laughingly refers to the author as his parsimonious friend, though he may very well consider anyone in search of such parsimony to be out of touch with the world around them.

As with our 'government to government' example, conditions starting out as challenges have given birth to opportunities that hopefully will make resource management effective and accepted by all parties. In our work, following the above middle path has led to a number of joint initiatives utilizing both forms of knowledge. A particularly interesting one is a film documentary done through the Torngat Secretariat and local communities, following fishers harvesting salmon in their subsistence fishery while communicating how their traditional knowledge leads them to employ the strategies they do with regard to where to fish, when to fish, how to fish, when to stop fishing, and a number of other considerations highly relevant to resource management.

Challenge 3: Adapting to rapid environmental change

Food security and continuity is an increasingly salient topic throughout the arctic as milder climate poses significant challenges to access, abundance and availability of several traditional wild foods. Both arctic char and Atlantic salmon are keystone species during climate moderation due to their morphology and behavior. Both are anadromous, though char migrate only to their adjacent saltwater estuaries or slightly beyond while multi sea winter salmon migrate thousands of kms to southwest Greenland. Habitat requirements are similar resulting in considerable overlap in distribution in arctic and northern boreal regions of the Northwest Atlantic. Atlantic salmon are more flexible, however, tolerating a wider range of water temperature at various life cycle stages. There is already some information suggesting char displacement in its southern range and salmon claiming these regions as fresh and saltwater temperatures increase, though char that remain appear to be larger due to increased prey availability associated with warmer (Reist *et al.*, 2006; Dempson *et al.*, 2007; DFO, 2007; Felt, 2008).

Climactic change is still not well understood at species-specific levels. Thus, both will provide critical information relevant to their own future range and population as well as more general consequences for Inuit harvesting practices and consumption. It is therefore critical that baseline information, drawing on both scientific evidence as well as *Inuit Qaujimajatuqangit,* be assembled to understand the magnitude and consequences of continued climate warming. In this way, local research can supplement ongoing scientific research focusing on arctic char in more northerly locations. To the extent that the opportunities associated with previous challenges are acted upon, Nunatsiavut, Nunavut and their respective wildlife and fisheries co-management boards are particularly well situated to utilize this important research opportunity.

Conclusion

Aboriginal people are important participants in small-scale fisheries for consumption, barter, exchange and wage income throughout the world. It is therefore imperative that they be included in any analysis of small-scale fisheries. It is likely that challenges and opportunities they face are shared with other similar fishers. Drawing on examples primarily from Labrador, we have reviewed the legal and legislative mandates entrenching their rights in these fisheries, as well as three important challenges as they acquire important management roles associated with these rights. To stress the positive side of these challenges, we suggest that each be seen as an opportunity. In our discussion, we outline these opportunities as well as various strategies underway to realize them. Challenges and opportunities are ultimately about capacity, the capacity to manage renewable resources to directly benefit aboriginal people in ways that are sustainable and consistent with the values and understanding that has guided them in resource utilization for hundreds and, in some cases, thousands of years. To the extent that experiences and challenges discussed here have resonance for other aboriginal fishing people beyond Canada and northern Europe, we hope this discussion will be useful.

References

Alaska Fish and Game (2010) http://www.subsistence.adfg.state.ak.us/geninfo/publctns/articles.cfm

Allain, J. (1996) Library of Parliament, Government of Canada. Parliamentary Information and Research Services. http://www2.parl.gc.ca/Content/LOP/ResearchPublications/bp428-e.htm

Armitage, P. (2009) Legalization of Aboriginal knowledge in Canada: methodological, ethical and property rights implications. Sami Rights in Coastal Landscapes and Seascapes Conference, April 22-24, 2009. Norway: University of Tromsø.

Australian Bureau of Statistics (2006) Census QuickStats.

Ben-Dor, S. (1966) *Makkovik: Eskimos and settlers in a Labrador Community.* Newfoundland Social and Economic Studies, 4. St. John's: ISER.

Berger, T. R. (2006) Conciliator's Final Report. http://www.ainc-inac.gc.ca/al/ldc/ccl/fagr/nuna/lca/nlc-eng.pdf.

Brice-Bennett, C. (1986) Renewable resource use and wage employment in the economy of Northern Labrador. Background Report for the Royal Commission on Employment and Unemployment.

Brice-Bennett, C. (1977) *Our Footprints are Everywhere: Inuit Land use and Occupancy in Labrador.* Nain: Labrador Inuit Association.

Caldwell, B. (1999) Fisheries Law Papers and Articles. http://www.admiralitylaw.com/fisheries/Papers/treatyrights.htm

Day, A. (2004) Fisheries in New Zealand: The Maori and the quota management system. Report prepared for First Nation Panel on Fisheries. http://www.fns.bc.ca/pdf/NewZealand.pdf.

DFO (2007) Can Arctic char adapt to climate change? DFO 2007-1321. Cat. No. Fs23-523/1-2007E. Ottawa: Department of Fisheries and Oceans.

Dempson, J. B., O'Connell, M. F. and Cochrane, N. M. (2001) Potential impact of climate warming on recreational fishing opportunities for Atlantic salmon, *Salmon salar* L., in Newfoundland, Canada. *Fisheries Management and Ecology,* 8, 69-82.

Dempson J. B., Shears, M., Furey, G. and Bloom, M. (2007) Resilience and stability of north Labrador Arctic char, *Salvelinus alpinus,* subject to exploitation and environmental variability. *Environmental Biology of Fishes,* 83(1), 57-67.

Ellis, S. C. (2005) Meaningful consideration? A review of traditional knowledge in environmental decision making. *Arctic,* 58(1), 66-77.

Felt, L. (2008) Importance historique et disponibilite future du saumon atlantique (*Salmo salar* L.) pour les populations de la cote du Labrador dans un context de changements climatiques. *Saumons Illimites,* 31(3), 20-24.

Felt, L. and Natcher, D. (2008) Effective governance in the changing political landscape of Labrador. *Newfoundland Quarterly,* 101(1), 32-35, 42.

Felt, L., Natcher, D., Procter, A., Sillitt, N., Winters, K., Gear, T., Winters, D., Andersen, S., Ford, R., Flowers, H., Rich, S. and Kemuksigak, R. (2011) The more things change . . .: patterns of country food harvesting by the Labrador Inuit on the North Labrador coast. In Natcher, D., Felt, L. and Procter, A., eds., *Settlement, Subsistence and Change Among the Nunatsiavummiut (Labrador Inuit)*. Winnipeg: University of Manitoba Press, pp. 176-227.

Fischer, J. (2000) Participatory research in ecological fieldwork: a Nicaraguan study. In Neis, B. and Felt, L., eds., *Finding Our Sea Legs: Linking Fishery People and Their Knowledge with Science and Management*. St. John's: ISER Books, pp. 41-54.

Government of Canada (2008) Aboriginal Peoples in Canada in 2006. http://www12.statcan.ca/census-recensement/2006/as-sa/97-558/p2-eng.cfm

Government of Newfoundland and Labrador (2006) Community Accounts. http://www.communityaccounts.ca/communityaccounts/onlinedata/getdata.asp

Hersoug, B. (2002) *Unfinished Business: New Zealand's experience with Rights-Based Fisheries Management*. Delft, Netherlands: Eburon.

Jorgenson, M., ed. (2007) *Rebuilding Native Nations: Strategies for Governance and Development*. Tucson: University of Arizona press.

Kennedy, J. (1977) Northern Labrador: an ethnohistorical account. In Paine, R., ed., *The White Arctic: Anthropological Essays on Tutelage and Ethnicity*. St. John's: ISER.

Kennedy, J. C. (2007) New foods, lost foods: local knowledge about changing resources in coastal Labrador. In Parrish, C., Turner, N. J. and Solberg, S. M., eds., *Resetting the Kitchen Table: Food Security, Culture Health and Resilience in Coastal Communities*. New York: Nova Science Publishers, pp. 87-98.

LILCA (Labrador Inuit Land Claims Agreement) (2003) Labrador Inuit Association.

Mackey, M. A. and Orr, R. (1988) The seasonal nutrient density of country food harvested in Makkovik, Labrador. *Arctic*, 41(2), 105-109.

Magdanz, J., Charles, S., Utermohle, J. and Wolfe, R. J. (2002) The production and distribution of wild food in Wales and Deering, Alaska. Technical Report 259. Juneau, Alaska: Division of Subsistence, Alaska Department of Fish and Game.

Menzies, C. (2006) *Traditional Ecological Knowledge and Natural Resource Management*. Lincoln, Nebraska: University of Nebraska Press.

Nadasdy, P. (2007) The gift in the animal: the ontology of hunting and human-animal sociality. *American Ethnologist*, 34(1), 25-43.

Naves, L. C. (2009) Alaska migratory bird subsistence harvest estimates, 2004-2007. Technical Paper 349. Anchorage, Alaska; Alaska Department of Fish and Game, Subsistence Division.

Neis, B. and Felt, L., eds. (2000). *Finding Our Sea Legs: Linking Fishery People and Their Knowledge with Science and Management.* St. John's: ISER Books.

NLCA (Nunavut Land Claims Agreement) (1993) Agreement between the Inuit of the Nunavut settlement area and Her Majesty the Queen in Right of Canada. http://www.tunngavik.com/documents/publications/1993-00-00-Nunavut-Land-Claims-Agreement-English.pdf

NWMB (Nunavut Wildlife Management Board) (2004) Nunavut Wildlife Harvest Study. Nunavut, Canada.

Procter, A., Felt, L. and Natcher, D., eds. (2011) Introduction to settlement, subsistence and change among the Nunatsiavummiut (Labrador Inuit). Winnipeg: University of Manitoba Press, forthcoming.

Qaujisarvingat Centre (Inuit Tapiriit Kanatami. Inuit Knowledge) (No date) http://www.inuitknowledge.ca/

Reist, J. D., Wrona, F. J., Prowse, T. D., Power, M., Dempson, J. B., King, J. R. and Beamish, R. J. (2006) An overview of effects of climate change on selected arctic freshwater and anadromous fishes. *AMBIO*, 7, 381-387.

Samson, C. (2003) *A Way of Life That Does Not Exist: Canada and the Extinguishment of the Innu.* St. John's: ISER.

Statistics New Zealand (2006) QuickStats about New Zealand.

Taylor, V. R. (1985) The early Atlantic salmon fishery in Newfoundland and Labrador. Canadian Special Publication of Fisheries and Aquatic Sciences 76. Ottawa: Department of Fisheries and Oceans.

Thorpe, N., Hakongak, N., Eyegetok, S. and the Kitikmeot Elders (2002). *Thunder on the Tundra: Inuit Qaujimajatuqangit of the Bathurst Caribou.* Tuktu and Nogak Project. Vancouver: Generation Printing.

U.S. Census Bureau (2006) Facts for Features.

Usher, P. (1982) Renewable Resources in the Future of Northern Labrador. Report for the Labrador Inuit Association, Nain.

Wenzel, G. (1999) Traditional ecological knowledge and Inuit: reflections on TEK research and ethics. *Arctic*, 52(2), 113-124.

Wenzel, G. (2004) From TEK to IQ: Inuit *Qaujimajatuqangit* and Inuit cultural ecology. *Arctic*, 41(2), 238-250.

White, G. (2006) Cultures in collision: traditional knowledge and Eurocanadian government processes in Northern land claims boards. *Arctic*, 59(4), 401-414.

Williamson, T. (1997) From Sina to Sikujâluk: our footprint. mapping Inuit environmental knowledge in the Nain District of Northern Labrador. Prepared for the Labrador Inuit Association. Nain, Labrador.

CHAPTER 9

Contemporary visions for small-scale aquaculture

Ben Belton[1] and David C. Little

This essay considers 'small-scale aquaculture' (SSA) from a variety of perspectives. We first examine the origins and usage of the term and offer a definition based on the social characteristics of those who practice it. We examine drivers of contemporary SSA, the various roles that it plays in supporting agrarian livelihoods and its relationship to wellbeing and poverty with references to examples drawn from across Asia. The challenges and opportunities presented to small-scale producers and culture systems are assessed, and the likely future of small-scale aquatic production systems is discussed.

Introduction

The literature on small-scale fisheries is highly multidisciplinary, with strong social scientific underpinnings. Although much of this work is concerned with implementing approaches to resource management that fulfil development goals (reducing poverty and improving the sustainability of natural resource use perhaps most pressing among them), considerable emphasis has also been devoted, only partly instrumentally, to understanding small-scale fisheries as social-ecological systems in their own right. This is in contrast to work on small-scale aquaculture (SSA) which has remained largely the domain of biologists, and which has more often than not assumed a rather normative tone, concerned principally with enabling the establishment of SSA as a productive activity assumed *a priori* to have positive implications for development. This has meant that detailed intellectually satisfying analyses of SSA and its implications with respect to poverty, livelihoods, food security and gender relations have been notable by their absence; since although most accounts of SSA are liberally peppered with references to these concepts,

1 Corresponding author: The WorldFish Center, Bangladesh, b.belton@cgiar.org

they are usually conceived of in unproblematic and rather mechanistic terms and largely without reference to any broader political-economic context or rigorous interpretive frameworks. Attempts to redress this balance have led to recent debate which has challenged conventional wisdom regarding the role of SSA in poverty reduction and how SSA should best be defined.

It is not our intent to revisit these questions extensively here. Rather we aim to move understanding of SSA forward by considering contemporary culture systems from a variety of perspectives. In doing, so we attempt to trace their origins, to re-evaluate the roles they play in supporting agrarian livelihoods, and to assess the challenges and opportunities presented to small-scale systems of aquaculture and small-scale aquaculture producers. We first provide a synopsis of the SSA narrative and a working definition of SSA to elucidate our treatment of the topic in this essay. The second section examines factors driving development of small-scale forms of aquaculture, particularly in Asia. The third section addresses a variety of pressures and incentives affecting SSA which determine choices regarding management and market orientation. A fourth section looks at some of the livelihood functions of SSA and its implications for wellbeing and vulnerability. Section five asks whether SSA will remain viable in the face of increasingly globalized trade in aquatic commodities and associated regulation of production. The conclusion looks to the likely future of small-scale aquatic production systems.

Origins and redefinition of SSA

'Small-scale aquaculture' has been championed by development institutions for well over three decades, with reference to a narrative which cites its potential to alleviate poverty, enhance food security and promote economic development. Emphasis on *small-scale* forms of aquaculture derives in part from their being positioned in implicit opposition to the 'industrial' shrimp production promoted by multi-lateral donors and national governments during the 1980s, which subsequently fell out of favor as a development strategy in the face of associated ecological degradation and social dislocation that occurred during this period. Nakamura (1985, p. 97) is instructive with regard to the dichotomy in approaches to aquaculture development that emerged as a result:

> There are two fundamental approaches to applying aquaculture in developing countries: improving small-scale, subsistence-level

> operations to meet immediate local needs or establishing large-scale, commercial industry based on the production of expensive species for export. The latter approach, geared toward increasing cash flow and therefore foreign exchange, may provide some employment for the poor, but mainly benefits only a small sector of society. The former strategy, however, directly benefits a larger number of people, especially the poor, by providing jobs and a modest income as well as a source of inexpensive protein. These two basic approaches are not mutually exclusive, but small-scale aquaculture is the more appropriate approach for rural communities[2].

As this excerpt indicates, emphasis on *small-scale* aquaculture is based on the assumption that the limited nature of the resources (land, labour and inputs) required for its practice places it within the grasp of poor, subsistence or semi-subsistence farming households that would otherwise be excluded from partaking in any benefits derived from larger commercial aquaculture operations. As Edwards *et al.* (2002, p. 223) put it, 'Rural aquaculture [a term used interchangeably with SSA] defies simple definition although the beneficiaries are clear: the poor'. In addition, SSA has been characterized variously as family owned and operated, reliant predominantly on family labor, utilizing small areas of land and/or water, and spanning a range of systems; from those involving limited investment in assets and operational costs and comprising but one segment of diverse livelihood portfolios, to others requiring more substantial investments in time, labour, infrastructure and capital.

These characterizations are problematic from a variety of standpoints. First, it is by no means certain that the poorer strata of any given community will necessarily be able to muster the resources required to engage in even low-investment, low-input forms of aquaculture. Second, equating scale to the physical area occupied by operations is a rather arbitrary given that culture systems are extremely diverse. This means that a one hectare pond might yield anywhere from a few tens of kilograms to several hundred tonnes of fish per annum and, correspondingly, require anywhere from no outlay to operating costs of several hundred thousand dollars. Third, as

2 In reality what has played out over subsequent decades is a 'third way' characterized by the emergence of an array of systems falling somewhere between these two rhetorical poles. An enormous diversity of systems exists even for shrimp, for which complete realization of the vision of industrial corporately-owned intensive production has been the exception rather than the rule.

Purcell and Brown (2005, p. 280) state, 'scales should be seen not as things in themselves with inherent qualities', nor are they external entities to which actors respond; an observation which implies that there is nothing intrinsic to aquaculture practiced on a 'small-scale' that lends it a greater capacity to deliver development outcomes than other forms of production. Indeed, growing recognition of this possibility has resulted in a number of recent publications from authors who have identified capital-intensive forms of aquaculture practiced on a large scale as making significant indirect contributions to poverty reduction. Finally, its rather vague boundaries and lack of theoretical content mean that SSA is an imprecise analytical concept. Recognition of this issue has led Belton *et al.* (forthcoming) to advocate a typology of aquaculture which utilizes relations of production as a heuristic with greater explanatory power than scale regarding the likely developmental outcomes associated with the different forms of aquaculture practiced in Bangladesh.

We will not dwell further on these issues as they have already received substantial attention elsewhere. Rather, for the purposes of this essay we typify SSA as an activity practiced predominantly by quasi-peasant producers, for whom aquaculture constitutes one element of a larger total livelihood portfolio. Ellis' (1993, p. 13) definition of peasants as, 'households which derive their livelihoods mainly from agriculture, utilise mainly family labour in farm production, and are characterised by partial engagement in input and output markets which are often imperfect or incomplete' is useful in clarifying this statement. We use the term 'quasi-peasant' rather than peasant in acknowledgement of the waning importance, eloquently chronicled by Jonathan Rigg (2006), of agriculture in rural livelihoods across wide swathes of Southeast Asia and China in the face of ever stronger urban-rural linkages, which means that classical representations of the peasantry may no longer hold to the extent that they once did. This caveat notwithstanding, Ellis' definition allows for identification of SSA as an activity practiced by households that derive a significant portion of their income from agricultural activities other than aquaculture, utilize mainly family labour in their productive activities, and are not completely integrated into input and output markets for aquatic produce (due, for instance, to consuming some or all of the fish they grow, or to procuring some portion of the inputs they use in fish culture from non-market sources).

Rural roots, urban drivers, multiple forms

The rapid rise of aquaculture has been very closely linked to concurrent economic growth and urbanization. This has been most impressive in the parts of Asia where fish has traditionally been a key dietary item, and in Egypt and Nigeria which also possess a cultural affinity for freshwater fish that can no longer be consistently met from wild stocks, rapid population growth and large urban populations. Typically, such areas have also experienced significant levels of agricultural intensification which have placed additional pressures on wild fish stocks through the modification of aquatic habitats, intensive use of agrochemicals, and exploitation of surface waters for irrigation. Urban income effects have been particularly important drivers of increased fish production in rural and peri-urban areas due to the high income elasticity of fish. This is vividly illustrated by the fact that urban consumers in Bangladesh eat approximately twice as much fish per capita as their poorer rural counterparts, despite the vast majority of production originating in rural districts. Seed availability has been another key driver. A reliance on seed produced off-farm has made fish culture far more dependent on informal seed supply networks than many other forms of agriculture. Until quite recently, dependence on seed captured from the wild limited the geographical scope and market orientation of aquaculture development. Networks of producers and traders of juvenile fish have, however, expanded massively in the last three decades, as private hatcheries have developed to supply them following key public sector investments in artificial reproduction.

Modification of floodplains (the areas which support the highest rural population densities) for purposes of flood prevention, irrigation and road construction has seriously impacted the productivity of inland fisheries but, at the same time, created numerous opportunities and resources for fish culture for large numbers of rural households including those with small landholdings. Much of this potential often remains unrealized however since for most households in these areas changes in access to both common-pool and household-managed water resources have occurred in tandem with dramatic shifts in the availability and scope of non-farm opportunities, particularly in manufacturing and services. These trends mean that many households possessing ponds and other water bodies, such as rice fields and ditches (collectively defined as household-managed aquatic systems) with potential for utilization in conventional aquaculture, choose not to follow this path or, after having attempted to do so, quickly abandon it. These decisions are often related to the low opportunity costs associated with non-stocked systems (which yield self-recruiting culturally preferred species such as snakehead, walking catfish and climbing perch), since these can continue to meet subsistence needs or even provide a saleable crop at low additional risk and investment. This tendency may be accentuated where, as in Northeast Thailand, off-farm activities reduce the availability of family labor required to manage systems more actively, and where there is a high risk of flood, theft or both. However, the ready availability of hatchery-derived seed means that mixed culture of stocked exotics and indigenous aquatic animals is increasingly defining the management of both household and common property aquatic systems.

Culture-based fisheries – capture fisheries in natural closed water bodies which are largely dependent on artificial stocking for recruitment (Lorenzen, 1995) – have expanded rapidly in recent years. The importance of such water bodies as safety nets for the poor, and the low input nature of fish culture within them, have made them an attractive domain for external interventions. Very large productivity gains coupled with maintained or improved biodiversity and protection of vulnerable beneficiaries have been achieved in some cases (see Garaway *et al.*, 2006 in relation to Laos; Arthur *et al.*, 2010). However, the institutional forms that define the governance of these systems, the variety of water bodies brought under culture, and the range of management options employed are highly variable, leading to outcomes which are also heterogeneous in terms of equity and impacts on biodiversity, and which in some cases have resulted in effective privatization, elite capture, and linked loss of access rights and exclusion of the poor, despite intentions to the contrary among project planners. It is also notable that, at least in Bangladesh, inland common-pool aquatic resources – including large perennial and smaller seasonal water bodies and sections of canals, rivers, streams and floodplains – are increasingly subject to enclosure for large-scale commercial aquaculture as natural stocks decline and the potential and incentives to profitably enhance them rise accordingly. This trend is particularly noteworthy because in the past most inland aquaculture development has occurred on privately owned agricultural lands and thus avoided many of the social and ecological problems associated with the tropical shrimp boom; many of which were a direct outcome of similar processes of enclosure.

Cage culture, which also frequently demands the *de facto* privatization of common pool resources – albeit usually to a more limited and less problematic degree – has expanded rapidly in a number of locations in response to opportunities offered by growing and increasingly differentiated high value urban markets, even where pond culture is already well developed[3]. While often occupying a limited physical area, intensively managed cages tend to be dependent on considerable environmental services related to feed supply and treatment of effluents which equate to a large ecological footprint. Relations of production vary significantly among cage-based systems, and include those operated by live-in workers on behalf of absentee owners in the Philippines and Indonesia, and the loosely vertically integrated semi-contract grow-out initiated by transnational feed companies (most notably Charoen Pokphand) in Thailand. In arrangements such as these, partial or complete separation of cage operators from the means of production and the effective or actual exploitation of the surplus value of their labour mean that they cannot be considered to engage in SSA defined as a quasi-peasant activity. However, where cage farmers retain privileged access to seed or feed from non-market sources and the activity remains a component of broader agrarian or 'aquarian' livelihood portfolios, it remains largely consistent with our characterisation of SSA. This is the case for producers as diverse as grouper fisher/farmers in southern Thailand (Sheriff *et al.*, 2008), snakehead producers in the Mekong Delta, and grass carp growers in Northern Vietnam.

Intensify or extensify?

One of the most important questions associated with contemporary SSA is whether extensive and semi-intensive forms can persist in the face of the seemingly inexorable rise of intensive, pellet-driven culture systems. Directly integrated semi-intensive forms of aquaculture (in which wastes and by-products from some agricultural sub-systems within the farm are reused as inputs into others) are extremely attractive in terms of their ecological sustainability. However, the most productive and highly developed of these, the pond-dyke systems of South China and the VAC systems of Vietnam (explained below), despite being very highly integrated through complex arrays of on-farm 'bio-resource flows', have never been entirely closed in terms of nutrient cycling; demanding the use of additional inputs imported

3 Cage culture is the production of fish in floating net pens. These are often installed in public water bodies. This is in contrast to pond culture which, as the name suggests, takes place in ponds. These are most commonly constructed on land with exclusive use rights.

from off-farm in order to achieve substantial levels of fish production. In their classic work on agriculture-aquaculture systems in South China during the early 1980s, Ruddle and Zhong (1988) reported that no sooner had a small feed factory opened in the commune where they conducted their study, than formulated pig and fish feed began to displace agricultural byproducts sourced both on and off-farm as pond inputs. This tendency has since become ever more pronounced due to the self-evident advantages which pellets offer in terms of feed conversion efficiency, fish growth and labor allocation.

The Northern Vietnamese VAC system, – 'V' stands for *vuon* (garden), 'A' for *ao* (pond) and 'C' for *chuong* (livestock quarters) – shares many features in common with the South Chinese pond-dyke system, having developed in a similar manner partly in response to extremely high population pressure on land and resources experienced in the Red River Delta provinces around Hanoi. Here too, the once highly integrated components of the homegarden system have become progressively delinked, the higher value pond element on many farms has been expanded and intensified at the expense of the others, and inputs such as grasses and manures produced on or near the farm have been largely replaced by commercial formulated feeds. In the process these operations have, according to our definition, ceased to be small-scale or quasi-peasant as a result of the more complete integration into input/output markets demanded by their specialization and intensification. In addition, these households are no longer oriented first and foremost towards satisfying subsistence needs and, in many cases, may have long since ceased to be dependent upon agriculture as a primary occupation as a result of the burgeoning non-farm opportunities available in and around the capital city.

However, it would be wrong to read this transition as indicative of a universal or imperative 'ladder of intensification', as a variety of other small-scale culture systems in locations where different agro-ecologies and agrarian political-economic realities hold sway have been rather more resistant even to quite limited intensification. Bangladesh is blessed with a vast array of water bodies, including some four million or so homestead ponds or 'borrow pits', most of which are less than 0.1 hectare in size and are excavated adjacent to rural homes for the purpose of elevating them above the level of monsoon flooding and providing water for domestic uses. Until the 1980s, fish entered these ponds during floods and were harvested after the waters receded, but the practice of deliberate stocking was rare. This gradually changed as the inland fishery declined and hatchery-produced seed became widely available; a process accelerated by a succession of major donor funded projects which promoted semi-intensive culture based on simple management practices such as appropriate stocking, harvesting, fertilization and supplementary feeding. At present the vast majority of homestead ponds are stocked with hatchery seed, but although the principles and benefits of semi-intensive pond management are widely understood in rural areas its uptake has been very patchy.

The division of homestead ponds between siblings upon inheritance was once widely recognized to discourage active management for fish culture due to the free riding and intra-familial disputes which often ensued. However, as their value as productive resources has increased, families unable to manage them collectively have tended to lease them to others. This trend (observed both in Bangladesh and West Bengal) has meant that abandoned ponds have become increasingly rare, while lease holders tend to manage them in a commercially oriented manner commensurate with their need to recoup lease costs. Clearly then, the drive to intensify is not uni-causal, nor necessarily universal, nor is the progression from small-scale quasi-peasant production to what we have referred to elsewhere as 'quasi-capitalist' aquaculture, despite what the Chinese and Vietnamese cases might seem to suggest when considered in isolation.

There are a variety of interlinked reasons for this. Vietnam and Bangladesh have superficially similar profiles in terms of the contribution of agriculture to GDP (around 20%), the proportion of the workforce employed in agriculture (approximately half) and, in the Red River Delta, population density (around 1000/km^2). However, agriculture in Vietnam has assumed a form which is more intensive, more market driven (including export markets) and, consequently, often of higher value than in Bangladesh, where attempting to ensure subsistence level rice production remains the fundamental goal of the great majority of rural households. The roots of these differences are complex, but reflect, among other things, the more equitable distribution of land and historically higher population density in North Vietnam, the extremely high incidence of landlessness or near-landlessness in Bangladesh, and the generally more dynamic performance of the economy over the last two decades in the former. As a result, a risk-averse 'subsistence ethic' (Scott, 1976), which prioritizes the pursuit of cultivating sufficient paddy to ensure a minimum level of household food security throughout the year above all else, continues to prevail amongst large portions of the rural population in Bangladesh. This outlook seems to impinge upon the uptake of even basic technologies (such as those required to upgrade from extensive to semi-intensive fish culture) which, viewed from the perspective of conventional economic rationality would appear to offer clear-cut benefits. This tendency is compounded by the low opportunity costs associated with extensive fish production in pre-existing ponds located close to the homestead which allow subsistence fish needs to be met with very little effort or expense.

This risk-averse outlook is reflected in the unwillingness of many households with borrow pits in areas subject to frequent flooding to invest the resources required to raise their dykes in order to protect them – a step which would make it possible to substantially increase their productivity. This subsistence-orientated risk aversion, along with the highly skewed distribution of capital and production factors, helps to explain why the rapid growth of intensive pellet-fed aquaculture experienced in recent years in Bangladesh has not emerged from the evolution of the activities of peasant producers, but from the entrepreneurial actions of an entirely separate class, comprised mainly of educated urban individuals, investing in new forms of aquaculture as purely capitalist enterprises. Although this pattern can be observed throughout Asia, the distinction between practitioners of small-scale peasant aquaculture and intensive capitalist aquaculture is generally less sharply defined in many parts of Southeast Asia, where land and wealth tend to be distributed somewhat more equitably and rural poverty is rather less crushing than on much of the Indian subcontinent. Thus, households which have shifted, over a period of three decades, from partly subsistence rice cultivation, to commercial semi-intensive aquaculture or intensive horticulture and, ultimately, into intensified forms of fish production are fairly commonplace in Central Thailand, but rare in Bangladesh (Belton and Little, 2008).

In the case of Bangladeshi and Vietnamese shrimp cultivation (much of which at least partially fulfils our criteria for consideration as small-scale given, despite strong forward integration into global export markets) there are also clear incentives which work in favor of the maintenance of extensive or diversified systems, despite apparent technical advantages offered by intensified and specialized alternatives. Many of these relate to the endemic nature of shrimp disease, particularly the virulent White Spot Syndrome Virus. The likelihood that stocked shrimp will contract the virus during grow-out, resulting in significant mortalities, is high given the very poor bio-security of extensive shrimp ponds. Under these circumstances, pursuing a strategy of stocking at higher densities and applying more inputs represents an unacceptable risk for all but a handful of wealthy individuals who can afford to invest in closed, semi-intensive systems.

In Bangladesh, most producers adopt the opposite strategy, relying on the extensive cultivation of wild shrimp post-larvae (PLs) which are trapped in ponds flooded by the tide and consume only naturally occurring food. These are usually augmented by multiple low level stockings of hatchery produced black tiger shrimp PLs. In these systems disease is probable, but low stocking densities and investment mean that it is unlikely to prove financially catastrophic. Whereas intensified systems tend towards specialized monoculture, in these extensive systems yields of fish (both deliberately stocked and naturally present) usually equal or exceed those of shrimp, and are sold on local markets or consumed on-farm, thereby offsetting disease-induced shrimp losses and contributing to local food supply. The effect of such extensification and diversification strategies is that these systems are quite resilient in the face of chronic disease. Other common diversification strategies such as alternate rotations of rice and shrimp have also proven less disease-prone than back-to-back shrimp cropping cycles.

An additional incentive to maintain small-scale shrimp farming in low-intensity states may ultimately derive from preferences for 'ethical' modes of food production among some sets of Northern consumers since, these farms produce shrimp which are effectively 'free-range' as compared, for instance, to the extremely intensive systems deployed in Thailand; the latter being in many ways analogous to feedlot livestock production and appropriating more ecosystem space and services than the former due to their reliance on fuel and manufactured feeds.

Wealth, welfare or well-being?

The issue of values brings us to consideration of SSA's contribution to well-being. Well-being is increasingly understood to include both a material dimension (the connotation of welfare or standards of living, suggesting a foundation in economic prosperity) and a 'subjective' dimension related to personal perceptions and levels of satisfaction (White, 2010). Not surprisingly, the material aspects of SSA (its ability to raise levels of household income and food intake) have been foregrounded in work on the topic to date. Far less attention has been paid to whether subjective factors unique to the individual – emotions, values, personal preferences – play a role in determining choices regarding SSA, or rank among the outcomes derived from participating in it.

Revisiting old evidence with this new conception of well-being in mind suggests that subjective dimensions may play an important role in mediating farmer involvement in SSA. Two excellent studies evaluating the circumstances of SAA development in the disparate contexts of Laos and Zambia reveal a number of interesting points of comparison. In Laos, fish farming is associated with modernity in both national policy discourse and the popular imagination, with the result that, 'fish ponds are an aspirational asset that is used as an indicator of wealth and class' (Bush, 2004, p. 231). In some cases, ponds constitute 'a space of power' where guests are invited to socialize with the senior members of village committees and are entertained with food and liquor, furnishing the pond owner with respect from community members as well as contributing to the maintenance of local power relations. As Bush has noted, this does not negate the fact that farmers may obtain income and nutritional benefits from their ponds. It does however indicate that 'the decision to adopt aquaculture... is embedded within wider social and political aspirational goals' (p. 232).

A comparable tendency is reported by Harrison (1996) who found that rural inhabitants in Luapula, Zambia, identified certain characteristics including fish pond ownership and cash crop production as symptomatic of being "developed", while development workers saw farmers who had constructed fish ponds as "progressive". As Harrison notes, 'for some, though not all, self-identification with such progressiveness is important', and 'may provide both status and material benefits through access to development assistance' (p. 273). In both cases, SSA introduced through external interventions is equated with modernity and prestige in the eyes of developers and those they seek to develop. The construction of a pond and production of fish thereby bestows a measure of status and, presumably, personal pride or satisfaction, upon the owners, and may have the additional effect of facilitating access to, or production of, other ostensibly entirely unrelated resources and non-economic capitals. Thus, although not couched in terms of well-being, these accounts appear at least partially amenable interpretation to this manner, and perhaps also hint at the existence of self-reinforcing connections between the subjective and material aspects of personal well-being.

Harrison has also shown that, in Zambia, practicing fish culture allows pond owners to meet the obligation to slaughter an animal in order to honor visitors; an outcome which has subjective as well as instrumental social outcomes. The ability to 'gift' fish to relatives and to feed household guests also featured in a list of 23 factors identified by Haque *et al.* (2010) as incentives for producing tilapia fingerlings in rice fields in Northwest Bangladesh, which contributed to the activity's widespread uptake despite rather modest monetary and caloric gains. Homestead ponds in Bangladesh are also often valued more for the regular supply of fresh fish that they provide for home consumption than for the cash incomes they generate, even when the market equivalent value of the two is approximately equal. This suggests a trade-off between economic imperatives and affective reasoning in a culture where fish is a highly prized food, whereby households that can afford to consume fish opt to do so for reasons of the emotional satisfaction that it provides, in preference to earning cash income. In other contexts, even the decision to culture fish may be the result of enjoyment derived from the activity itself, as the increasingly common phenomenon of hobby fish farming in Thailand suggests.

The material aspects of SSA also deserve some attention, since, as noted in the introduction, its poverty reducing capacity is often uncritically cited by proponents. However, as Harrison has observed in Zambia:

> Most fish farmers, although not from a recognizable "elite" have certain characteristics in common. Adopters are more likely to be men, to be slightly better off, to be slightly better educated, and much more likely to be active participants in social and political activity than others in the community. These facts are closely connected with one another and to some extent causally associated (1996, p. 271).

This, we would argue, represents a fairly accurate reflection of pond-based SSA in a great many other locations, although the degree to which women are involved varies considerably from place to place, and active participation of pond owners in social and political activities may be less of a defining feature where the activity is very widespread and less closely linked to externally inserted development efforts. That individuals or households practicing SSA are likely to be in a better than average position relative to others in their communities might seem to imply that the activity is unlikely to bring about much poverty *reduction*. Such a conclusion overlooks the potentially important and rather less obvious role that SSA may play in poverty *prevention* – a distinction which Béné *et al.* (2010) have also made with reference to small-scale fisheries – particularly given that in locations such as Bangladesh the situation of even those who are well-off relative to their neighbours may remain perilous at best.

For households such as these, fish produced in the homestead ponds, rice fields, trap ponds and other connected resources which make up farmer-managed aquatic systems represent fungible assets which can be strategically liquidated to reduce or avoid high interest debt burdens associated with rice cultivation, the purchase of rice for home consumption or other essential expenses. These systems may therefore provide a buffer against transient poverty, with possible long-term implications contingent on the manner in which the debt is to be repaid and the ability of the household to repay it. A number of convincing studies in South Asia have found that even quite well-off families frequently slide into poverty in the long-term as a result of the capriciousness of circumstance and that, had these numbers been smaller, net reductions in overall poverty would have been much larger; suggesting that the poverty prevention role of SSA may be important at the aggregate level also. Rather than conceptualizing SSA as a simple economic mechanism for reducing poverty in the medium term, it may therefore be more appropriate to view it as a household insurance policy against future hardship which also enhances the subjective well-being of its members in the present.

Co-existing or competing?

Another question pertinent to SSA is whether it will continue to prove viable or attractive as the specialized aquaculture which already accounts for the bulk of output in most Asian countries continues its remarkable expansion. Seen from one perspective, the story of Vietnamese pangasius appears to suggest not. In the space of scarcely a decade, pangasius has been transformed from a fish virtually unknown outside Southeast Asia (grown in the Mekong Delta for partly subsistence purposes in tiny household ponds fed with human excreta), into a global commodity worth more than $1 billion per year, sold in over 100 countries worldwide, and rivalling Atlantic salmon in terms of volumes traded. This unprecedented developmental trajectory, particularly over the last five years, has been matched by an equally precipitous decline in the numbers of very small producers. This is the result of rapidly rising costs, ever tighter margins and, increasingly, greater regulation and traceability requirements; all of which translate into an inability to compete with far less numerous but cumulatively far more productive large capitalist farms. On the face of it, this appears to be a classic case of industrial consolidation, mirroring that which took place, albeit over a considerably longer timeframe, in the global salmon industry and for a range of other non-traditional agricultural export crops.

However, as Khiem *et al.* (2010) have shown, this 'adverse incorporation' of small producers into a newly established global value chain has not resulted in widespread dispossession and proletarianization. That this did not occur is in part a function of Vietnam's voracious domestic demand for aquatic products. This has meant that many producers who found it necessary to abandon catfish grow-out have been able to utilise their ponds to either 'outgrade' into farming a range of other fish species with lower production costs or 'downgrade' into nursing pangasius, red tilapia and the seed of other species[4]. The fact that producers at this lower end of the spectrum often continued with other agricultural activities such as rice cultivation and orchards whilst they engaged in pangasius production also contributed to their resilience. The value that ponds retain as productive assets even after the failure of particular culture systems is also apparent in Central Thailand, where numerous unsuccessful inland shrimp producers have taken up culture of tilapia and other fish species.

It may be that successive waves of entry, maturation, failure, adaptation and diversification are symptomatic of aquaculture development played out over the long run, leading to gradual increases in the number of species cultured and the variety of systems in which they are grown. While in some cases (such as the VAC system) this will tend to result in a shift away from quasi-peasant forms of production, it may also result in the creation of sufficient space for the co-existence of a variety of systems, including those characterized as SSA. An indication of the likely persistence of certain forms of SSA is provided if one refocuses attention so that it rests not on the aquaculture component of the farming system alone, but views the pond as an indivisible component of the homegarden.

There is a substantial literature detailing the composition and functions of homegardens throughout the world. In Bangladesh,

> A typical homegarden... occupies a small space adjacent to the house within the farmstead compound. It is very intensively cultivated to grow and raise a wide variety of plant and animal species utilizing primarily household female and child labor. The primary production goal here is to meet household consumption demand (Ali, 2005, p. 266).

Such homegardens produce fruits, timber, firewood, litters, vegetables, livestock fodder, as well as meat, dairy, fish, and organic fertilizers, based on cultivation practices which 'maximize output at a minimum risk since the cost of input is very low compared to output' (Ali, 2005, p. 266). They are there-

4 'Downgrading' is the opposite of upgrading, corresponding to a reduction in the functions of an activity within a value chain. 'Outgrading' refers to the exit, either strategic or forced, from a value chain or a 'strand' within it, and entry into an alternative lower level activity (Khiem et al, 2010).

fore resilient in social-ecological terms, and are reported to contribute half of farm income in Bangladesh on average; a figure which rises to more than 80% among near landless households. The integral role which these systems play in supporting rural households means that, considered not as a standalone enterprise but as a single component of the larger homegarden, SSA is unlikely to decline even in the face of huge expansion of intensive capitalist aquaculture. Indeed, there is evidence that markets and products are becoming increasingly differentiated along rural/urban lines in Bangladesh, with the carps typical of quasi-peasant production continuing to dominate fish consumption in rural areas while pangasius and tilapia from large commercial systems gain an increasing share of urban markets. The persistence of actively managed ponds in homegarden systems is also in evident in Northeast Thailand despite the ever greater importance of the non-farm economy in rural livelihoods.

The continuation of certain forms of SSA aimed at semi-subsistence or local consumption is thus likely, even as the character of agrarian livelihoods continues to undergo rapid transformation. It should also be noted, however, that where overwhelming incentives to commercialize exist, as in the case of the VAC system (which is essentially also a form of homegarden), modes of production may change rapidly. Other markedly different dynamics exist where quasi-peasant producers have been incorporated into the value chains of global export commodities. Mounting traceability requirements imposed with the aim of ensuring food safety in the Global North, and the proliferation of private third party certification schemes into which these have been integrated, mean that the position of this type of producer may become increasingly difficult to maintain over time, given logistical difficulties in implementing product traceability and standards compliance among large numbers of resource poor producers. As the example of Vietnamese catfish indicates, the effect of any exclusion of small producers from export markets will depend in large part upon the range of alternatives available to them.

Future visions for SSA

Urbanization and economic growth in many of the rice-fish societies where SSA has become established suggests that the expansion of quasi-capitalist and capitalist production systems will continue in coming decades. The tendency for these to become more intensive and uniform in their production processes and output will, on the one hand, satisfy the growing demand of mass markets for safe animal-source products but, on the other, is unlikely to meet the strong cultural attachment to the diverse and local that is still prevalent in rapidly growing areas of Asia, where a high cultural (and monetary) value is attached to traditional and often 'wild' foods. Such is the resilience of food cultures, particularly in Asia, that it is possible to envisage the continued existence of considerable market space for small-scale producers (perhaps seeking to out-grade from the production of mainstream species) to meet these aspirations, which are already tangible in the consumption habits of the trend-setting middle class. Such development would in some respects mirror, or pre-empt, the renewed interest in local food and culture in Northern economies, where a parallel economy in local food and drink has emerged, or been revived, alongside modern agri-food systems dependent on global supply chains.

The nature of the development experienced in emerging Southern economies is crucial in determining the future of SSA. Improved transportation infrastructure and the expansion of urban regional and district centres appears to be facilitating more local and temporary migration from rural to urban contexts, as compared to earlier longer-term and permanent migration from rural areas to a handful of megacities. By allowing families to continue straddling a rural-urban divide, this trend may encourage retention of cultural memory and indigenous rural knowledge, and facilitate the persistence of agrarian elements in livelihoods which would otherwise be lost following wholesale relocation to distant cities. The rapidly increasing growth in demand for aquatic products relative to local supply is also likely to provide continued opportunities for domestic expansion by Southern producers, which may well prove preferable to the more stringent and unpredictable demands of ultimately much smaller European, North American and Japanese markets, particularly for those producers at the smaller end of the spectrum. South-to-South trade in cultured aquatic produce is also increasingly prevalent for high volume products such as Chinese tilapia, which now features strongly in African markets, and Vietnamese pangasius, which is making inroads in Latin America and is also widely consumed in neighboring Cambodia. Producers associations and other groups with protectionist leanings frequently claim that such developments will hinder the development of infant aquaculture of all stripes in importing countries, but it remains to be seen whether these fears prove valid.

The rapid growth of aquaculture and broader development are occurring at a critical time in human history, during which increasing unpredictability associated with climate change and greater volatility in food prices and food security are already a reality. Climate change is expected to bring particularly severe impacts to the densely settled deltas of Asia where SSA is most prevalent and where, if maintained or further developed, it might play an important part in adaptation approaches intended to enhance social-ecological resilience. The incorporation of wetlands into urban and peri-urban landscape planning and management to reduce impacts of flooding is becoming mainstreamed and provides opportunities for fish culture. Diverse SSA strategies from aquatic vegetable production to floating cage-based aquaculture have also been trialled and promoted as a basis for easing future disruptions in food supplies in flood affected communities. Studies which have documented aquaculture in high flood risk contexts also point to numerous indigenous adaptive modifications in site selection and design. In line with the strong linkages between SSA, and the persistence of traditional homegardens mentioned above, the development of contemporary models that serve local recreational and aesthetic as well as local food security and flood mitigation functions are to be expected.

Such evolution is likely to occur in parallel to a more general industrialization of food production, but the greater inherent energy and nutrient efficiencies of low-intensity household level systems may provide further incentives for their maintenance. Resource pressures make hikes in the relative costs of energy and limiting nutrients such as phosphorous inevitable in the medium term. These tendencies may mean that production systems in which energy and nutrient efficiency is high, especially those based on polycultures that exploit aquatic food webs *in situ* and encourage closer linkages between the effluents from fish production and crop fertilization requirements, ultimately gain a renewed competitive advantage as compared to more energy consumptive intensive alternatives, and could favor their persistence as forms of SSA which meet the food fish needs of very large numbers of poorer consumers. Evidence for this is the continuance and indeed growth in demand for small carps and tilapias, raised semi-intensively throughout Asia for poor urban and rural consumers, which has occurred alongside expansion of markets for both high- and low-value pellet-fed fish.

The likely future of SSA is therefore by no means homogeneous, nor fixed, being indivisible from the broader historical context in which it is situated. This landscape represents the sum of: interactions between local and global economic forces and transformations; the interplay between agrarian change and urbanization and its accompanying restructuring of livelihoods and modes of production; global environmental change and shifting resource availability; and cultural and individual subjective preferences. The inherent complexity associated with these elements and their interactions mean that it is impossible to predict a single overriding fate for SSA. Rather, we must anticipate multiple futures in which a huge diversity of aquatic production systems – at least some of which are small-scale – persist, emerge, adapt or disappear in response to the specificities and imperatives presented by diverse configurations of development in different locales.

References

Ali, A. M. S. (2005) Homegardens in smallholder farming systems: examples from Bangladesh. *Human Ecology*, 33(2), 245-270.

Arthur, R. I., Lorenzen, K., Homekingkeo, P., Sidavong, K., Sengvilaikham, B. and Garaway, C. J. (2010) Assessing impacts of introduced aquaculture species on native fish communities: Nile tilapia and major carps in SE Asian freshwaters. *Aquaculture*, 299, 81-88.

Belton, B. and Little, D. (2008) The development of aquaculture in central Thailand: domestic demand versus export-led production. *Journal of Agrarian Change*, 8(1), 123-143.

Belton, B., Haque, M. M. and Little, D. C. (Forthcoming) Does size matter? Reassessing the relationship between aquaculture and poverty in Bangladesh. *Journal of Development Studies.*

Béné, C., Hersoug, B. and Allison, E. (2010) Not by rent alone: analysing the pro-poor functions of small-scale fisheries in developing countries. *Development Policy Review*, 28(3), 325-358.

Bush, S. R. (2004) *A Political Ecology of Aquatic Living Resources in Lao PDR.* PhD Thesis. Sydney: University of Sydney.

Edwards, P., Little, D. C. and Demaine, H. (2002) Issues in rural aquaculture. In Edwards, P., Little, D. C. and Demaine, H., eds., *Rural Aquaculture.* Wallingford: CAB International.

Ellis, F. (1993) *Peasant Economics.* UK: Cambridge University Press.

Garaway, C. J., Arthur, R. I., Chamsingh, B., Homekingkeo, P., Lorenzen, K., Saengvilaikham, B. and Sidavong, K. (2006) A social science perspective on stock enhancement outcomes: lessons learned from inland fisheries in southern Lao PDR. *Fisheries Research*, 80, 37-45.

Haque, M. M., Little, D. C., Barman, B. and Wahab, M. A. (2010) The adoption process of ricefield based fish seed production in Northwest Bangladesh: an understanding through quantitative and qualitative investigation. *Journal of Agricultural Extension and Education*, 16(2), 161-177.

Harrison, E. (1996) Digging Fish ponds: perspectives on motivation in Luapula Province, Zambia. *Human Organization*, 55(3), 270-278.

Kheim, N. T., Bush, S. R., Chau, N. M. and Loc, V. T. T. (2010) Upgrading Smallholders in the Vietnamese Pangasius Chain. ODI grant number RO334. Long Xuyen: International Development Research Centre.

Lorenzen, K. (1995) Population dynamics and management of culture-based fisheries. *Fisheries Management and Ecology*, 2, 61-73.

Nakamura, R. (1985) Aquaculture development in India: a model. *BioScience*, 35(2), 96-100.

Purcell, M. and Brown, J. C. (2005) Against the local trap: scale and the study of environment and development. *Progress in Development Studies*, 5(4), 279-297.

Rigg, J. (2006) Land, farming, livelihoods, and poverty: rethinking the links in the rural South. *World Development*, 34(1), 180-202.

Ruddle, K. and Zhong, G. (1988) *Agriculture-Aquaculture in South China: The Dike-Pond System of the Zhujiang Delta*. UK: Cambridge University Press.

Scott, J. C. (1976) *The Moral Economy of the Peasant: Rebellion and Subsistence in Southeast Asia*. USA: Yale University Press.

Sheriff, N., Little, D. C. and Tamtikamton, K. (2008) Aquaculture and the poor – is the culture of high-value fish a viable livelihood option for the poor? *Marine Policy*, 32, 1094-1102.

White, S. (2010) Analyzing wellbeing: a framework for development practice. *Development in Practice*, 20(2), 158-172.

Section 4:

Instruments and reforms

CHAPTER 10

Whose fish is it anyway? Iceland's cod fishery rights

Gabriela Sabau[1]

Fish is a gift of the regenerative power of nature which can support and sustain peoples' livelihoods. This valuable gift has been depleted since the middle of the nineteenth century due to lack of clearly defined property rights. The theory explaining how open access fisheries are overused is well known, as are the scientific tools that have been used to control overfishing, such as maximum sustainable yield (MSY), total allowable catch (TAC) and individual transferable quotas (ITQs). This essay will demonstrate that the design and implementation of a successful fishery is more an art than a science, by discussing Iceland's efforts to manage its cod fishery in the last 35 years by using a property system based on ITQs.

Introduction

Fish is a gift of nature endowed with the power to renew itself and thus sustain peoples' livelihoods (Shiva, 2010). As a gift, it belongs to the commons; it is open to everyone, but must be shared and treated gently to give it a chance to renew itself. This characterization of fish has been lost in the last 150 years in Iceland's drive for industrialization and economic growth supported by increasingly powerful technologies. Fish has become a natural resource which is economically exploited and "developed in a planned manner, to facilitate the generation of revenues and growth of capital" (Shiva, 2010, p. 229), in spite of the fluctuations naturally occurring in fisheries. Fish has also become a source of unearned economic profit (scarcity rent) which it was hoped could be perpetually maintained with proper management. The main tool for fisheries management is the maximum sustainable yield (MSY), a

1 Corresponding author: Sir Wilfred Grenfell campus, Memorial University, Canada, gsabau@grenfell.mun.ca

theoretical concept based on Milner B. Schaefer's empirical observation of the biological pattern of fish stock growth, and his over-optimistic idea that there was a harvestable surplus of the fish stocks that scientists could estimate accurately (Finley, 2007). The objective of MSY is to harvest the maximum number of fish at their maximum weight (Schaefer, 1957). Studies have shown that MSY in open-access fisheries does not prevent overfishing, as each fisher tries to individually maximize harvest (Pauly and Thia-Eng, 1988). The model explaining how overfishing of an unregulated fishery occurs has been developed by Gordon H. Scott, who applied economic principles to the biological population models. It demonstrates that the maximum economic yield (MEY) is achieved when less than the MSY is harvested, but neither level is chosen by the fisher. Each individual harvester will try to fish as much as he can, as for him "the fish in the sea are valueless...because there is no assurance that they will be there for him tomorrow if they are left behind today" (Gordon, 1954, p. 135). Since the 1950s there have been numerous attempts by various fishing nations to regulate open-access fisheries. The efforts range from raising the marginal cost of fishing (by limiting annual catches – with the target typically being the MSY, restrictions on fishing gears, temporary closure of fishing grounds, imposition of limited seasons) to setting a tax/landing fee on fish harvests (when the scarcity rent is collected by the government), or providing stewardship incentives for individual harvesters in the form of rights-based catch shares. The fisheries which opted for this latter solution hoped to increase their economic efficiency since the "quotas in the fishery will flow to those that gain the most net benefit from them due to lower costs.....but rents are retained by the fishing industry" (Stavins, 2011, p. 91). Some 150 major fisheries in countries such as New Zealand, Australia, Canada, Iceland, the Netherlands and USA have established systems of individual transferable quotas (ITQs) since the mid-1970s. It seems that besides increasing quota owners' profits, the implementation of a well-designed ITQ system may prevent fisheries stocks "collapse across diverse taxa and ecosystems" (Costello *et al.*, 2008). Other economists worry that, in the presence of recurrent ecological uncertainty, fishing quotas may cause "an unjust and disturbingly large government-engineered alteration of income distribution", due to increased windfall scarcity rents (Weitzman, 2002, p. 325), and consider correctly chosen landing fees superior to harvest quotas.

This essay considers Iceland's experience with the ITQ fisheries management system, to demonstrate that assigning property rights to fishers based on a scientifically-determined quota does not automatically secure fisheries' success. Iceland's saga with ITQs implementation is a long process of trial and error which implies continuous broad-based dialogue and adaptation to accommodate not only economic efficiency but also ecological dynamics and social needs and aspirations. The design and implementation of a successful fishery is more an art than a science.

Iceland's slowly emerging fishing rights

Iceland is an island in the North Atlantic Ocean blessed with some of the best fishing banks in the world. The most important of the Icelandic fisheries is the demersal or groundfish fishery which in recent years has generated over 80% of the total fish harvest (Eythorsson, 2000). For 1,000 years the fish stocks around Iceland were harvested by foreign fleets (English, German, Danish, and Norwegian), and specialized Icelandic fishing industry only emerged around 1870 (Eggertsson, 1996). Icelandic fisheries remained international for another century, with large foreign fishing fleets still "taking almost half of the demersal catch as late as 1975" (Arnason, 1996). The extension of Iceland's exclusive economic zone (EEZ) from four nautical miles (1952) to 50 nautical miles (1972) and to 200 nautical miles (1976) led to the "nationalization" of the fish resources. According to Act 3/1988 of the Icelandic Parliament, "The fish stocks around Iceland are the property of the Icelandic people" (Matthiasson, 2003). With foreign competition removed, the Icelandic fishing fleet was rapidly modernized, growing from no trawlers in 1970 to 80 stern trawlers in 1980 (Eythorsson, 2000). Yet, bad news concerning the imminent collapse of the cod fish stocks in a 1975 report by the Marine Research Institute (MRI) in Reykjavik led to imposition of an overall catch quota in the demersal fisheries. The precautionary annual quotas recommended by MRI scientists were small but were systematically exceeded by a growing Icelandic fishing fleet. For instance, in 1976 the total allowable catch (TAC) for cod was almost halved, from 400,000 to 230,000 tonnes, yet 346,253 tonnes were actually landed (ICES, 2010). New measures (such as limited fishing days) aiming to limit vessels' fishing effort for cod were introduced in 1977 without much success, and the cod stock continued to decline rapidly. This was the background for the introduction of individual vessel quotas in the demersal fisheries in 1984. All vessels over ten gross registered tonnage (GRT) were allotted a quota share based on their average catch during the three previous years. Smaller vessels were to catch from a common pool. All vessels that were involved in fishing had to have a permit issued by the Ministry of Fishery.

A complex fishing rights regulation

The system of ITQ was extended to all Icelandic fisheries in the Fisheries Management Act of 1990. All vessels above six GRT and all major commercial stocks became subject to quotas. The ITQ system works like this: the Directorate of Fisheries (DoF) issues annual catch quotas (in kgs) to individual vessels as a share of the TAC. The TAC is based on MSY and the precautionary approach. It is set every fishing year for each species (25 species representing approximately 95-97% of the total annual catch value) by the Minister of Fisheries upon recommendation by the MRI. The annual catch entitlement (ACE) is a simple multiple of the TAC for that fishery and the vessel's TAC share.

The owners of small vessels (below six GRT) were initially offered a choice between ITQs and a limited effort management system. As very few chose the quota system, in 1996 they were allotted a TAC for cod and a given number of sea-days. They were also allowed free hook and line fishing for cod for 23 days between April 1st and October 31st. This loophole greatly expanded the fleet of small vessels and their operations, making fishing effort reduction measures observed by those vessels operating under the ITQ system ineffective (Haraldsson, 2008). In 2004 the small vessel loophole was closed in order to restrict entry into the fishery and the small boat owners were allotted quotas in other valuable species (except cod). The Fisheries Management Act provides that both quota shares and ACEs can be fully or partially transferred between vessels. The quota shares are allocated permanently and are divisible, "they can be separated from vessels and transferred as an independent commodity to other vessel owners, either by transfer of permanent quota shares or by leasing for one year only" (Eythorsson, 1996, p. 272). To prevent quota concentration, the Act provides that not more than 50% of the annual catch quota is freely transferable between different owners, with a single firm not allowed to hold more than 12% quota shares of cod and not control more than 12% of the value of all TACs.

Since September 1998, all quota leasing transactions are to be made anonymously through a stock market structure, the Quota Exchange Market (QEM). A Directorate of Fresh Fish Prices was also set up in 1998 to control landing prices, resolve disputes between boat owners and crew, and control leasing transactions. Since 2004, all fishing rights are subject to a fishing fee calculated as a percentage of the profits (9.5% in 2009).

A comprehensive, transparent monitoring scheme

Iceland has a comprehensive system for monitoring and controlling individual vessels. "Even good science is not going to help in the absence of comprehensive monitoring schemes (including on-board observer programmes) to suppress quota-busting, high grading and other nefarious practices" (Pauly, 1996, p. 111). All fishing vessels are obliged to keep detailed catch logs about the fishing area, quantities fished and catch composition. Landed catch is weighed by inspectors from the DoF and then registered in the central data base. Vessels' logs and landing reports are published on the DoF website. If inspectors observe that the catches contain too many small fish (more than 25% cod smaller than 55 cm) the fishing grounds can be closed. Minimum mesh size for cod nets is 135 mm. Fishing with bottom trawl is prohibited 6-12 miles from the coast. In order to stop discards, a general discard prohibition clause is in place. If a vessel overfishes or brings in vast amounts of by-catch, the owner has the option to obtain additional quota within a certain period of time after landing. Thus, fishers have an incentive to land all the fish they catch. The system has other flexibility features. A provision allows the use of the catch quota for one species to count against a limited catch amount of another species. It is permitted for the year's catch to exceed the year's quota by 5% in some species. The excess is then deducted from the following year's quota. It is permitted to postpone fishing for part of the quota and to transfer up to 20% of the year's quota to the following fishing year; catches of undersized fish in some cases (e.g., cod <50 cm) count only as half their weight against quota. This is to discourage discards; it is also permitted to land a catch exceeding the year's quota by maximum 5%, as long as the catch is auctioned and the bulk of the value of the catch goes to the MRI.

The DoF can issue reprimands and suspend the commercial fishing licences of vessels which violate the management legislation. The lawbreakers are subject to fines or, in some cases, imprisonment for up to six years. The responsibility for proper conduct is placed on the fishers, who must show evidence that they have quotas for the species they fish and land.

How successful is the Icelandic ITQ system?

In December 2008, just before the first effects of the financial meltdown became obvious, an article in *The Economist* presented Iceland's ITQ system as a success story, a "model way to catch and keep fish", made possible by giving fishers rights that offer a reasonable expectation of profitable long-term fishing by encouraging the conservation of stocks: "The system is clear, open and fairly simple, and it is well policed. It thus enjoys the respect of fishermen. And it is based, crucially, upon scientists' assessments of stocks, not politicians' calculations of electoral advantage" (Economist, 2008, p. 15). How successful is the Icelandic ITQ system?

This question can be answered by considering the economic, ecologic and social impacts of this property rights system over the short- and the long-run. In the short run, the success can be assessed by remembering the main reasons for introducing the ITQ system in Iceland's fisheries. According to Arnason (1996, p. 71), "the introduction of the ITQ system in the various fisheries has been motivated by two major concerns. The primary one is the desire to improve the economics of the fishing activity. The second, actually closely related to the first, is the desire to conserve the fish stocks". The first reason is justified by the Icelandic economy's heavy dependence on its fishing industry, which provides 40 % of export revenues and employs 5.2 % of the labor force (Statistics Iceland, 2010). "Icelanders simply cannot afford to run an inefficient fishing industry" (Arnason, 1996, p. 72). Indeed, the fishing industry is economically efficient. It generates substantial economic (scarcity) rents which have secured Iceland's relative prosperity in recent times. The net profit of harvesting and processing demersal species increased from 24.6% to 25.1% in the period 2008-2009 (Statistics Iceland, 2010). The cod industry is highly profitable due to improved harvesting and processing technology and highly reliable transportation systems. Iceland is known for the quality and freshness of its fish products, which are Marine Stewardship Council (MSC) certified. In addition, in 2010 Iceland established its own FAO-ISO certification system, the Icelandic Responsible Fisheries Certification Programme, "to increase the fishers' responsibilities" (Johnsen and Eliasen, 2011).

Concerning the second reason, stock conservation, things are more complex. It may be true that dedicated access provides quota owners with incentives for stock conservation or rebuilding of depleted stocks (Breen and Kendrick, 1997). However, it is worth remembering that "stock conservation is achieved by setting the total allowable catch conservatively while individual transferable quotas mainly enhance economic efficiency" (Hannesson, 2005, p. 237). Since 1955, assessments of the Icelandic cod stock by the International Council for the Exploration of the Sea (ICES) show that the spawning stock biomass is slowly decreasing from 938,351 tonnes in 1955 to 231,743 tonnes in 2005 (ICES, 2010). At the same time, fishing mortality rates exhibit "a clear increasing trend from 0.3 in 1955 to approximately 0.6 in 2005" (ICES, 2010). A harvest control rule for calculating TACs was established in 1995.

The annual TAC for cod was set at 25% of the fishable biomass of four-year and older harvestable cod. In 2009, in the Icelandic government's five-year management plan for the cod stock, the exploitation rate was lowered to 20% of the fishable biomass. The main objective of the management plan is to ensure an increase in the size of the cod stock that, according to the ICES MSY Framework, can generate a spawning stock biomass above 220, 000 tonnes by 2015 (ICES, 2010). Only time will tell if the Icelandic quota ownership system has been beneficial to cod stock recovery or not. This is where the long-run dimension of fisheries' success comes in.

It is also in this time frame that the social aspects of a successful fishery need to be considered. According to Icelandic fisheries expert Jon Kristjansson, "the value of the fishery is not how much money you can make on it but how many people can make a living out of it" (World Fishing, 2009). The ITQ system is generating substantial economic rents but concentrates them in a few hands; at the exclusion of those whose livelihoods used to directly depend on fishing activity. This is the perverse effect of the ITQ system which hurts both small-scale fishers and the fish stocks. "To this day, around the world, fishing is primarily a small-scale subsistence activity, subjected to the rhythms of season and tide, subservient to the nature of fish and the ability of fishermen" (Bavington, 2010, p. 130). Has the Icelandic ITQ system protected fishing as a "small-scale subsistence activity"? Twelve years after the introduction of the ITQ system in the cod fishery, it was assessed that "the most significant result from the reform is a rather massive re-distribution of wealth and income" (Eythorsson, 1996, p. 281). The system has produced winners and losers. The theory of ITQs predicts that "in most ITQ fisheries the number of vessels and jobs drops significantly and the holders of the ITQs acquire valuable assets" (Hilborn, 2007, p. 155). In Iceland, the winners are vessel owners who were allotted quota shares in 1984 and 1991, or those who acquired quota shares when the prices were still low. They have experienced a nice increase in their capital assets as the increase of capital value of quota shares was substantial. They have also benefitted from the increase in the price of annual catch quotas. For instance, in 1996 the annual quotas were traded at prices between 10% and 25% of permanent quotas, depending on species (Arnason, 1996). The quota owners have also increased their revenues by offering their quota for lease for annual returns which, for cod, have always been high; in the 20%-27% range (Eythorsson, 1996). "The losers are the fishermen, or *the fishing crews*, who have been thrown into a market where eventually, only the lowest bidder gets the chance to catch the fish" (Eythorsson, 1996, p. 281).

At least three other groups in Iceland consider the ITQ system unfair: fishers whose vessels have reduced their quota holdings and need to participate in the costly quota purchases; community groups outside the fishing industry not involved in the initial allocation of quota rights in the 1990s (they ask for sharing/reallocation of quota rents by means of taxes); and inhabitants of fishing villages and towns from which quotas have been permanently transferred. These are critical of the initial allocation of quotas and often suggest that their community should receive (or should have received) its own local quota and thus participate in the distribution of the fisheries rent. All these are justified claims that, sooner or later, need to be addressed by the Icelandic nation, as owners of the fish stocks, if we do not overlook "the fact that dominion over things is also imperium over our fellow human beings" (Cohen, 1927-28).

Whose fish is it anyway?

According to current legislation, "the exploitable marine stocks of the Icelandic fishing banks are the common property of the Icelandic nation........ The allocation of harvest rights provided for by this Act neither endows individual parties with the right of ownership nor irrevocable control over harvest rights" (Fisheries Management Act, 2006). Being harvesting rights (even if permanently allocated for the time being), and not ownership of the fish stock, Iceland's ITQs are imperfect property rights. They are reminiscent of the property relations brought about by the emergence of large scale publicly traded corporations characterized by the separation of ownership and control. This separation means that the control function is not allocated on the basis of 'clearly defined property rights'. The ownership of shares, like the ownership of bonds, is indeed clearly defined: the shareholder can buy, sell, or hold those rights. But those rights do not 'add up', to a real ownership-based control of the company when the shareholders are atomized and dispersed (Stiglitz, 1999). Yet, the strength of the Icelandic ITQ fisheries management system lies in the "poorly" defined property rights over fishing quotas and in the separation between the fish owners and the quota owners (those in control). The "poorly" defined property rights can be revised anytime, if need be, while the owners and those in control tend to check and balance one another and promote their mutual responsibility and accountability. According to the Fisheries Management Act (2006), the owners have the responsibility to promote marine stock "conservation and efficient utilisation, thereby ensuring stable employment and settlement throughout Iceland". Quota–owners as the beneficiaries of economic rents are accountable to more than this group of owners. "New responsibilities towards the owners, the workers, the consumers, and the State thus rest upon the shoulders of those in control" (Berle and Means, 1991, p. 8). This doesn't mean that the design and implementation of the Icelandic fisheries' property rights system is easy and straightforward. In fact, the system has been evolving for more than 35 years (Arnason, 1996) as a slow and painful process of trial and error, of permanent redefinition of the fusion between private rights and public order that is characteristic of Iceland's development path. Since its adoption in 1991, the Fisheries Management Act has been amended several times (in 1992, 1994 and 2006) to reflect numerous changes in the ITQ management system.

The Icelandic system of property rights for fishing is not a technical formula or a clear-cut legal system as their allocation involves subtle economic, social and political choices. As such they are rather a "historical record of winners, losers and social accommodation in economic and political struggles over a nation's direction" (Kennedy, 2011, p. 20). The implementation of the ITQ system is a work in progress that needs continuous adjustment, decision-making and institutional change. Simply following recommendations concerning MSY or TAC values is not sufficient for making the system work efficiently and equitably. "Positioning fisheries to conserve ecosystems and livelihoods requires much more than preventing overfishing. It involves institutional change that encompasses the public good of the oceans (biodiversity, ecosystem integrity, and sustainability) and societal values (existence, aesthetic and amenity)" (Grafton *et al.*, 2008, p. 631). It also involves a correct understanding of the concept of property rights.

Property rights are not the rights of individuals over things but "complex relations of reciprocal rights and duties among people with respect to things" (Kennedy, 2011, p. 26). The "things" over which the property rights are exercised, in the case of fisheries, are living, moving creatures with complex life and reproductive patterns which require specific habitats and careful human interference. It is wrong to simply conceptualize fish as "living property or swimming inventories that can be owned long before they are actually captured" (Bavington, 2010, p. 11).

Due to the dynamic complexity of the relationship between self-interested humans and unpredictable fish stocks which is not always easy to grasp, governance of fisheries has been identified as a "wicked problem" (Jentoft and Chuenpagdee, 2009). "Decisions about how to deal with it must be made on a routine basis, but governors would also have to be prepared for sudden changes that occur for reasons that may or may not have anything to do with how they actually try to govern them" (Jentoft and Chuenpagdee, 2009, p. 556). The collapse of the Newfoundland fisheries in Canada has demonstrated that even the richest fishing grounds can disappear almost overnight with chances of the fish stocks returning being marred by huge uncertainties. It seems like the fish stocks have their own agenda and travel destinations which are independent of all human efforts. Maybe they are not owned by humans at all.

It has been said that the "implementation of ITQs is an irreversible social experiment" (Eythorsson, 1996, p. 281). I beg to differ. The design and implementation of the individual transferable quotas system in the Icelandic fisheries is indeed a very interesting social experiment but it is not an irreversible one. Like any human construct, it is transient in nature and can be reversed at any time by a subsequent human choice, or simply by the disappearance of the fish stocks.

References

Anonymous (2009) A survey of the sea: an Icelandic success: a model way to catch and keep fish. *The Economist,* 390(8612), 15-17.

Anonymous (2009) *World Fishing and Aquaculture,* 1. Iceland.

Arnason, R. (1996) On the ITQ fisheries management system in Iceland. *Reviews in Fish Biology and Fisheries,* 6, 63-90.

Bavington, D. (2010) *Managed Annihilation: An Unnatural History of the Newfoundland Cod Collapse.* Vancouver: UBC Press.

Berle, A. A. and Means, G. C. (1991) *The Modern Corporation and Private Property.* Rutgers: Transaction Publishers.

Breen, P. A. and Kendrick, T. H. (1997) A fisheries management success story: the Gisborne, New Zealand, fishery for red rock lobsters (*Jasus edwardsii*). *Marine and Freshwater Research,* 48, 1103-1110.

Cohen, M. R. (1927) Property and sovereignty. *Cornell Law Quarterly,* 13, 8-30.

Costello, C., Gaines, D. D. and Lynham, J. (2008) Can catch shares prevent fisheries collapse? *Science,* 321, 1678-1681.

Eggertsson, T. (1996) No experiments, monumental disasters: why it took a thousand years to develop a specialized fishing industry in Iceland. *Journal of Economic Behavior and Organisation,* 30, 1-23.

Eythorsson, E. (1996) Theory and practice of ITQs in Iceland. Privatization of common fishing rights. *Marine Policy,* 20, 269-281.

Eythorsson, E. (2000) A decade of ITQ-management in Icelandic fisheries: consolidation without consensus. *Marine Policy,* 24, 483-492.

Finley, C. (2008) A Political History of Maximum Sustained Yield, 1945-1955. In Starkey, D. J., Holm, P. and Barnard, M., eds., *Oceans Past: Management insights from the history of marine animal populations.* London: Earthscan.

Gordon, H. S. (1954) The economic theory of a common-property resource: the fishery. *Journal of Political Economy*, 62(2), 124-142.

Grafton, R. Q., Hilborn, R., Ridgeway, L., Squires, D., Williams, M., Garcia, S., Groves, T., Joseph, J., Kelleher, K., Kompas, T., Libecap, G., Lundin, C. G., Makino, M., Matthiasson, T., McLoughlin R., Parma, A., San Martin, G., Satia, B., Schmidt, C-C., Tait, M., and Zhang, L. X. (2008) Positioning fisheries in a changing world. *Marine Policy*, 32, 630-634.

Hannesson, R. (2005) Rights based fishing: use rights versus property rights to fish. *Reviews in Fish Biology and Fisheries,* 15, 231-241.

Haraldsson, G. (2008) Impact of the Icelandic ITQ system on outsiders. *Aquatic Living Resources,* 21, 239-245.

Hilborn, R. (2007) Defining success in fisheries and conflict in objectives. *Marine Policy,* 31, 153-158.

ICES (2010) Report of the North Western Working Group (NWWG) 2.4.2. ICES Advice, pp. 18-23.

Jentoft, S., Chenpagdee, R. (2009) Fisheries and coastal governance as a wicked problem. *Marine Policy,* 33, 553-560.

Johnsen, J. P. and Eliasen, S. (2011) Solving complex fisheries management problems: what the EU can learn from the Nordic experiences of reduction of discards. *Marine Policy*, 35, 130-139.

Kennedy, D. (2011) Some caution about property rights as a recipe for economic development. *Accounting, Economics, and Law*, 1(1), Article 3.

Matthiasson, T. (2003) Closing the open sea: development of fishery management in four Icelandic fisheries. *Natural Resources Forum, 27,* 1-18.

Pauly, D. and Thia-Eng, C. (1988) The overfishing of marine resources: socioeconomic background in Southeast Asia. *Ambio,* 17(3), 200-206.

Pauly, D. (1996) ITQ: the assumptions behind a meme. *Reviews in Fish Biology and Fisheries,* 6,109-112.

Scott, A. (1955) The fishery: the objectives of sole ownership. *Journal of Political Economy,* 63(2), 116-124.

Shiva, V. (2010) Resources. In Sachs, W., ed., *The Development Dictionary: A Guide to Knowledge as Power*. London: Zed Books Ltd.

Schaefer, M. B. (1957) Some considerations of population dynamics and economics in relation to the management of the commercial marine fisheries. *Journal of the Fisheries Research Board Canada,* 14(5), 669–681.

Statistics Iceland (2010) Labour market, employed persons by economic activity, sex and regions 2008-2010 (NACE REV 2). http://www.statice.is/?PageID=1191&src=/temp_en/Dialog/varval.asp?ma=VIN01103%26ti=Employed+persons+by+economic+activity%2C+sex+and+regions+2008%2D2010+%28NACE+REV+2%2E%29+%26path=./Database/vinnumarkadur/rannsoknir/%26lang=1%26units=Fjöldi

Stavins, R. N. (2011) The problem of the commons: still unsettled after 100 years. *American Economic Review* 101, 81-108.

Stiglitz, J. (1999) Whither Reform? Washington, D.C.: The World Bank. http://siteresources.worldbank.org/DEC/Resources/84797-1251813753820/6415739-1251814010799/stiglitz.pdf

Weitzman, M. L. (2002) Landing fees vs. harvest quotas with uncertain fish stocks. *Journal of Environmental Economics and Management,* 43(2), 325-338.

CHAPTER 11

Do all answers lie within (the community)? Fishing rights and marine conservation

Maria A. Gasalla[1]

The role of small-scale fisheries in developing countries has been widely discussed in the fisheries and "commons" literature. However, in the conservationist debate, even small-scale fishing has often been seen as a potential evil that indeed threatens ecosystem health and should be broadly and rapidly eliminated. Besides the global commitments to increase the protection of world's marine and coastal regions, several international *fora* have focused on the issue of how to conciliate fisheries with conservation, with both human and ecological dimensions taken into account in more contemporary governance systems (i.e., ecosystem approach to fisheries, or even fisher's self-governance). This essay draws attention to some examples from Brazil in order to consider alternatives towards community-based processes in fisheries governance and in the political economy of marine resource use. Promising pathways of potentially participative initiatives for small-scale fisheries illustrate the opportunities and challenges of current options where power imbalance between players often undermines socially-inclusive marine conservation.

Introduction

The role of small-scale fisheries in developing countries has been widely shown and discussed in the fisheries and "commons" literature (Berkes, 1989; Ostrom, 1990; Ostrom *et al.*, 1999; Allison and Ellis, 2001; Berkes *et al.*, 2001). A pro-poor function with a capacity to provide labor and cash income to resource-poor households has also been frequently associated with overexploitation (Bené *et al.*, 2010), in that there are too many fishers chasing

1 Corresponding author: Fisheries Ecosystems Lab, University of São Paulo, São Paulo, Brazil, mgasalla@usp.br

too few fish and thus making too little income (Sumaila, 2008). However, the usual economic assumption that small-scale fisheries is an economic dead-end was deemed incorrect because such fisheries serve other needs in developing countries, such as human and socio-economic development, poverty reduction, employment, food security, dignity, labor buffers, adaptive markets, protein source, and social redistribution (Bené *et al.* 2007; 2010).

In some conservationist debates, in contrast, even small-scale fishing has often been seen as a potential evil that threatens ecosystems and their biodiversity health and should be broadly and rapidly eliminated (e.g. in serveral e-discussion gorups). Besides the global commitment to protect and conserve at least 10 percent of the world's marine and coastal regions by 2012 (CDB, 2006), several international *fora* have focused on how to conciliate fisheries with conservation, with both human and ecological dimensions taken into account in more contemporary governance systems (i.e., EAF, ecosystem approach to fisheries, or even fisher's self-governance).

In the real world, in the struggle to access marine resources industry leaders have the ability to push for the perennial use of the public wealth. Small firms and small communities are often seen as unwelcome impediments to the attainment of their objectives. Industrial operations in coastal and marine ecosystems have frequently been enthusiastic to claims and policies working in the direction of eliminating fishers, and biodiversity conservation initiatives may help in such a purpose even with the noble goal of environmental protection.

Marine protected areas (MPAs) have been promoted around the world as a key management tool or an effective means of protecting marine and coastal resources and biodiversity (Agardy, 1997; Allison *et al.*, 1998; Hilborn *et al.*, 2004), but agendas have shown little concern about their impact on the livelihoods, culture and survival of small-scale and traditional fishing and coastal communities that can certainly contribute to the success of its implementation (Diegues, 1995; 2002; Hilborn *et al.*, 2005).

In some cases, when such a concern apparently exists a masked negotiation can be in place driven by veiled economic interests and political power furthermore the desired ecological concern that may have motivated or justified their original formation (Degnbol et al., 2006). Jentoft et al (2011) give pertinent examples in respect to that the perspective of MPAs as open, complex governance systems consisting of diverse stakeholder coalitions raises the issues of definitional power: whose goals are becoming the real goals of the MPAs, whose interests count more when the MPAs are designed and mandated, who wins or loses as a consequence of the way the MPA is being formed, is marine biodiversity conservation really the issue?.

There is considerable room for policy debate about the objectives: on where we want to be in the well known trade-off between the environmental impact of fishing and fishery yields, but also between marine conservation and fishing rights. There is no denying that fishing can modify marine ecosystems producing impacts to biodiversity and also that different societies make different choices about how much environmental change might be accepted in return for sustainable food production. Considering the outcomes, as suggested by Berkes and Folke (1998), by recognizing the interconnected nature of the social and ecological subsystems, and using ideas of complex adaptive systems and institutional development (Ostrom, 1990; Berkes, 1989; 2002) one can understand why relationships between actors involved in marine conservation may evolve differently in distinct contexts, territories and countries.

In one place, a stewardship role can be attributed to local small-scale communities as a sense of ownership and responsibility amongst natural resource users to manage and protect natural resources for their long-term sustainability and cultural heritage values. In another place, a project that starts with particularly acrimonious relations among the actors has little chance of success in building trust and cooperation (Chuenpagdee and Jentoft, 2007). However, in order to ensure that the known mistakes of large-scale fisheries are not repeated, it seems important to think about the shared way: while the state would have the fundamental responsibility to ensure that access to resources is fair and equitable, local communities would also have a commitment to sustainable use and protection of marine and fisheries resources (Kurien, 2002). The sharing of these mutual concerns seems to be the only way forward (Jentoft, 2003; Kurien, 2010) if honestly set, requiring a partnership accord between the interest groups involved in the fisheries sector to negotiate a trustworthy process of sharing the rights and authority for responsible stewardship (Gutierrez *et al.* 2010; Kurien, 2010).

Negotiating marine fisheries co-management in Brazil can be considered a process that is still in its infancy. The main advantage of this is that innovative governance systems can still arise from local territorial programs for small-scale fisheries and sustainable use reserves when locals and emerging leaders have a clear and open mind about the possibilities and potential of responsible fishing. The main disadvantage is that there is a long way to go to develop awareness, ethics, capacity building, organization and networking, especially in terms of fishery conservation and marine protected areas.

Nevertheless, if it is possible that marine conservation can become effective and socially inclusive and responsible, it is worth examining some available tools from the Brazilian experience. Alternatives and perspectives will be shown by illustrating the opportunities and challenges of current options.

The Brazilian experience

Are you aware of a place where social scientists talk about conservation of biodiversity, and natural scientists are interested in community-based approaches? One can say that both the awareness of the social movements (i.e., a need for social justice) and concerns of biodiversity loss have been reaching at least a part of the Brazilian scientific community in the last decade, allowing a sort of interchanging "transdisciplinarity". If it can go reasonably in science then again in the policy debate the "social-environmental" discourse attempts to create mechanisms to implement such ideas but in practice it is difficult to transcend the political speech with ethical commitments. Here some promising pathways of potentially participative initiatives for small-scale fisheries in Brazil will illustrate the strengths and weaknesses of current available options. They seem to raise the question: Can Brazil actually hold the potential of being at the forefront of more socially inclusive marine conservation?

Country's current situation and its coastal small-scale fisheries

Brazil currently has a potentially growing economy, a multifaceted and creative culture, and notably the world's largest biodiversity wealth. Although the country still faces a difficult prospect in trying to curb illiteracy, violence and corruption in several areas, it has proven to be a stable democracy in the last couple of decades and faces no imminent social, political or economic unrest; neither is it naturally exposed to extreme events or dramatic disasters. With a coastline of around 8,000 km, which stretches over diverse ecological systems, Brazil has almost one million small-scale artisanal fishers and three million people depend on its economy (Diegues, 2008). However, these "artisanal fisheries systems" are highly vulnerable to a variety of external factors and pressures such as large-scale tourism, pollution, population pressure in the coastal zone, uncontrolled industrial development operating at the land-sea interface, oil and gas drilling, the expansion of modernized inshore fisheries and competing large-scale fleets. Nevertheless, the traditions of small-scale, inshore fishing have continued to prevail and remain the economic backbone of Brazil's coastal poor (Diegues, 2008).

Following the decline and collapse of important industrial fisheries, small-scale artisanal fishers have reoccupied some coastal waters in a process that was described as "re-artisanalization" of fishing activities. At present, catches from artisanal fisheries represent 60% of the country's total fishery landings (Diegues, 2008). However, in the most industrialized areas of the country, the maritime territoriality of small-scale fishers has been partially or completely lost, which contributes to marginalization and poverty aggregation.

Even if Brazil still seems far from either ecologically, economically, or socially efficient science-based management programs or sustainable harvests in terms of fisheries management, some experiences and perspectives regarding community-based options for small-scale fisheries should be highlighted. When there is not a strong central government control of fishing, community-based co-management seems to be a promising alternative. In this case, territorial fishing rights can be assured in the context of marine conservation.

On the territorial concept, fishing rights and conservation

As we will see in the following sections, the concepts of territory, "territorial idea", and fishing territoriality are quite important as: (1) the local productive structure of fisheries including resources and space, (2) a notion of ownership and rights-based access in a particular living-and-working maritime space, (3) unity of development planning, (4) unity for biodiversity, natural capital, or ecosystem treatment, consequently useful for: (5) "ecosystem-based fishery management" (EBFM) which is mainly area-based, (6) "marine protected areas" (MPAs) which are, by definition, focused on a delimited space to be preserved, as well as for (7) the "large marine ecosystem" (LME) context.

This conceptualization is not trivial, since fishing territories are important for setting social objectives regarding small-scale fishing rights and conservation. As shown by Hasbaert (2005) in "The Myth of no territorialization", the territorial idea is very ancient, used in biology two centuries ago, subsequently incorporated in human sciences through geography, and reaching public policies in the last quarter of the 20th century. Territorial delimitation is not often a consequence of public policies but rather of a very peculiar articulation based on a local productive structure (Santos, 1996). Fishing territoriality is thus based on the knowledge used to locate productive fishing grounds, and its legitimacy occurs through formal or informal institutions in order to maintain long-term access through time (Malberg, 1985; Begossi, 1998). Then, fishing rights (as well as most marine conservation spatial planning) are based on a maritime territoriality idea. Territories have been used to restrict and control the way resources are exploited, and the use and transfer of commons rights. Therefore, by defining where (and by whom) fishing is allowed, both EBFM and MPAs may affect power relationships related to territorial use rights.

MPAs refer to any coastal or marine area in which certain uses are regulated to conserve natural resources and biodiversity but also historical and cultural features. While numerous studies have examined the potential ecological and biological impacts of MPAs (Agardy, 1997; Allison *et al.*, 1998; Roberts and Hawkins, 2000; Hilborn *et al.*, 2004), few have focused on their social implications for communities and stakeholders that heavily depend on fisheries for their livelihood. Thus, a particular MPA may be both a "biological success" and a "social failure" (Christie *et al*, 2003; Diegues, 2008).

In Brazil, several MPAs have been established opportunistically and not defined either by scientific (biodiversity or cultural values) or social (e.g., aiming to preserve the livelihoods and survival of traditional fishing cultures) criteria. The implementation of no-take sanctuaries or marine parks with restricted human interventions were constrained and challenged not only by the absence of administrative infrastructure and proper capacity for compliance and surveillance (Moura *et al.*, 2007; Gerhardinger *et al.*, 2010), but also by the social justice of territorial trade-offs in the face of industrial pressure.

Genuine Brazilian MPAs: Extractive and sustainable use reserves

In the late 1980s, a new sort of protected area came about in Brazil: the so-called "sustainable use reserves" (SUR), which combined the idea of sustainable exploitation with the aim of conserving natural resources and biodiversity (Diegues, 2008). By definition, SUR are "protected areas aimed at sustainable use and conservation of natural renewable resources by traditional extractive populations". They can also be described as areas of social and ecological interest because they are based on the premise of a real capacity for exploiting natural resources in a responsible way without jeopardizing the objectives of conservation. Unlike no-take reserves, which were created by the government and supported mainly by non-governmental organizations (NGOs), SUR were an outcome of pressure from traditional communities, such as rubber tappers in the Amazon, who wanted to save the resources that afforded them a livelihood before they were completely destroyed. Later on, the concept of "extractive reserves" – a formal and legally recognized co-management system successfully instituted in the western Amazonian forest economies – evolved to the coastal and marine domains of traditional fishing communities, which turned into Marine Extractive Reserves (MER) (Diegues, 2008). Acknowledging that the environment and society stand to benefit from helping the coastal poor secure access to their traditional sea territories and livelihood resources, MERs were a radical departure from setting up and managing no-take MPAs (Pinto da Silva, 2004). At present there are 20 coastal MER and 19 legally recognized SUR in Brazil, three of them on the coast (ICMBio, 2010).

A MER requires that certain biological, social and cultural criteria be satisfied before it is established. In MERs, fishing or aquaculture is the main, and often the only, economic activity. So a MER is a community-based, site-specific, multi-use, land and sea resource management approach based on claims of culturally distinct groups with longstanding livelihood ties to "artisan-scale" production territories (Cordell, 2006). MERs can also be considered "new commons" that are being built by coastal communities, particularly fishworkers, to protect their fishing territory from encroachment by other economic activities such as large-scale tourism, industrial fisheries, port construction, and oil and gas drilling (Diegues, 2008; Seixas *et al.*, 2009). In contrast to no-take reserves created without consulting the local people (e.g., Environmental Protection Areas off the coast of São Paulo State) and disrespecting traditional fishing rights, a MER is established mainly in response to a formal demand by local communities, fishers' associations and cooperatives. The application, addressed to the Ministry of Environment, should also indicate the approximate area traditionally used by local fishers. Assessments are required to identify fishery resources and biodiversity concerns, economic and social issues, fishing technology, markets, social organizations and conflicts, among other issues. A study by Diegues *et al.* (2008) concluded that (1) restrictions and threats of eviction (as in the case of integral protection units), are the main causes of conflicts between traditional fishers and managers, where fish resources and traditional fishers are the main losers, (2) fishers' associations must be encouraged to establish no-take zones inside the MERs to protect their natural resources, (3) the more successful reserves are those that, in addition to the sustainable use of natural resources, are able to improve fishers' incomes and provide adequate social benefits, particularly educational and health services, (4) women's participation in local associations greatly contributes to the MERs success.

It seems clear that the idea in Brazil that biodiversity conservation can be achieved only through no-take protected areas is based mainly on the fact that institutions are inefficient and incapable of managing natural resources and developing rational control rules or user consensus. Therefore, it is very important to include and spread knowledge on the available portfolio of options in the current conservation debate. On the other side, capacity that reinforces the power of traditional communities, who depend on natural resources, to make decisions needs to be strengthened (e.g., see the Recife Letter in Pedrosa, 2011).

"Territories of citizenship": Is there a potential route to strengthen fishers' self-governance?

"Territories of Citizenship" is an innovative Brazilian public policy for poverty and social inequalities reduction in the neediest regions of the country, especially in rural areas, and in support of family agriculture (Territories of Citizenship, 2009). This country-based territorial concept and policy has also migrated to the fisheries arena, considering that small-scale fishing communities are analogous to rural communities and concentrate several low-income and poor populations. Basically, the conceptual purpose of the "Territories of Citizenship" are economic development and the unification of basic citizenship programs through a strategy of sustainable territorial development, where social participation and integration of actions among the federal government, states, and municipalities are crucial to represent the reality of social groups and institutions. This approach, led by the Ministry of Agricultural Development, aims to facilitate the planning of governmental actions for the betterment of these selected regions with low human development. In 2009, the Brazilian Ministry of Fisheries and Aquaculture created the analogous program "Territories of Fisheries and Aquaculture" (MPA, 2010), to identify fishing and aquaculture territories that are threatened, with the aim of improving their capacity for self-governance by supporting their collective interests. This seems to be a promising route for strengthening community-based management in territories that deeply require collective participation for self-development and survival. This can also be the case in areas close to urban centers that are marginalized by urban societies, competing initiatives, and conservation.

However, the policy is still in its infancy and it needs a proper fund allocation, sound participatory evaluations and locally-engaged projects adapted to local ecosystems, culture and identity. The territorial approach to fisheries co-management can be an important pathway for dealing with both social exclusion and environmental degradation in Brazil. Nevertheless, it will only occur when present "territories" become legitimate with nominated and representative councils and targeted funds, focused on a more ecosystem-oriented approach. In some places, it may eventually happen independently of government institutions.

Coastal fishing agreements

Both the Ministry of Environment (through the National System of Protected Areas) (Prates and Blanc, 2007) and the Ministry of Fisheries and Aquaculture (MPA/FIPERJ, 2009) have been using another sort of community-based management tool, locally called "Fishing Agreements".

The former institution applied this new legal instrument (IBAMA IN nº 29/2002) through projects in the Amazonian region (such as the ProVárzea Project in Santarem and Mamiraua in Tefé), as a way of formalizing use and access rules defined by fishing communities in processes triggered by environmental conflicts and competition by natural resources (Oviedo *et al.*, 2003). Between 2002 and 2007, around 20 fishery management projects were undertaken involving 150,000 people, promoting the strength of institutions, raising local leadership, improving management, and making the fishing economy in Amazonia more environmentally sound (Vidal, 2010).

While more than 50 fishing agreements currently exist in the Northern region of the country, their implementation in the coastal zone is still developing. A new attempt has been made recently in a coastal community of the Southeastern region, in the State of Rio de Janeiro (Ilha Grande Bay), where conflicts between fishers and local marine protected areas are common (MPA/FIPERJ, 2009). However, the way this mechanism will be defined is still in the planning phase, subjected to the organization of co-management arrangements and to the definition of marine conservation priority goals.

The participatory aspect of "Fishing Agreements" in communities that lacked public attention and organization can be a promising differential agenda that will also require capacity building to be properly and effectively implemented.

Marine environmental compensation and fishing communities: May it flow to a healthy direction?

Compensation means recompense, i.e., the loss suffered by an owner or resource user as a consequence of being deprived of his property or resource. It can be monetary or indirect and, in the former case, the amount of compensation payable cannot be less than the monetary value into which that person/community might have converted this property/resource. Restrictive measures in marine conservation or stock recovery programs may trigger compensation processes to socio-economically vulnerable fishing communities (e.g., the cod moratorium in Eastern Canada). On the other hand, indirect compensation to fishers can occur in the case of environmental damage or disasters, such as oil spills or tsunamis, but also through the licensing of projects that will potentially impact the fisheries ecosystem, such as drilling.

In Brazil, projects that cause a significant environmental impact pay a portion (more than 0.5%) of their total cost into an environmental fund as compensation (Federal Law No. 9985/2000). This process is particularly pertinent to the issue of marine conservation and fishing rights. Each environmental office determines which conservation area(s) is eligible to receive compensation as a result of the project's impacts. If compensation is warranted, the office also decides the exact amount of the compensation, based on the degree of impact. When a project is expected to harm a specific conservation area, the environmental office is also responsible for approving the license for the project. When compensation is warranted, the monies must be spent on the conservation area that is harmed or on creating a new conservation area, depending on the criteria of the office. Recently, negotiations led to the creation of an Environmental Compensation Fund, which will be an alternative to developers and other entities that are required to pay environmental compensation but do not want to get involved in a bureaucratic process (UNEP, 2010). This is particularly interesting because the fund that is financed by monies paid as compensation will be applied to the creation, management, and implementation of protected areas such as parks and reserves.

Moreover, increasing investments in Brazilian pre-salt and oil and gas reserves can potentially result in further funds that will provide compensation to both the marine environment and vulnerable communities, and small-scale fishers are important actors in this scene. The recent discovery of large reserves of gas along the Santos Basin off the more industrialized coast of the country has also seen increased investments in parallel activities. It is to be emphasized that benefits from related compensation funds should be applied to societal and conservation needs in a way that reaches vulnerable communities such as traditional fishers in an ethical and respectfully way .

However, processes disconnected from fishing communities can be very inefficient in making compensation work in practice. This is particularly the case when the officers in charge of fishery-related compensation programs and the paying companies are not able to monitor, verify and evaluate middlepersons reporting so that appropriate compensation occurs benefiting a particular fishing community in a corruption-free positive way based on their social-ecological reality. It seems important to alert that transparent and equal-opportunity bidding procedures are still required. Additionally, it is important that professionals engaged in participatory diagnosis do not participate in both the steps before as well as those after the approval of licensing projects (impact assessment and mitigation steps). Otherwise, the process will be biased, and real benefits will not accrue to local people. In this sense, the fishing communities' degree of organization may have a positive or negative impact on this sort of compensatory process when the definition and setting of local demands are essential steps towards beneficial pay-backs.

The development of compensation schemes that contribute to management and to local and independent responsible fishing agendas should be recommended (Maramar, 2010). Overall, the Brazilian environmental compensation process seems to be a potentially promising way to trigger community-based approaches to both small-scale fisheries management and marine conservation. However, *de facto* real benefits have not been seen so far in several affected fishing communities while it is expected that Brazilian democratic society will evolve through fairer and stronger institutions able to set these procedures in a cleaner, technical, and more solid framework.

Other ways forward: Solidarity trade networks for fisheries

The social solidarity economy is one of the responses to the current economic crisis, both for the South and the North, and should be mentioned considering that it adds value to sustainable resource use based on collective approaches and social-ecological justice. Within a global context, such an economy can only develop by establishing the link between economy and society, local and global, labor and investment, as well as production, consumption and the environment. In the fisheries world, networks of solidarity economy can enhance and trigger a virtuous cycle where market- and trade- imposed benefits of harvests are based on social well-being, economic growth and ecological health (i.e., reduction of impacts on ecosystems, proper by-catch management, and maintenance of target species population under sustainable levels). In 2010, Brazil hosted the first System of Equitable and Solidarity Trade in the world that is recognized and supported by the state (through the National Council of Solidarity Economy). The system set a series of parameters to follow during the execution of public policies aimed at creating employment and revenues through promotion of the solidarity economy and of fair trade. Its objectives include supporting consumer education in order to adopt sustainable habits, and the organization of consumers to purchase products and services from fair and solidarity trade; reinforcing the national identity of equitable and solidarity trade by publicizing the concept and by adopting practices that are inherent to it; and favoring practices of equitable prices by those who produce, commercialize and consume (NCSE, 2011).

It is expected that such a system may possibly extend to the fisheries sector in Brazil, adding value to sustainable products. In this sense, certification processes for small-scale fisheries at both the local and global scale should involve promoting these initiatives for healthy and sustainable supply-chains that include both conservation of fishery resources and fishing rights. They might promote the reduction of illegal, unregulated, and unreported fishing while supporting artisanal fishers' access to resources, territories, and markets. Besides capture fisheries, this will also encourage market access for community aquaculture and marine fish cultivation.

On "divorced" public policies and institutional weakness

Although several instruments and mechanisms are potentially available for reconciling marine conservation with fishing rights in Brazil, a clear disconnect between different policies may drive institutional weakness that impede proper implementation of community-based fisheries management in the coastal and marine realms. The evidence raises concern about the lack of integration of the different policies, especially between those mentioned before and coastal zone management programs.

The main problems include poor inter-institutional coordination of coastal and ocean governance with fisheries and aquaculture related issues, administrative and bureaucratic systems, and structural problems for MPA delivery. A recent critical analysis (Gerhardinger *et al.*, 2010) showed that the National System of Marine Protected Areas in Brazil was perceived as weak, with few recognizable marine conservation outcomes on the ground. The following major flaws were identified: institutional crisis faced by the national conservation agency; poor management within individual MPAs; an overly bureaucratic management and administrative system; financial shortages creating structural problems, and a disconnection between MPAs policy and their delivery. Furthermore, the need for a better understanding of the role of 'leaderships' in the performance of MPA networks, more effective official peer-review mechanisms, and more localized audits (and reforms if necessary) of the federal biodiversity conservation agency were also identified. It is important to highlight that a continuing focus on the designation of more MPAs while not addressing these issues on the present ones will achieve little beyond fulfilling biodiversity commitments on paper. On the other hand, an updated inclusion of fishing rights into the definition of access rules for marine conservation units will be more than pertinent.

Overall coastal zone management, which involves multiple stakeholders at local, state and national levels that in many cases overlap and conflict, and the agendas for natural resource use (living and non-living), seem to be quite incompatible. Also, bias and strong influence of other sectors that lack an environmental commitment often undermine local initiatives based on participatory processes of vulnerable communities such as fishers. For instance, there is a claim from NGOs that while they are invited to participate in public debate aiming to legitimate several decision-making processes in the coastal zone, they are not allowed to bring "real ground-based participatory agendas from the local communities". On the other hand, the fact that different NGOs are competing for funds and influence at international and national levels seems to undermine their common goals. In addition, the lack of organization within fishing communities and the poor participation of academic institutions in public *fora* also result in weak influence of these sectors in the (non-integrated) coastal zone management as a whole.

Finally, and probably the most important dilemma that still persists without a proper answer: if the fishing communities are not organized, capable mature and capacitated to evaluate and elaborate on the benefits and damage/loss of marine conservation and fishing rights, how can they not remain confined to the eternal delivery of disconnected public policies?

A future perspective: the "science" behind fisher's knowledge

Fishers often rely on an exhaustively knowledge of the natural milieu for their livelihoods. Case studies around the globe show clearly that this knowledge can be joined in collaboration with scientists, government managers and NGOs to become effective and appropriate counterparts in fisheries science and management (Johannes *et al.*, 2000; Haggan *et al.*, 2006). In Brazil, the potential of application and incorporation of fisher's local knowledge in small-scale fisheries management has been highlighted as a promising perspective through a development more linked or rooted in the local culture. Its potential applicative role to better manage tropical fisheries (Silvano *et al.*, 2009) and also in marine high-level science programs (Gasalla and Diegues, 2011) has been discussed.

Also, new methodologies based on its application particularly directed through a policy and decision-making framework have also been locally developed. In terms of science-based global change programs, for instance, Gasalla and Diegues (2011) recently developed the so-called "ethno-oceanography" conceptual framework which has been considered promising to address environmental knowledge on global change issues by interdisciplinary modeling that combines fishers (bottom-up) with scientific knowledge (top-down).

Leite and Gasalla (2010)[2] proposed a new method to incorporate fisher's knowledge in EBFM issues in Southeastern Brazil, seeking consensus in the identification of essential fish habitats, through the spatial location of both spawning and recruits grounds of fishery resources and their seasonal patterns. Such new methods seem somehow promising for future application in participatory approaches that can contribute to reconciling fish-

2 World Small-Scale Fisheries Congress 2010 Best Student Paper Award

ing with marine conservation. It should be emphasized that the development of an "ecosystem approach to small-scale fisheries" should respect the characteristics of this type of activity. A proper approach must be locally developed to deal with its particularities and needs, very distinct from industrial-oriented fisheries policies.

Conclusion

Marine fisheries ecosystem services must be guaranteed for future generations, and both biodiversity conservation and poverty alleviation should be high-level objectives related to this goal. The political economy of natural resource use is a subject that often lacks depth in the conservationist debate. Therefore, potential instruments available for community-based and territorial initiatives in Brazil should be taken into consideration in dialogues between marine conservation and fishing rights.

The conservationist debate in developed countries plays a central role in the way the settings of no-take zones for marine conservation have evolved. However, it should be considered that wealthy and well-structure societies may be capable of appropriately compensating those vulnerable actors negatively affected by conservation policies, which may not be the case in developing countries where small-scale fishing also provides a unique source of protein and food. Moreover, some developed societies have the potential to absorb and redirect human capital to work with sectors other than fishing, which may not be possible in some developing countries. This should by no means deny the needs for more efficient marine biodiversity conservation programs in developing countries, including Brazil. However, policy integration, and local community involvement are essential for the success of such conservation goals especially when the government role on fishing is not strong or solid.

There can be still doubts about the capacity of Brazil to establish socially responsive, economically realistic and environmentally sound MPAs, but as a giant social laboratory (according to Sachs, 2006) it seems to be able to come closer to that, in regard the tropical country context. However, for this to become reality, in-depth technical capacity and experience in developing particular ways of finding solutions for good practices, as well as the support of international agencies for socially-inclusive marine conservation projects are needed.

A new paradigm should allow that marine conservation include the perspective of the people whose lives are closely intertwined with the ocean and those who essentially make a living from fisheries. In this sense, it is likely that Brazil can become at the forefront of the construction of community-based alternatives for marine conservation that can interest fishers in environmental preservation and make local people the stewards of their territory.

Acknowledgments

I thank the fishers who always kindly collaborate in our field research, and Antonio Carlos Diegues, Fabricio Gandini, Marta Collier Ferreira Leite, Fikret Berkes, Amanda Rodrigues, Carolina Minte-Vera, Cristiana Seixas, Fikret Berkes, Joachim Carolsfeld, René Scherer, Ana Parma and Ratana Chuenpagdee for fruitful discussion and collaboration.

References

Agardy, M. T. (1997) *Marine protected areas and ocean conservation.* San Diego, California: Academic Press.

Allison G. W., Lubchenco, J. and Carr, M. H. (1998) Marine reserves are necessary but not sufficient for marine conservation. *Ecological Applications,* 8(S79-S92), 76.

Allison, E. H. and Ellis, F. (2001) The livelihoods approach and management of small-scale fisheries. *Marine Policy,* 25, 377-388.

Begossi, A. (1998) Property rights for fisheries at different scales: applications for conservation in Brazil. *Fisheries Research,* 34, 269-278.

Bené, C., Macfayden, G. and Allison, E. (2007) Increasing the Contribution of Small-scale Fisheries to Poverty Alleviation and Food Security. FAO Fisheries Technical Paper, 481. Rome: Food and Agriculture Organization.

Bené, C., Hersoug, B. and Allison, E. (2010) Not by rent alone: analysing the pro-poor functions of small-scale fisheries in developing countries. *Development Policy Review,* 28(3), 325-358.

Berkes, F. (1989) Common property resources. Ecology and community based sustainable development. London: Belhaven Press.

Berkes, F. and Folke, C. (1998) Linking social and ecological systems: Management practices and social mechanisms for building resilience. UK: Cambridge University Press.

Berkes, F., Mahon, R., McConney, P., Pollnac, R. and Pomeroy, R. (2001) *Managing small-scale fisheries: Alternative Directions and Methods.* Ottawa: International Development Research Centre.

Berkes, F. (2002) Cross-scale institutional linkages: perspectives from the bottom up. In Ostrom, E., Dietz, T., Dolsak, N., Stern, P. C., Stonich, S. and Weber, E.U., eds., *The Drama of the Commons.* Washington: National Academy Press, pp. 293-321.

Chuenpagdee, R. and Jentoft, S. (2007) Step zero for fisheries co-management: what precedes implementation. *Marine Policy,* 31, 657-668.

Cordell, J. (2006) Dynamics and challenges of MPA development and coastal protection. In *Taking Marine Management to Scale: Connecting Societies, Coastal Landscapes and the Sea.* Washington: World Bank.

Christie, P., McCay, B. J., Miller, M. L., Lowe, C., White, A. T., Stoffle, R., Fluharty, D. L., McManus, L. T., Chuenpagdee, R., Pomeroy, C., Suman, D. O., Blount, B. G., Huppert, D., Eisma, R-L, Oracion, E., Lowry, K. and Pollnac, R. B. (2003) Toward developing a complete understanding: a social science research agenda for marine protected areas. *Fisheries,* 28(12), 22-26.

Degnbol, P., Gislason, H., Hanna, S., Jentoft, S., Raakjaer Nielsen, J., Sverdrup-Jensen, S. and Wilson, D. C. (2006) Painting the floor with a hammer: Technical fixes in fisheries management. *Marine Policy,* 30(5), 534-543.

Diegues, A. C. S. (1995) Repensando e recriando as formas de apropriação comum dos espaços e recursos naturais. In Diegues, A. C. S., ed., *Povos e Mares: leituras em sócio-antropologia marítima.* São Paulo: NUPAUB-USP, pp. 209 -236.

Diegues, A. C. S. (2002) The myth of wilderness and the fate of traditional communities in the Brazilian Amazon. The Commons in an Age of Globalization, the Ninth Conference of the International Association for the Study of Common Property, Victoria Falls, Zimbabwe. http://dlc.dlib.indiana.edu/documents/dir0/00/00/08/12/

Diegues, A. C. S. (2008) *Marine protected areas and artisanal fisheries in Brazil.* India: SAMUDRA Monograph, International Collective in Support of Fishworkers.

Gasalla, M. A. and Diegues, A. C. S. (2011) People's seas: "ethno-oceanography" as an interdisciplinary means to approach marine ecosystem change. In Ommer, R. E., Perry, R. I., Cochrane, K. and Curry, P., eds., *World fisheries: A social-ecological analysis.* Oxford, UK: Wiley-Blackwell, pp. 120-136.

Gerhardinger, L. C., Godoy, E. A. S., Jones P. J. S., Sales, G. and Ferreira, B. P. (2010) Marine protected dramas: the flaws of the Brazilian National System of marine protected areas. *Environmental Management*, Online First, doi:10.1007/s00267-010-9554-7

Gutiérrez, N. L., Hilborn, R. and Defeo, O. (2011) Leadership, social capital and incentives promote successful fisheries. *Nature*, 470, 386-389.

Haesbaert, R. (2005) *O Mito da desterritorialização*. Brazil: Bertrand Brasil.

Haggan, N., Neis, B. and Baird, I. G., eds. (2006) Fishers' Knowledge in Fisheries Science. Coastal Management Sourcebooks 4. Paris: UNESCO.

Hilborn, R., Stokes, K., Maguire, J. J., Smith, T., Botsford, L. W., Mangel, M., Orensanz, J., Parma, A., Rice, J., Bell, J., Cochrane, K. L., Garcia, S., Hall, S. J., Kirkwood, G. P., Sainsbury, K., Stefansson, G. and Walters, C. (2004) When can marine reserves improve fisheries management? *Ocean and Coastal Management*, 47(3-4), 197-205.

Hilborn, R., Orensanz, J. M. and Parma, A. (2005) Institutions, incentives and the future of fisheries. *Philosophical Transactions: Biological Sciences*, 360(1453), 1471-2970.

ICMBio (2011) Instituto Chico Mendes de Conservação da Biodiversidade - Dados Gerais das Unidades. Dados Gerais das Unidades de Conservação - outubro de 2010. http://www.icmbio.gov.br/comunicacao/downloads (accessed April 13, 2011).

Jentoft, S. (2003) Co-management: the way forward. In Wilson, D. C., Nielsen, J. R. and Dengbol, P. eds., *The fisheries co-management experience: Accomplishments, challenges and prospects*. London: Kluwer Academic Publishers, pp. 1-13.

Jentoft, S, Chuenpagdee, R. and Pascual-Fernandez, J. J (2011) What are MPAs for: on goal formation and displacement. *Ocean and Coastal Management*, 54, 75-83.

Johannes, R. E., Freeman, M. M. R. and Hamilton, R. J. (2000) Ignore fishers' knowledge and miss the boat. *Fish and Fisheries*, 1(3), 257-271.

Kurien, J. (2001) Forging unity: the agenda. Indian Ocean Conference. Chennai, India: International Collective in Support of Fishworkers (ICSF) and International Ocean Institute (IOI).

Kurien, J. (2002) Some thoughts on moving from unsustainability to sustainability of tropical multi-specie eco-system and livelihood-oriented, small-scale, marine fisheries in developing countries. In Gréboval, D., ed., Report and documentation of the international workshop on factors contributing to unsustainability and overexploitation in fisheries. FAO Fisheries Report, 672. Rome: FAO, pp.161-163.

Kurien, J. (2010) Negotiating fisheries co-management in Aceh Province, Indonesia. Notes on process. OSRO/INS/601/ARC Fisheries and Aquaculture Project. Rome: FAO.

Leite, M. C. F. and Gasalla, M. A. (2010) A method for assessing fek/lek as a practical tool for ecosystem-based fisheries management: seeking consensus in southeastern brazil. *World Small-Scale Fisheries Congress,* Bangkok, Thailand.

Malberg, T. (1985) Territoriality at sea: preliminary reflections on marine behavioral territories in view of recent planning. *Man-Environment Systems,* 15, 15-18.

Maramar (2010). Maramar Institute for Responsible Management of Coastal Zone, Brazil. Environmental Governance Initiative. http://www.maramar.org.br/iniciativas/inga (accessed November 05, 2010).

Moura, R. L., Dutra, G. F., Francini-Filho, R. B., Minte-Vera, C. V., Curado, I. B., Guimarães, F. J., Oliveira, R. F., and Alves, D. C. (2007) Gestão do uso de recursos pesqueiros na reserva extrativista marinha do Corumbau, Bahia. In Prates, A. P. and Blanc, D. eds., *Áreas aquáticas protegidas como instrumento de gestão pesqueira.* Brasília: MMA/SBF, p. 179.

MPA /FIPERJ (2009) Relatorio do Acordo de Pesca: uma possibilidade para a Baía da Ilha Grande/RJ. Brasília: Ministry of Fisheries and Aquaculture Report.

MPA. (2010) Política Territorial da Pesca e Aquicultura. Brasília: MPA. http://www.mpa.gov.br/mpa/seap/Jonathan/mpa3/planos_e_politicas/docs/politica_territorial.pdf

NCSE (2011) *National Council of Solidarity Economy.* http://www.mte.gov.br/ecosolidaria/cons_default.asp (accessed April 13, 2011).

Ostrom, E. (1990) Governing the commons: The evolution of institutions for collective action. Cambridge, UK: Cambridge University Press.

Ostrom, E., Burger, J., Field, C. B., Norgaard, R. B. and Policansky, D. (1999) Revisiting the commons: local lessons, global challenges. *Science,* 284, 278-282.

Oviedo, A., Vasquez, R. and Ruffino, M. L. (2003) Acordos de pesca: A comunidade é quem faz. Brasília: WWF Brasil/IBAMA/ProVárzea.

Pedrosa, B. M. J. (2011) Uniting for Change. *Samudra report,* 58, 36-40.

Pinto da Silva, P. (2004) From common property to co-management; lessons from Brazil's first Maritime Extractive Reserve. *Marine Policy,* 28(5), 419-428.

Prates, A. P. and Blanc, D., eds. (2007) *Áreas aquáticas protegidas como instrumento de gestão pesqueira.* Brasília: Ministério do Meio Ambiente, Secretaria de Biodiversidade e Florestas.

Roberts, C. M. and Hawkins, J. P. (2000) Fully-protected marine reserves: a guide. Washington, D.C.: WWF Endangered Seas Campaign.

Sachs, I. (2006) Desenvolvimento includente, sutentável e sustentado. São Paulo: Garamond.

Santos, M. A. (1996) *A natureza do espaço.* São Paulo: HUCITEC.

Seixas, C. S., Minte-Vera, C. V., Ferreira, R. G., Moura, R. L., Curado, I. B., Pezzuti, J., Thé, A. P. G. and Filho, R. F. B. (2009) Co-managing commons: advancing aquatic resources management in Brazil. In Lopes, P. and Begossi, A., eds., *Current Trends in Human Ecology.* Newcastle upon Tyne: Cambridge Scholars Publishing, pp.156-182.

Silvano, R., Gasalla, M. A., and Pacheco, S. (2009) Application of fishers' local ecological knowledge to better understand and manage tropical fisheries. In Lopes, P. and Begossi, A. eds., *Current trends in Human Ecology.* Newcastle upon Tyne, UK: Cambridge Scholars Publishing, pp. 76-100.

Sumaila, R. (2008) Heath based fisheries management for development. International Institute of Fisheries Economics and Trade's Conference, July 22-25. Nha Trang: Vietnam.

Territories of Citizenship (2009). Territories of citizenship: integration of public policies to reduce inequalities. Brazil: Brazil Government. http://www.territoriosdecidadania.gov.br (accessed February 02, 2010).

UNEP (2010) United Nations Environmental Program. http://www.unep.org/dec/onlinemanual/Enforcement/InstitutionalFrameworks/EconomicInstruments/Resource/tabid/1016/Default.aspx (accessed April 02, 2011).

Vidal, M. D. (2010) Manejo participativo da pesca na Amazônia: a experiência do ProVárzea. *Ciência e Natura, UFSM,* 32(2), 97-120.

CHAPTER 12

Conflicting gears, contested territories: MPAs as a solution?

José J. Pascual-Fernández[1] and Raquel de la Cruz-Modino

Fishing gear conflicts have a long tradition in coastal areas. These conflicts have increased recently, influenced in many cases by the entrance of new stakeholders like recreational fishers. Marine protected areas (MPAs) are territorial institutional arrangements that risk reinforcing these conflicts but, given the right conditions, may also be a means of alleviating them. In Spain, artisanal fishers have obtained a privileged position in the development of MPAs. In some cases, MPAs have provided them with increased control over their traditional fishing areas, at the same time assuring increased governability while maintaining ecosystem health. In this essay, we analyze the elements and processes that have helped or hindered these developments and their implications for conflict resolution and coastal governability.

Introduction

Human societies have developed a diversity of adaptations in order to make a living from aquatic environments. To make it possible, humans needed to overcome the fact that these environments are not well suited to our natural capacities. Hominid evolution was developed in terrestrial habitats, and our capacities have been built through a slow process of evolution in a diversity of land ecosystems. This is not to say that hunters and gatherers, the most long-term adaptation of human societies, did not develop strategies to use abundant aquatic resources where they had such opportunities, but they usually needed relatively sophisticated cultural tools to be able to assure their subsistence. Gathering shellfish escaped these requirements, as this

1 Corresponding author: Institute of Social and Political Sciences, University of La Laguna, Tenerife, Spain jpascual@ull.es

activity could be developed on beaches or intertidal areas with relatively simple technology or just with bare hands. Sometimes, fishing from the shore or rivers could also be developed with relatively simple technology and environmental knowledge. But human societies soon expanded their activities into more distant water environments; then, the complexity of the technology, knowledge and abilities needed for successful adaptation increased rapidly. Gears, knowledge and skills allow human societies to control the harvest of fishing resources, but these three factors are not evenly distributed in any human societies, and even less if we compare human groups with varied access to capital or technology. European history in fisheries is plagued with conflicts related to access to resources, and to the introduction of new technology; especially when this technology may introduce new actors in the activity, or expel others from it. In coastal areas, cultural and socio-economic changes may also introduce new actors, resulting in conflicts for territories and resources, as well as clashes in images about ecosystem management and rights over resources.

To understand these conflicts, we need to take into account that in this realm natural resource management is beyond the capacity of individual users or productive units (McCay, 1981), and need to be tackled in much wider terms. While in agriculture technology used on one farm may not affect others in its vicinity, in fisheries, a new technology may deplete the resources for all in a relatively short time. As a consequence, we frequently find resistance to technological innovation in fisheries, especially when these are not accessible to all, or may be understood as a menace for the survival of activities and populations already dependent on the resources.

Fishing technology such as vessels and gears are deeply related to environmental knowledge that has constituted a tool for territorial appropriation in fisheries for centuries. On land, the control of territories for human populations has been developed using many different strategies. Frequently, the control of territories was related to the defense of the perimeter, where exclusion implied continuous surveillance of the area in order to impede new entrants' access to resources. However, in some cases the measures devised to control the territories were more related to the management of knowledge inside the boundaries of the social group, as knowledge accumulated for generations made possible the adaptation to ample territories with a low density of resources, which are unpredictable to find and use without that knowledge (Cashdan, 1983).[2] In fisheries this is usually the tool that human populations have used for centuries to gain

2 See also the debate between Dyson-Hudson and Smith (1978) and Cashdan (1983).

some level of control over large fishing areas. As we will discuss, these features have been shaped differently in the last decades as a consequence of technological change.

Technology is a key factor in explaining how human populations have been able to earn a living in fisheries. Fishing technology implies the use of gears and tools developed jointly with the specific skills that made its utilization possible. This is also dependent on knowledge about the prey and the environment. Gears may be differentiated based on such factors as capital needed to obtain and use the gear, knowledge and skills necessary to find and catch fish, areas of operation, periods of the year to be employed, capacity of capture, by-catch, needs of workforce, side-effects on environment, etc. Choosing one gear versus another, or substituting gears through innovation, is therefore not an easy process. For instance, new gear may have consequences on the marketing of the produce, influencing prices, or even expelling workers from the activity. Introducing new gear usually does not occur without resistance and conflict. In some cases, arguments against the gear may be related to resource conservation. In other instances, resistance may be related to the effort of local populations to maintain control over resources. As we will see, however, the process can move beyond opposing new technology, to devising new institutional arrangements focused on the same purpose: to obtain some level of security over access to local resources and, hence, livelihoods.

A Spanish history of conflicts over gears

In Spain, we find numerous cases of conflict linked to innovation in fisheries, probably as in any other country. A good example is the conflict associated with the use of purse seines in Galicia at the turn of the 19th-20th century, as analyzed by Giráldez Rivero (1993). The conflict was between fishers who used *xeitos*, traditional net gear for the capture of sardines, and others using purse seines (*traiña*) that were several times more efficient.[3] The *xeito* owners could not easily access the new technology because of its cost. At the same time, they were being displaced from the market due to the lower prices of sardines as a consequence of the increased productivity of purse seining.

3 In the evaluation of Giráldez Rivero (1993), while the *xeito* could capture 7,000 sardines, and could be used once a day, the purse seine could obtain more than 50,000, and be employed several times a day. For the best description of these gears and their use at the beginning of the 20th century see Rodriguez Santamaría (1923).

Traditional fishers viewed the innovation as a menace to their way of life and to their independence, as only owners of canning factories or other capitalists of land-based activities could afford this technology. This conflict is a good example of how small-scale fishers tried to resist technological and capitalist development that was radically transforming their way of life. Small-scale fishers advocated a natural order where they could maintain relative independence from capital, as these new gears meant a total change in social relations in the fishery. Their opposition was also argued with precautionary concerns, claiming that purse seines were a destructive gear that compromised the future of the sardine fishery due to impacts on the sea bottom because of trawling, or because catches of small fish compromised sardine survival. Neither of those charges was founded, but they were indeed used.

These claims were not new, as they had been similarly used one century earlier to denounce the use of new gear in the sardine fishery. It was very difficult to prove these charges, but the opponents of purse seining used extensively scientific conservationist arguments, mixed with others related to moral behavior, against their implementation. In a few years the conflict was all around Galicia, with strong surges of violence in some cases; e.g., purse seine boats destroyed with dynamite (Giráldez Rivero, 1993). The solution devised by the authorities was to divide fishing areas, first banning purse seines within three miles from the coast, and later from the inner area of the Galician estuaries (*rías*). However, this did not stop the expansion of the new gears, as the canning businessmen continued to support them by lending money to fishers so that they too could acquire the new equipment. This reduced disputes and relaxed the former's image as greedy monopolists (Giráldez Rivero, 1993).

The same tendency of conflicts over technological innovation occurred around the same time with the development of trawling in Northern Spain, as Ortega Valcárcel (1996) explains. The first trawling steamer appeared in San Sebastian in 1878, and expanded to other areas like Cantabria, also in northern Spain. The steamers increased the efficiency of trawling, and conflicts between *cofradías*[4] (fisher guilds) and the new capitalist ship-owners

---

4 *Cofradías* are different from cooperatives in that they are created and supported by the state as local non-profit corporations with public rights. They represent the interests of the whole fishing sector by acting "as consultative and cooperative bodies for the administration, undertaking economic, administrative and commercial management tasks" and with ability to "cooperate in matters of regulating access to the resources and informing over infractions occurring in their territory" (J. Pascual Fernández, 1999, p. 71). *Cofradías*, as consultative bodies, may propose specific regulations to the

quickly aroused. The latter were again related to canning companies and to other land-based activities. In this case, also, the key argument against the new technology was the preservation of marine resources. From our current perspective, the claims of *cofradías* vis-à-vis trawling were in some sense premonitory, but the confrontation was also related to how *cofradías* and local populations lost control over resources, markets and power. Resistance against the prospects of losing a way of life, or at least losing some of the independence previously enjoyed, was one of the key elements behind the conflict (Ortega Valcárcel, 1996).

Canarian Archipelago was no exception to this pattern of gear conflicts in fisheries during the old regime. Later, around 1917-1930 conflicts surged between traditional users of beach seines – with a long history in the archipelago since the Spanish conquest in the XV century – and the purse seines introduced by canning companies and capitalist entrepreneurs. Once again, the reasons used to deny legitimacy for the new technology were related to resource conservation, but the more clear consequences drive to the expulsion of many fishers, although they strongly resisted the process.

Decades later, we find similar discourses in the *trammel war* developed between different fishing communities in Tenerife in the 1980s, as described by Pascual-Fernández (1991). The conflict involved communities where fishers defended the use of hook and line as the only permitted gears versus those communities where the use of trammels was extensive. Once again, resource exhaustion was a central argument against trammels, but the conflict also had a territorial significance: with trammels it was much easier to develop the activity in large areas, entering informal territories of other fishing communities, taking their resources and placing cheaper fish on the market. It was also simpler to intensify the activity by increasing the length of the trammels. The use of trammels was legal, but not the sizes or the method employed. As a consequence, fishers of other communities in the area opposed this gear, finally obtaining support from the administration, which banned trammels in the North of Tenerife and in almost all the Canary Islands (Pascual-Fernández, 1991).

Similar processes have repeatedly occurred in the Canary Islands in the last decades. During the 1980s and 90s, there were conflicts between fishers from different islands, and some *cofradías* lobbied to ban pots or longlines in their waters, claiming that these gears endangered resources.

---

administration (for instance regulations about gears), develop activities of management and paperwork for fishers, and manage the first sale of catches. See also (Alegret, 1998, 1999; Symes, Steins, & Alegret, 2003).

For instance, there was conflict between *cofradías* and fishers from less populated islands, against those from Tenerife and Gran Canaria who were much more numerous and used more productive gear. In some sense, by banning specific gears they tried to prevent fishers from other islands entering their waters, thus preserving resources and markets. Once more the position of the *cofradías* was justified with arguments related to conservation. In the previous examples, fishers and their organizations used biological-conservationist claims, more or less based on science and research, to restrain, for instance, the use of purse seines or trawling in the mainland, or trammels, pots and longlines in the Canary Islands.

At the turn of the 20th century, reasoning related to social order – that it was unacceptable to give advantages to a few capitalists because it compromised traditional livelihoods, transforming the fishery and expelling fishers from the activity – was especially used to restrain technological innovation. This logic, of course, was not considered relevant at the time because of its strong liberal context, but clearly influenced the position of fishers and their organizations. The arguments in the background rejected imposed changes, but also implicitly denied the rights of these new entrants to exploit the common resources used for centuries, and conceived as theirs, by local fishers. The measures devised for the appropriation of these resources – acquiring and storing environmental knowledge and skills fitted to traditional technology – became almost useless with the innovations introduced by new entrants.

In the conflicts described above, we see a clash of images between advocates of different gears. Images can be understood as judgments and assumptions on fundamental matters such as the relation between society and nature or the role of government (Kooiman *et al.*, 2005, p. 20). They are relevant because people tend to act based on their images of how the world works, or should work. Frequently, "it is the assumptions on how the system should be that determines what are perceived as the problems" (Jentoft *et al.*, 2010, p. 1315), and consequently shapes approaches to solutions. In the conflicts noted, we can distinguish one image embedded in traditional livelihoods and customs, with emphasis on maintaining the *status quo*, marking precaution and conservation of resources against the advent of new technology and outsiders. This image is also based on an understanding that local people had special rights over the resources and the areas where they had been fishing for decades or centuries. The competing image justifies technological and capitalist development as a source of progress for society, but disregards the risks associated with new technology for resources and environment. In this image, precaution was almost absent, with strong confidence in the capacities of science to solve any problems in the future. Until very recently, connections to the political and economic power of the second image were clear, in a context where developmentalism constituted a central element of the dominant liberal ideology, relevant since the end of the 19th century.

Small-scale fisheries: new menaces?

The conflicts described earlier usually ended with the implementation of the new technologies. Here we focus on the Canary Islands to analyze some of its specificities and how local fishers have responded to these new challenges, adapting and rephrasing images and measures frequently proposed with very different objectives in mind.

In the Canary Islands the challenges for small-scale fisheries derive from different sources. First of all, we find occupation of space by tourism and its infrastructure, due to the development of "sun and beach" resorts, spread along the coast of many islands. Second, there is a change in housing patterns and population growth across the coast. Thirdly, and this is particularly relevant for our purpose, there has been a change in the habits of the population and the activities performed on the coast. Many of these activities are now intensified. Recreational fishing in its many forms (for example angling from the shore, spear fishing, but especially fishing with boats) has generated unprecedented pressure on coastal and marine resources, previously used almost exclusively by artisanal fishers. From 2007 to 2009, the number of recreational fishing licenses in the Canary Islands was doubled from less than 20,000 to almost 40,000, while number of artisanal fishing vessels decreased strongly[5]. The contrasting tendencies between recreational fishing licenses on the rise and small-scale fisheries in decline clearly illustrate the extent of the challenges faced by coastal fishers in the Canary Islands.

A key element in this process is that small-scale fishers no longer have the same capacities to control their territories. Traditionally, Canarian small-scale fishers had some control and appropriation of territories and resources through knowledge that only they had. Knowledge and intellectual assets of marine habitats have a key role here, as was the case in hunter-gatherer societies with scarce and unpredictable resources, where knowledge about where, when and how to obtain basic resources was a vital element for survival

5 Source: Department of Fisheries, Canarian Government.

(Cashdan, 1983). For artisanal fishers, defending the social boundaries of the group was essential in order to preserve their accumulated knowledge. This "secrecy" allowed them to transmit precise knowledge from father to son about the configuration of the seabed, the ecological niches, the points on the coast used to locate fishing spots at sea by triangulation (*las marcas*), the best periods of the year to develop a fishery or to visit a spot depending on the tides, or many other variables relevant to locating and capturing marine species (Jorion, 1978; Pascual-Fernández, 1991).

Such secrecy served to limit competition by restricting the flow of data about where or how the fish could be located (McCay, 1978). In small-scale fisheries this constituted a true information economy, where the proper management of information acquisition, storage, retrieval, transmission and use were essential to the success of the production units (Andersen and Wadel, 1972; Pascual-Fernández, 1991). In this sense, it was usual to find as many strategies for "espionage" and "theft" as those devised to safeguard the information. Knowledge accumulated within the family was transmitted to new generations in a process that was always costly in time and dedication for young generations (Pascual-Fernández, 1991). The cost of obtaining this information was very high for new users, and thus acted as a deterrent to entry. However, in the last years the massive emergence of new technology make it much easier to obtain the necessary information to develop fisheries activities in the area, compromising the capacity of artisanal fishers to maintain traditional forms of ownership. For instance, the use of global positioning system (GPS) and echo sounder has removed much of the advantage enjoyed by professional fishers. Now, recreational fishers may get information about good fishing spots, and store the 'knowledge', using the GPS, simply by passing close to the boats of artisanal fishers while they fish. Sharing this information is also very easy with all the facilities internet provides. Many recreational boat fishers have access to this technology, and to other sophisticated fishing gear. Also, poaching with different gears and selling these catches – illegally – has increased in the last years. As a consequence, their competition with regard to resources and markets has increased, putting pressure on small-scale fishers to adapt to the new circumstances.

New solutions? Marine reserves in Spain

In the Canary Islands and other coastal areas of Spain, the survival of small-scale fisheries is threatened in a new era of changes that compromise resource sustainability. We suggest that an adjustment in institutional arrangements and new patterns of governance are necessary to achieve clear rules about resource use and rights. As we will illustrate, marine protected areas (MPAs) can potentially be a good way of meeting these demands. However, a careful analysis of its goals, the step zero of its implementation and the images stakeholders held about them are badly needed in a research agenda (Jentoft, *et al.*, 2010; Jentoft *et al.*, 2011). The standard perception, based on a myriad of cases from around the world, is that MPAs are frequently implemented in ways that further marginalize small-scale fishers and local communities. As a consequence, many voices – from social scientists to community activists – ask for caution in their implementation, so as not to destroy valuable livelihoods and local communities. Our experiences demonstrate that in some cases protected areas may actually help to preserve not only endangered species but also endangered livelihoods, supporting the small-scale fishing activity.

Under the Spanish Constitution of 1978, Article 148, fisheries administration in Spain is shared between the State (territorial waters) and the Autonomous Regions (interior waters) (Suárez de Vivero and Frieyro de Lara, 1994). This characteristic is especially relevant for marine and fisheries governance, as both national and autonomous regions legislate on the protection of the marine environment under their jurisdiction, creating a necessity for coordination that has not always been successful.

In Spain, concerns about the sustainability of fishing activities gave rise to the start of the legislation for 'marine reserve' in 1982, in the form of "*Ministerial Order of Maritime Restocking*" administered by the Ministry of Agriculture and Fisheries. The first reserve created under this Ministerial Order was the *Marine Reserve of Tabarca* in Alicante, on the Spanish Mediterranean Coast, in 1986. Since then, the Spanish ministry in charge of these affairs[6] has created several reserves, where the "*main goal is the sustainability of artisanal fisheries*" (Revenga, 2003). As specified in the 1982 order, the State must consult with the National Federation of *Cofradías* and with the Spanish Institute of Oceanography, prior to the reserve establishment.

Today, besides marine reserves, marine protected areas (MPAs) with different legal status, designations, and implications for local fisheries, have been created in Spain. In 2011 there are 28 MPAs, ten created by the state

6 This Ministry has been renamed several times through its history. In 2008 it was merged with the Ministry of Environment, and is now referred to as *Ministry of Environment and Rural and Marine Affairs.*

only on marine areas and two considering marine and terrestrial areas, and sixteen created by the autonomous governments on marine areas, while some of them are under shared responsibilities (De la Cruz-Modino, 2008; Pascual-Fernández and De la Cruz-Modino, 2008).

An MPA is not an easy technical fix. It is also a social institution, created not only to secure the non-human side of ecosystems, but also devised to allocate rights, preserve uses and/or exclude users, although many scientist may only focus on its capacity to protect ecosystems (Degnbol, *et al.*, 2006). It may also facilitate the maintenance of maritime heritage and coastal cultures or their demise, or create a tourist attraction, with all their advantages or troubles. For instance, the Marine Reserves of Fishing Interest created by the Galician Regional Government have helped small-scale fishers to maintain control of their traditional fishing areas and resources. The legal framework used for their establishment requires some degree of involvement of *cofradías*, and fishers have taken a leading role in creating some of them. These marine reserves are well financed to undertake surveillance, making poaching very difficult. While recreational boat fishing is banned or restricted, small-scale fishing with traditional gears is usually permitted. One of the main requirements for allowing fishing inside an MPA is the evidence that a specific small-scale fishing boat has been fishing there for some time, and joined in a local *cofradía*. The *Patron Mayor* (president) of the *cofradía* is usually a member of the governing bodies, and in some cases can maneuver to maintain a high degree of control on anything that happens in the area. So MPAs in this case are also a way of retaining control of resources in the hands of local small-scale fishers, at the same time excluding any new entrant considered to threaten the health of the ecosystem or fishers livelihood.

To a certain extent, these protected areas constitute perimeter defended territories, previously unfeasible because of the high cost of surveillance and enforcement. This is linked, on one side, to the images created and spread around the world – mainly by academics and conservation organizations – about the necessity to preserve marine ecosystems. On the other side, MPAs have been rethought and reinterpreted by local fishers and their *cofradías* in Spain, as a way to emphasize the traditional activities and resource friendly technology performed by small-scale fishers, contributing thus to preserving the ecosystem at the same time as their livelihoods. Through this device, they have gained clear support from conservation organizations. We present here two examples to illustrate this point.

The marine reserve of Punta La Restinga and the Sea of the Calms, El Hierro, Canary Islands

This marine reserve was suggested in 1989 by biologists from the University of La Laguna who were trying to create a Network of Marine Reserves (MR) in the Canary Islands. Initially the idea was not well received by the local fishers of La Restinga because they lack confidence in the researchers, and the fact that outsiders suggested this possibility made fishers doubt their intentions. There was no precedent of MRs on the Canary Islands, and the position of state and local authorities, despite being interested in the network idea, was not in favor of the proposal. However, the discrepancy between what was implied by the development of the reserve and the conservation measures already implemented by the local fishers was not huge, and years later the same fishers began to view the idea as a possibility. Several reasons contributed to this gradual change. For instance, there had been some struggles in the community about the use of pots and longlines. A majority of fishers had decided to ban these gears, and the marine reserve proposal reinforced this position. The menaces of new entrants, especially spear fishers, were also present. Local artisanal fishers detected the need to protect their resources from non-compliance with rules agreed inside the community, and from threats by new entrants.

Around mid nineties fishers negotiated the design of a new marine reserve within their *cofradía* in collaboration with the university researchers. Their continued presence in the village and involvement by a son of a prominent local fisher – a marine biology student at the time –, contributed to a better understanding of what the reserve meant, and to the process of image convergence. The revised design for the MR assumed the majority of conservation measures already adopted locally. This second proposal was viewed as an extension of the fishers' decisions to manage their fishing area. The main feature of the proposal in relation to the previous situation was the creation of a no-take zone, which generated controversy. But the exclusive allowance to continue the tuna fishery inside that area for fishers from La Restinga was the key to its final acceptance. Fishers led the final stage in the process by voting in the *cofradía* in 1995 the final project for the establishment of the MR, and submitting their decision to the state, that published the Order in January 1996. Fishers and the *cofradía* obtained institutional support for the conservation rules they had already adopted, increasing the surveillance in their fishing territories, and at the same time avoiding the danger of new entrants like spear fishers taking their resources. In this case the process was clearly one of converging images between researchers and local fishers, and giving time for the process to progress (more than five years since the first suggestions until the real discussions for the possibility of the implementation) was important for achieving this convergence.

The national legislation that authorizes marine reserves, even when management is shared between national and regional governments, gave artisanal fishers the role of key stakeholders. Also, as they promoted the final petition, they could influence the design of its governing bodies. Since its inception, local fishers have promoted new rules inside the protected area; for instance restricting some tourist uses, like the number of scuba-divers in specific spots. In this way they have also obtained the capacity to place limits on tourist development in the area.

Marine reserve of Fishing Interest of Os Miñarzos, in Lira, Galicia

The first proposals related to the possibility of implementing a Marine Reserve of Fishing Interest (MRFI) in Lira came from a research group at the University of La Coruña, who had been working with local fishers and other coastal populations for many years. The disaster of the Prestige oil tanker polluting the area and creating huge problems for fishing activity produced an increased awareness about environmental fragility and the need for conservation.

The initial suggestion by the researchers of considering an MRFI in the first months of 2002 led to a process of careful analysis of what this could possibly mean for people in the area. The creation of this MRFI in Lira clearly took into account the experience of La Restinga and the Sea of the Calms Marine Reserve in El Hierro, which served as a model for its design. Fishers from El Hierro travelled to Galicia to share their experience, and the commission in Lira travelled to La Restinga to observe and discuss the process in El Hierro. The La Restinga marine reserve was considered a success in terms of community empowerment; the fact that it is being used, promoted and exported as a model for MPA initiatives in other regions of Spain means that it also initiated a collaborative process to create a new image of MPAs in Spain. In Lira the image of what the MPA could be was built interactively, after a long debate. In this phase there was no formal involvement of the administration, although a public servant of the fisheries administration in Galicia provided advice as a frequent visitor to the area who was well respected by local fishers. Scientific advice from university researchers was relevant in this process. Taking into account all this information, local fishers devised a model of MRFI to fit their needs, and ensured that they were strongly represented in the governing bodies (half of the board). With the MRFI in force, conflicts between local fishers and poachers and new entrants have also clearly diminished.

In both La Restinga and Lira, the final outcome of the protection process was increased control over the traditional territories used by local fishers, avoiding the possibility that new entrants may diminish the relevance of small-scale fishers in the management and use of the area. In both cases, legislation by the ministry in charge of fisheries management, and the presence of long-standing institutions like *cofradías*, helped to ensure that artisanal fishers would retain a clear level of control of the resources used traditionally.

Conclusions

Gear conflicts are not only related to specific gears that some groups support or reject. Usually these conflicts have much wider implications. They are also conflicts over territories and resources controlled over the time by some populations that become threatened by technological innovations. These conflicts can also be understood as conflicts about livelihoods, because these innovations are frequently introduced by specific actors external to the traditional user groups. These conflicts can also be related to huge numbers of new entrants using technology previously reserved only for some traditional user groups. For instance, the conflicts between small-scale fishers and recreational boat fishers in several areas of Spain relate to the technologies the latter are allowed to use, which the former try to influence the authorities to restrain.

In this essay, we illustrate how institutions like MPAs can be a device to enable control of small-scale fishers of their fishing areas and its resources. Both La Restinga and Lira reserves show how MPAs may help to secure small scale-fisheries and livelihoods, and their dominance over certain coastal areas, increasing the public visibility of the activity. They have reinforced local fishing identities and a sense of ownership and responsibility over marine territories. They have been successful in conceiving how to adapt MPAs to preserve their traditional way of living. In this sense, they have rebuilt MPA arrangements in new ways. Fishers are not passive actors in these processes, and in some cases they even built alliances with conservation organizations in this endeavor. This pattern, as we commented earlier, is uncommon in many other areas of the world. The reason this experience is so different from the general rule is the way power works at the level of the local fishing community in Spain, and in Spanish fisheries as a whole. The role of *cofradías* as long-standing institutions that represent the small-scale fishing sector cannot be minimized. If fishers were not as well organized and therefore able to speak with one voice, and by law have a right to be heard, the outcome might have been very different. Small-scale, artisanal fishers need to be organized and thus empowered, particularly those who have the most to lose because their livelihood depends on fragile resources. With organization, as demonstrated in La Restinga and Lira, they may be in the driver's seat of MPAs, making sure that their interests in maintaining their livelihood and their communities are recognized.

Small-scale fishers usually need associations like *cofradías* to be able to stand against those who have more power than they have – like tourist developers, government agency representatives, scientists etc. Therefore, organizing small-scale fishers is important, and should be part of the initial process of implementing MPAs, if such organizations do not yet exist. Not only will it make MPAs work in the interest of small-scale fishers and local communities; it will also help make MPAs more successful, because local people will be less hostile toward them. Getting small-scale fishers organized is not an impossible endeavor (Frangoudes *et al.*, 2008). It requires a change of focus. In many cases perhaps a little bit less effort should be placed on analysis and monitoring of the non-human side of ecosystems before and after the implementation. Instead, greater efforts should be placed on making it easier to build fishing-related institutions in coastal areas, and getting people organized. This takes some time, but so does the establishment of MPAs. Without these conditions, building an MPA can be Russian roulette, not only for small-scale fishers but also for other stakeholders.

Acknowledgments

This research is part of the project "Governability analysis applied to the creation process of Marine Protected Areas (GOBAMP, CSO2009-09802)," supported by the Ministry of Science and Innovation of Spain and the European Regional Development Fund. We acknowledge the collaboration from the Network of Marine Reserves of the General Secretary of the Sea, Ministry of Environment and Rural and Marine Affairs of Spain. We are grateful to the comments provided by Álvaro Díaz de la Paz, Ratana Chuenpagdee and Svein Jentoft to an earlier draft of this paper, and to the collaboration by Manuel González Ramallal.

References

Alegret, J. L. (1998) Property rights, regulatory measures and strategic responses among the fishermen of Catalonia. In Symes, D., ed., *Property Rigths and Regulatory Systems in Fisheries.* Oxford: Blackwell Science, Fishing New Books, pp. 175-187.

Alegret, J. L. (1999) Space, resources and historicity: the social dimension of fisheries in the Northwestern Mediterranean. In Symes, D. ed., *Southern Waters: management issues and practice.* London: Blackwell Science, Fishing New Books, pp. 55-65

Andersen, R. and Wadel, C. (1972) Comparative problems in fishing adaptations. In Andersen, R. and Wadel, C., eds., *North Atlantic fishermen: Anthropological essays on modern fishing.* St. Johns, Newfoundland: I.S.E.R.Memorial University of Newfoundland, pp. 141-165.

Cashdan, E. A. (1983) Territoriality among human foragers: ecological models and an application to four Bushman groups. *Current Anthropology*, 24(1), 47-66.

De la Cruz-Modino, R. (2008) *Turismo, pesca y gestión de recursos en la Reserva Marina Punta de La Restinga- Mar de Las Calmas (El Hierro-Islas Canarias) y el Área Natural Protegida de las Islas Medas (Girona, Cataluña)*. Unpublished PhD Tesis. La Laguna: Universidad de La Laguna.

Degnbol, P., Gislason, H., Hanna, S., Jentoft, S., Raakjaer Nielsen, J., Sverdrup-Jensen, S. And Wilson, D. C. (2006) Painting the floor with a hammer: technical fixes in fisheries management. *Marine Policy*, 30(5), 534-543.

Dyson-Hudson, R. and Smith, E. A. (1978) Human territoriality: an ecological reassessment. *American Anthropologist*, 80(1), 21-41.

Frangoudes, K., Marugán-Pintos, B. and Pascual-Fernández, J. J. (2008) From open access to co-governance and conservation: the case of women shellfish collectors in Galicia (Spain). *Marine Policy*, 32(2), 223-232.

Giráldez Rivero, J. (1993) El conflicto por los nuevos artes: conservacionismo o conservadurismo en la pesca gallega de comienzos del siglo XX. *Ayer*, (11), 233-251.

Jentoft, S., Chuenpagdee, R., Bundy, A. and Mahon, R. (2010). Pyramids and roses: alternative images for the governance of fisheries systems. *Marine Policy*, 34(6), 1315-1321.

Jentoft, S., Chuenpagdee, R. and Pascual-Fernandez, J. J. (2011) What are MPAs for: on goal formation and displacement. *Ocean and Coastal Management*, 54, 75-83.

Jorion, P. (1978) Marks and rabbits furs: location and sharing of grounds in coastal fishing. *Peasant Studies*, 7(2), 86-100.

Kooiman, J., Bavinck, M., Jentoft, S. and Pullin, R., eds. (2005) *Fish for Life: Interactive Governance for Fisheries.* Amsterdam: Amsterdam University Press.

McCay, B. J. (1978) Systems ecology, people ecology and the anthropology of fishing communities. *Human Ecology*, 6(4), 397-422.

McCay, B. J. (1981) Development issues in fisheries as agrarian systems. *Culture and Agriculture*, 11, 1-8.

Ortega Valcárcel, J. (1996) *Gentes de mar en Cantabria.* Santander: Universidad de Cantabria, Banco Santander.

Pascual-Fernández, J. (1991) *Entre el mar y la tierra. Los pescadores artesanales canarios.* Santa Cruz de Tenerife: Ministerio de Cultura - Interinsular Canaria.

Pascual-Fernández, J. (1999) Participative management of artisanal fisheries in the Canary Islands. In Symes, D., ed., *Southern Waters: Issues of management and practice.* London: Blackwell's Science, Fishing New Books, pp. 66-77.

Pascual-Fernández, J. J. and De la Cruz-Modino, R. (2008) Los espacios marinos protegidos en España: ¿nuevas formas institucionales para las estrategias de apropiación? In Beltrán Costa, O., Pascual-Fernández, J. and Vaccaro, I., eds., *Patrimonialización de la naturaleza: el marco social de las políticas ambientales* . Donosti: Ankulegi Antropologia Elkartea, pp. 199-221.

Revenga, S. (2003) Las Reservas Marinas Canarias (España). In Moreno, D. and Frías, A., eds., *Actas de las I Jornadas sobre Reservas Marinas y I Reunión de la Red Iberoamericana de Reservas Marinas (RIRM). Cabo de Gata, Almería 17-23 de Septiembre de 2001.* Madrid: Publicaciones del MAPA, Secretaría Técnica, pp. 101-111.

Rodríguez Santamaría, B. (1923) *Diccionario de artes de pesca de España y sus posesiones.* Madrid: Sucesores de Rivadeneyra.

Suárez de Vivero, J. L. and Frieyro de Lara, M. (1994) Spanish marine policy - role of marine protected areas. *Marine Policy*, 18(4), 345-352.

Symes, D., Steins, N. and Alegret, J. L. (2003) Experiences with fisheries co-management in Europe. In Wilson, D. C., Nielsen, J. R. and Degnbol, P., eds., *The fisheries co-management experience : accomplishments, challenges, and prospects.* Dordrecht: Kluwer Academic Publishers, pp. 119-132.

CHAPTER 13

Governance reforms to develop a small-scale fisheries policy for South Africa

Moenieba Isaacs[1]

The Marine Living Resources Act 18 of 1998 classified and legalized artisanal and small-scale fishers in South Africa as subsistence fishers with access to marine resources only for consumption. This classification was in conflict with the livelihood needs of small-scale fishers, who then challenged the fisheries policy through a litigation process to recognize their right to practice their livelihoods. This essay presents the governance processes of pre- and post-apartheid regimes to address coastal poverty mainly through an Individual Transferable Quota system of rights allocation. Then, it considers the governance processes artisanal fishers were involved in to develop a new small-scale fisheries policy for South Africa, and the challenges of implementing this policy – limited resources, identification of small-scale fishers, and of communities and representative institutions necessary to manage small-scale fishing.

Introduction

The history of subsistence fishing dates back to the coastal indigenous "strandloper" people of South Africa. Later, a limited amount of subsistence fishing was allowed under provincial government control in the Western Cape Province and under traditional authorities in the Eastern Cape and KwaZulu Natal Provinces. Subsistence fishers were characterized by extreme poverty, limited livelihood opportunities, and whose livelihoods are dependent on marine resources. This group mainly harvest low value species and do not compete with the established fishing industry. Despite the important contribution of subsistence o the livelihoods of many households in coastal communities, most fishers were never formally recognized. Small-scale

1 Corresponding author: University of Western Cape, South Africa, misaacs@uwc.ac.za

fishers were categorized as subsistence, traditional, informal and artisanal who used recreational permits to harvest limited amounts for personal consumption and for sale. Although it was illegal to sell their catches under the recreational permits, most of them made a living through this means.

Since 1992, the discourse on South African subsistence fishing has changed dramatically; primarily through the social surveys of the Human Science Research Council in 1993 initiated by the National Party government (apartheid regime) and the Subsistence Fishers Task Group – SFTG (2000) initiated by the African National Congress (ANC) government (post-apartheid regime). In 2004, the Artisanal Fishers Association, Masifundise, and the Legal Resources Centre, with the support of academics, launched a class action lawsuit against the Minister of the Department of Environmental Affairs and Tourism (DEAT). This case, Kenneth George & Others v. the Minister, used the Constitution and the Equality Act to litigate on the social and economic impacts of the reform process (allocation of fishing rights). This case argued for the right to be recognized as a category of fishers and the right to practice their livelihoods. In May 2007, just before the inquiry was due to begin in the Equality Court, the Minister of DEAT agreed, with the consent of all parties, to find a long-term policy solution for more than 1,000 artisanal fishers. Claimants Kenneth George & Others concurred with putting the case on hold, and the agreement later became an order of the court. The agreement also provided relief mechanisms to allow fishers to subsist, and reserved 120-140 tonnes of west coast lobster (*Janus lalandii)* for poor fishers. These fishers are unemployed, and possess few or no assets, and livelihoods that are dependent on cash. Coupled to this is a high prevalence of alcohol and drug abuse, physical abuse of women, single parent families and teenage pregnancy. As a result of these problems, the rate of illiteracy in the community is high. It is also stated that the small-scale fisheries and subsistence policy would be reassessed, with broader participation and input from various stakeholders (Isaacs, 2011).

This essay presents the governance processes of pre- and post-apartheid regimes to address coastal poverty through an Individual Transferable Quota (ITQ) system of rights allocation. The policy makers viewed the ITQ as a means of job creation and security, sound working conditions, and improving health and safety of workers. New employment opportunities are drastically needed to cater for the unemployed, the underemployed, and the thousands of new entrants to the labor market. Hence, there is an overarching focus on creating new jobs, not by the state but by entrepreneurs, be they large-scale, small-scale or even micro-businesses. The fisheries department argues that this fits into the government's broader macro-economic policy to reduce poverty.

Through the ITQ system of allocating fishing rights, the established fishing companies were encouraged to enter into equity ownership arrangements with black-owned companies, to sell shares to employees, and to enter into harvesting and processing joint ventures with small-scale enterprises. The poor, marginalized fishers in coastal communities were encouraged to make use of the new opportunities created by the reform process by forming small enterprises or by applying for access rights as subsistence fishers.

Although the new fisheries policy recognized subsistence fishers and gave them legal rights to access marine resources, they were not allowed to sell any of their harvest. This was in conflict with their livelihood needs and contravened the policy and definition of subsistence fishers. The essay examines the governance processes subsistence fishers engaged in to develop a new small-scale fisheries policy for South Africa, and the challenges of this policy – including limited resources, identifying who the small-scale fishers are, and the communities and representative institutions necessary to govern small-scale fishing.

Governance transition of small-scale fishing in South Africa

South Africa has many coastal settlements that depend on the harvesting of marine resources for human consumption sale and commercial purpose. The planned establishment of coastal settlements was based on a common model, but 'physical, historical, economic, social and political factors have ensured continuing differentiation, posing different problems which will require distinctive approaches to change' (Lemon, 1991). van Sittert (2003) asserts that the concept of 'fishing community' is situated within the industrialization of the fishing industry between the 1930s and 1960s. This is especially the case on the West Coast of South Africa. The company-established fishing towns have had difficulties in operating as a unit, although some common interests have been managed through democratic or representative organs. Although the fishing communities along the west coast do not have the qualities of small, spatially defined geographic units with a homogenous social structure, they consider themselves fishing communities who derive a livelihood from marine resources.

Pre-democracy governance reforms

In May 1992, the Minister of Environmental Affairs and Tourism requested an inquiry into the socio-economic conditions of the fishing communities along the west coast of South Africa. De Wet Schutte's (1994) findings highlighted that poverty, lack of housing, alcoholism, unemployment and illiteracy were pertinent features of most communities along that coast. In addition, the report highlighted the low level of organizational capacity to administer complicated tasks of managing a fishing quota. Accordingly, the Minister requested the Quota Board to allocate 3,000 tonnes of hake (*Merluccius paradoxus* and *M. capensis)* along the west and south coast of South Africa to alleviate poverty and proposed setting up Fishermen's Community Trusts (FCTs) to provide support for those members of coastal communities who were dependent on fishing, rather than covering the whole community. A large number of communities established FCTs, and the Minister realized that the original amount set aside for poverty alleviation was insufficient. The FTCs were not involved in managing this allocation, but instead relying on the established industry to harvest, process and market the quota on behalf of the FTCs, and the income from the 3,000 tonnes was distributed to respective FTCs for re-allocation to fishers as part of the government's social relief to fishing communities. By 1993, 34 FCTs had been formed and the quota was increased to 10,000 tonnes of hake. The following year, the community quotas for hake and anchovy were raised even higher. The same principle of the established companies managing the quotas and the cash income distributed to the trust applied. The fishing trusts created local conflict over the legitimacy of present rights holders, corruption and mismanagement of organizations funds. A committee reviewed the community quota system and recommended a more coherent management framework, removal of cash payment, and the establishment of an umbrella trust for all communities, if this system of allocation of rights was to continue. Before political action could be taken, the Cape Supreme Court ruled that trusts of this type were not considered legitimate receivers of quotas according to the Sea Fisheries Act of 1988, which was still in force. Consequently, most trusts were dissolved (De Wet Schutte, 1994).

Poverty alleviation in fisheries during the apartheid era was mainly focused in the Western Cape (west coast) and in white and colored coastal settlements; it was thus regionalized and ethnicized[2] (van Sittert, 2002). The rationale for community quotas in the pre-democracy reforms was primarily a social security measure for the government and was never intended to develop the small-scale fisheries in coastal communities. Fishers

2 In the mid-1930s, the issue of poverty in fisheries was ethnicized as the national Carnegie investigation into white poverty and the liberals in the United Party and the Coloured Development Council looked into the effects of colored poverty (van Sittert, 2002).

in communities were not involved in harvesting, processing or marketing the quotas. The quotas were fished by large vessels which required heavy capital investment as well as competence for running the operations. The whole community trust system was based on selling the quotas back to the established operators for a relatively low price, and using the income as relief support to poor fishers within the respective fishing communities.

Democracy governance reforms

With the new democracy and constitution, access rights in coastal communities was strongly motivated by restitution and entitlement through livelihood based activities, while the new policy saw access to coastal communities as a means to keep people alive, on condition that they were not allowed to sell their catches. According to the new fisheries policy, Marine Living Resources Act 18 of 1998 (MLRA), subsistence fisher is:

> a natural person who regularly catches fish for personal consumption or for the consumption of his or her dependants, including one who engages from time to time in the local sale or barter of excess catch, but does not include a person who engages on a substantial scale in the sale of fish on a commercial basis (MLRA 18 of 1998, p. 12).

This new fisheries policy failed to respond to the fundamentally heterogeneous social, political and economic nature of fishing communities in South Africa, particularly neglecting the importance of creating institutional structures with poor communities. The definition of subsistence fisher was not applicable to many fishing communities, as their main intention was to acquire fishing rights and sell their catches. Many fishers in coastal communities on the west, south and east coast of South Africa strive to obtain formal fishing rights. Also, there was no data available on the number of fishers and the species suitable for subsistence fishing. During the policy formulation process, political and administrative efforts mainly concentrated on commercial reallocation, as subsistence fishing was still considered a marginal problem. But as the transformation of the commercial sector created litigations, bad publicity, and corruption, charges regarding reallocation for the subsistence fishing option reappeared on the political agenda. How to allocate rights and manage the subsistence fisheries was left to a task group (SFTG) established in 1999 to advise Marine and Coastal Management. According to the detailed survey of the SFTG, there were 143 fishing communities along the entire coast, comprising approximately 20,000 households with some 30,000 subsistence fishers. Of these, only

30 communities were located in the Western Cape, with the majority in KwaZulu Natal and Eastern Cape. However, the most contentious resources were mainly found in Western Cape waters, in particular West Coast rock lobster and abalone, in which both the commercial and the recreational sectors were heavily involved. According to the SFTG:

> Subsistence fishers are poor people who personally harvest marine resources as a source of food or to sell them to meet the basic needs of food security; they operate on or near to the shore or estuaries, live in close proximity to the resource, consume or sell the resources locally, use low-technology gear (often as part of long-standing community-based or cultural practice) and the kinds of resources they harvest generate only sufficient returns to meet the needs of food security (SFTG 2000, III).

At this stage it was clear that not all small-scale fishers fit into the strict interpretation of the SFTG but it was broader than the MLRA definition The task team acknowledged that the definition "excludes an important group of people who might previously have been considered *subsistence fishers* or *artisanal fishers*, but who would prefer to gain commercial rights" (SFTG, 2000). For this reason, it was recommended that a separate sub-category should be established to accommodate small-scale commercial fishers. The arguments in favor of such an arrangement were based on the high value of species such as abalone and West Coast rock lobster, as it would be more economically beneficial to sell and export these resources, rather than using them for household consumption. The task group also argued that subsistence fishers should be allowed to *graduate* to small-scale fishers to reduce poverty in coastal communities (SFTG, 2000).

Since the MLRA 18 of 1998 did not make provision for small-scale fishers, they were then categorized as 'limited commercial' in 2002 to fit into the ITQ system. Fishers who were allocated subsistence permits had to reapply for limited commercial quotas and this led to confusion at the local level. The response of the fisheries department to the excluded fishers was interim relief permits. The permits were allocated annually to poor fishers who did not obtain ITQs. These interim relief measures to fishers fitted into the ITQ rationale of allocating rights and were the ITQ system of social relief. Subsistence fishers and interim rights holders who wanted to continue fishing had to commercialize their entities or apply as individuals. As a result, many community welfare organizations were transformed into commercial enterprises (Isaacs, 2006).

In 2005/2006, the Long-Term Fishing Rights Policy allocated long-term rights (between 8 and 15 years) in 22 of the commercial species. The majority of the poor *'bonafide'* fishers were again excluded from fishing rights due to the language (only in English), cost, and complicated nature of the application forms. In addition, they had to form companies and compete with established companies harvesting the same resource in the inshore zone. The insufficient ITQs allocated to fishers and small enterprises in fishing communities, the lack of technical and management skills, the lack of start-up or investment capital, and the monopolistic business systems and structures of the established companies perpetuated their competitive advantage over these community entities. Many found it difficult to survive without entering into harvesting, processing and marketing agreements with established companies and, once the costs were deducted, their income was less than the basic income level of R30, 000 per annum[3] in South Africa (Isaacs, 2006). On a whole, most of the small enterprises formed and allocated fishing rights under the ITQ system thus became more vulnerable as they started to feel the cumulative pressures of the lack of infrastructure to exercise their fishing rights, the business skills needed to manage the quota, and the start-up capital to acquire the necessary equipment.

The arguments of the fisheries department with regard to interim relief measures, and defining subsistence fishers as limited commercial fishers, are part of the ITQ rationale of allocating rights and a wealth-based approach to redistribution. This also fits in with the broader macroeconomic agenda of creating small enterprises in communities to address poverty alleviation. Many fishers applied for these limited commercial rights, but only a small number were successful. Those who did get rights were allocated insufficient quotas, and many *bonafide* fishers were left out of the system completely and hence no longer had access to the sea (Isaacs, 2003).

Towards a new small-scale fisheries policy for South Africa

Throughout the new democratic government's allocation regime, subsistence fishers argued for better living standards, the right to be recognized as a category of fishers and the right to practice their livelihoods. They see themselves as fishers in the artisanal fishing sector who operate independently as a fishing unit with a boat, equipment, and in a labor intensive manner. The size of their boat is small to medium and their equipment is generally

3 This is approximately US$5,000 per annum.

traditional. Although their catch is obtained with a view to a cash sale, its disposal is usually through informal arrangements, with middle persons and local informal buyers who then sell on the informal markets. These fishers argued that the ITQ system placed a disproportionate emphasis on commercial imperatives at the exclusion of social and cultural issues, and the importance of fish as sources of food security to coastal communities. Furthermore, the policy is not based on a human rights approach, as the right to livelihood is compromised when the policy does not recognize the right to fish and to artisanal activities. Hence, the struggle for recognition as small-scale fishers started in 2004 when court papers were launched in both the High Court and the Equality Court. The Equality Court is a new court introduced in South Africa following the first democratic Constitution in 1996. The Equality Court aims specifically to give effect to the Equality Clause in the Constitution which states that 'everyone is equal before the law and has the right to equal protection and benefit of the law' (Section 1). This case argues that all of the above alleged violations of the rights of artisanal fishers arise because the State, through the Minister, failed to treat small-scale fishers equitably in comparison to the other fishing sectors. In failing to do so, the law is inequitable and discriminatory and hence violates the central tenet of the Constitution, that of the Equality Clause (Sunde, 2006).

In November 2007, the Minister of DEAT hosted a national summit on small-scale fisheries. At this meeting, a group of representatives from fishing communities in four coastal provinces were elected and mandated to oversee and develop a policy for small-scale fisheries in South Africa. A National Task Team, representing fishing communities, non-governmental organizations (NGOs), academics and government officials (The Department of Agriculture, Forestry and Fisheries – DAFF) have since developed a new draft policy for the small-scale fisheries sector (DAFF 2010, p. 7-8) which carefully differentiate between small-scale fishers, as well as small-scale community and small-scale sector from the commercial ITQ allocation. In addition to meeting their basic livelihood needs, small-scale fishers can engage in harvesting, selling and semi-commercial activities. With regard to small-scale fishing communities, the policy considers the complex nature and structural impacts of forced removals on coastal communities and includes the traditional and customary practices in harvesting, processing and selling of marine resources. The policy also considers those in coastal communities who engage in ancillary activities such as net-making and boat-building, which provide additional fishery-related employment and income opportunities to these communities (DAFF, 2010). Basically, the draft small-scale fisheries policy is informed by human rights principles.

It recognizes the right of small-scale fishers to secure a livelihood in the small-scale sector. The draft policy promotes gender equity and social security for fishers with a strong developmental focus to relieve poverty and ensure food security. Moreover, the policy promotes establishment of community organizations (legal entities) and encourages fishers to join such entities. Through the establishment of the legal entities the following are promoted: self-governance to co-manage marine resources; implementation of the new small-scale fisheries policy regulations; economic development through value-added post-harvest activities and market on the domestic and international markets. The draft small-scale fisheries policy's support of collective rights, multi-species allocation, a preferential zone, and of co-management systems is a paradigm shift from the mainstream ITQ model used in allocating long-term fishing rights in 2006 reorganize to further

> The policy proposes a shift away from past management approaches to one which emphasises community orientation and establishes mechanisms and structures for a community-based approach to harvesting and managing marine living resources by the sector and to the allocation of fishing rights to a legal entity closely associated with small-scale fishers. This shift gives preference to the fishers and communities that can demonstrate their historical involvement in the sector and the use of traditional fishing practices (DAFF 2010, p. 34-35).

Challenges of the draft small-scale fisheries policy

The challenges facing the draft small-scale fisheries policy include creating legal entities that are representative of those fishers who were left outside the formal rights allocation process. Isaacs (2011) shows that fishers who were allocated interim relief permits through the court agreement are becoming the new emerging elite, at the exclusion of marginalized fishers with relevant small-scale policy information. In addition, the nature and structure of the legal entities is crucial, as many fishers in the draft policy meetings stressed the difference between urban and rural fishing communities. They also mentioned that some fishers were not residing in the geographical location of the identified community, so the policy should consider how these legal entities would benefit or prejudice them. The challenge for small-scale fishers who actively engage in the informal space is to now participate in the formal structures. The history and custom of self-governance systems is absent in fishing communities in South Africa, and fishers do not have the infrastructure to harvest, market and process their catches.

Further, the policy does not mention what species will be allocated to legal entities. This is crucial as many of the inshore resources are under threat (abalone, West Coast rock lobster and line fish species). Poaching for high-value species is occurring at an alarming rate (see Hauck, 2008; Isaacs, 2011). Also, many of the existing rights holders harvesting in the inshore zone are competing for the same resources as the small-scale fishers the policy is meant to protect.

Another challenge for this policy is the capacity of DAFF staff to provide extension services to the legal entities. Most of the staff is not trained for this task and serious recruitment in this field is necessary. The most important aspect is the budget to roll out the implementation and development of management plans to reach the goals of the policy. It is unclear who, where, and which funding is committed to these tasks.

Most existing rights holders are opposed to the draft small-scale fisheries policy, especially with regard to allocating rights to legal entities (collectives) rather than to individuals. In draft policy meetings, many rights holders articulated their sense of fear with the collective allocation and stated clearly they wanted to remain individual rights holders at a small-scale level. They do not want to form part of any legal entity or community structures and feel the Equality Court order is imposing a new policy, with a new allocation system, forcing them to form new governance structure that they were not part of the drafting process.

What, then, is the solution for small-scale fisheries in South Africa? Can the new small-scale fisheries policy combine welfare and wealth-based functions when allocating fishing rights? The system of wealth-based allocation of rights, which mainstreams small-scale enterprises, will result in capital accumulation and wealth generation in fisher households and will eventually reduce their poverty (see Béné, 2003). Contrary to the expectation, this system of allocation failed because many rights holders were allocated unviable fishing rights, most of them were vulnerable, many sold their rights to established companies, and some went deeper into poverty. The welfare-based approach seems to fit into the collective allocation systems in preferential access zones, with multispecies allocation (according to the draft small-scale fisheries policy). This system could provide safety nets, reduce vulnerabilities and prevent fishers from going deeper into poverty. However, many of the existing rights holders do not agree with a collective allocation system and want to retain individual rights (Isaacs, forthcoming).

Béné (2003; 2004) and Béné *et al.* (2010) have made an important contribution in understanding poverty in fisheries and relating it to macroeconomic policies in developing countries, global economic systems and pro-poor models. In South Africa, macroeconomic policy and fisheries policy (e.g., MLRA) favor the wealth-based model of fishing rights allocation. The draft small-scale fisheries policy of 2010 shifts the system to collective allocation of fishing rights to encourage a welfare-based approach to fisher households. In reality, many fishers would prefer individual rights allocation. A phased-in approach towards a welfare-based, community-oriented, collective allocation is recommended by some. Clarity on the species, links to local economic development, the nature and structure of the legal entity, and extension support to fishing communities are keys to developing the small-scale fisheries in South Africa. Where necessary, the state, NGOs, and fishers need to be involved in setting up legal entities in communities to avoid a situation where existing rights holders reorganize and establish legal entities to claim and capture benefits designated for marginalized fishers (see Isaacs, 2003, 2006). Furthermore, there needs to be directed funding for an infrastructure to harvest, process and market marine resources for the local, regional and, where possible, international market.

Conclusion

The case in South Africa indicates that the protection of the Constitution and democratic freedoms, abiding by the human rights approach are all necessary, but insufficient, steps to ensure that the rights of poor people are protected. In fisheries, hard choices need to be made when addressing poverty, development and sustainability of the resource. This requires the political will to change the allocation paradigm and to support collective allocations to promote a welfare-based approach to address poverty in fishing communities. The critical issues of social and economic justice were the basis of the arguments in the litigation process for the recognition of small-scale fishers and their livelihoods. To protect the livelihoods of fishers, the rules of exclusion and 'substractibility' (a set of mechanisms designed by the community to regulate resource use among its members and to set rules for self-governance) should be applied. Clear rules are needed on how the resource should be regulated, who should participate in the harvesting, and who should benefit from the post-harvesting activities. In reality, there is currently a lack of experience of self-governance systems, a lack of infrastructure to harvest, market and process, and a lack of community organisation to manage such a system.

References

Béné, C. (2003) When fishery rhymes with poverty: a first step beyond the old paradigm of poverty in small-scale fisheries. *World Development*, 31(6), 949-975.

Béné, C. (2004) Small-scale fisheries: assessing their contribution to rural livelihoods in developing countries. FAO Fisheries Circular 1008. Rome: FAO.

Béné, C., Hersoug, B. and Allison, E. (2010) Not by rent alone: analysing the pro-poor functions of small-scale fisheries in developing countries. *Development Policy Review*, 28(3), 325-358.

DAFF (Department of Agriculture, Forestry and Fisheries). (2010) Draft policy for small-scale fisheries sector for South Africa. Government Gazette, Vol. 543, September 3, 2010, No. 33530. Pretoria: South Africa Government.

De Wet Schutte (1994) Report of the committee of inquiry into fishermen's community trusts. Commissioned by the Human Science Research Council. Cape Town: Ministry of Environment Affairs and Tourism.

Hauck, M. (2008) Re-thinking small-scale fisheries compliance. *Marine Policy*, 32(4), 635-642.

Isaacs, M. (2003) Understanding the social processes and politics of implementing a new fisheries policy, the Marine Living Resources Act 18 of 1998, in South Africa. Unpublished PhD thesis, University of WesternCape.

Isaacs, M. (2006) Small-scale fisheries reform: expectations, hopes and dreams of 'a better life for all'. *Marine Policy*, 30, 51-59.

Isaacs, M. (2011) Creating action space: small-scale fisheries policy reform in South Africa. In Jentoft, S. and Eide, A., eds., *Poverty Mosaics: Realities and Prospects in Small-Scale Fisheries*. Dordrecht, The Netherlands: Springer.

Isaacs, M (forthcoming) Paradigm shift: From Individual Transferable Quotas (ITQs) to collective allocations – A struggle for small-scale fishers in South Africa. Special Edition on Human Rights and Fisheries. *MAST*

Lemon, A. (1991) The apartheid city. In Lemon, A., ed., *Homes apart: South Africa's segregated cities*. Cape Town: David Phillip, pp. 1-25.

MLRA (Marine Living Resources Act) 18, Vol. 395, No. 747. (1998) Pretoria: South Africa Government.

SFTG (Subsistence Fisheries Task Group). (2000) Draft recommendations for subsistence fisheries management in South Africa. Cape Town: Chief Director, Marine and Coastal Management.

Sunde, J. (2006) Emerging concerns of fishing communities: sharing on organising in South Africa. International Collective in Support of Fishworkers (ICSF) Conference, Forteleza, July 4-6, 2006. Cape Town: Masifundise Development Trust.

van Sittert, L. (2002) "Those who cannot remember the past are condemned to repeat it": comparing fisheries reforms in South Africa. *Marine Policy*, 26(4), 295-305.

van Sittert, L. (2003) The tyranny of the past: why local histories matter in the South African fisheries. *Ocean and Coastal Management*, 46, 199-219.

CHAPTER 14

"Fishing is our gold mine": lessons learned from participatory fisheries management in Malawi

Steve J. Donda[1]

Malawi is one of the small countries in Southern Africa that does not have a gold mine but has the third largest lake in Africa, and is endowed with a diverse fish population in water bodies that include lakes Malawi, Chilwa, Malombe and Chiuta. Fishing activities in these lakes are mainly small-scale using a wide range of fishing gears that target different fish species. Until the early 1990s, management of the fishery resources in Malawi was based on government-centered biological conservation approaches. After participatory fisheries management was introduced in Lake Malombe on a pilot basis in 1993, fishing communities in lakes Chilwa, Chiuta and some parts of Lake Malawi adopted approaches that varied according to their specific environments. Following the success of this management approach, the Mbenje Island fishing community on Lake Malawi registered sustainable and profitable fishing that equates incomes with those of gold mines at the local level. This essay highlights lessons learned from participatory fisheries management initiatives in the four lakes of Malawi.

Introduction

Malawi is a landlocked country of about 13 million people (according to the 2008 census), with a total area of about 120,000 km^2, 20% of which is water (FAO, 1994). The water body is dominated by Lake Malawi but includes the very productive lakes of Malombe, Chilwa, Chiuta, the Shire River and associated marshes.

Malawi's economy is predominantly agricultural, and its economic performance is largely dependent on weather conditions and international

1 Corresponding author: Department of Fisheries, Malawi, sdonda@sdnp.org.mw

commodity prices (Europa, 1993). About 56% of the land area is arable and suitable for agricultural production, which accounts for about 90% of Malawi's exports (NSO, 1994). The fisheries sector contribution to the Gross Domestic Product (GDP) is relatively small compared to that of agriculture (about 4%), and provides employment to about 40,000 people and a livelihood to about 400,000 who work in ancillary activities. Annual fish catches in Malawi fluctuate between 60,000 and 80,000 tonnes.

The fisheries sector is sub-divided into two, the capture fisheries and aquaculture sub-sectors. This essay will mainly concentrate on the capture fisheries sub-sector. The capture fisheries sub-sector in Malawi is an important source of employment, rural income, food security, import substitution and biodiversity. Living aquatic resources provide somewhere between 60-70% of the nation's animal protein supply and nearly 40% of its total protein supply (GOM, 1994). Much of the fish is consumed in rural areas and thus contributes to the nutritional and household needs of some of the poorest people in the country.

The fisheries are multi-species and multi-gear, involving a number of harvesting techniques. They are divided into artisanal or traditional fisheries and commercial fisheries. The artisanal fisheries account for 85-95% of the total fish landings in Malawi (GOM, 1994) and have a big impact on the livelihoods of the majority that depend on fisheries. The commercial fisheries are comprised of a few stern trawls and a number of pair trawls. The target species for the artisanal fisheries, depending on the fishing gear, are *chambo* (*Oreochromis*), *kambuzi* (*Haplochromis*), *usipa* (*Engraulicypris sardella*), *utaka* (*Copadichromis*), *kampango* (*Bargrus meridionalis*) and *mlamba* (*Clariid gariepinus*). The main fishing gears are gillnets, *chilimira, nkacha,* longlines, handlines, fish traps, and species-specific seine nets are used for *chambo* and *kambuzi.*

General characteristics of major water bodies and their fisheries

Lake Malawi is the largest lake in Malawi, with an area of about 118,000 km^2. It is situated in a south-easterly section of the East African Rift Valley system and has a mean depth of about 264m. The lake has unique fish diversity and contains the highest number of freshwater fish species in the world. The fish species are believed to have hybridised within the lake, during its 4 – to 5 – million year history (Joyce, 2011).

The Lake Malawi fisheries are multi-species, with distinct fisheries named after fish species or gears. The fishing gears involved in the fishery range from pair trawlers and ring nets to less sophisticated gears like beach seines, gill nets and fish traps. *Chambo*, the local name for the *Oreochromis* species, is one of the six most commercially important species that are commonly fished in Lake Malawi. The artisanal fisheries are comprised of a wide range of fishing units, from traditional fishing gears and crafts – such as fish traps and handlines operated from dugout canoes – to relatively modern gears and craft – like seine nets operated from planked boats powered by outboard motors – and employ a lot of people. In contrast, the commercial fisheries comprise a few single or pair trawl units – currently about 20 – and employ only a small number of people.

Lake Malombe, just like Lake Malawi, is situated in a south-easterly section of the East African Rift Valley system, about 10 km from Lake Malawi. The two lakes are connected by the Shire River which is the only outlet of Lake Malawi. Lake Malombe is a natural impoundment of the Shire River, the outflow of Lake Malawi. The lake covers a surface area of about 390 km^2 with a maximum length of 29 km and a width of 17 km. The overall depth of the lake is between 5 and 9m. In 1924, Lake Malombe was so dry that it was possible to walk across it (National Archives, 1931). As a result, large sections of the lake bed were cultivated.

Lake Malombe is highly productive because the water column mixes freely and nutrients are easily recycled from the bottom. It achieved a maximum yield of 15,500 t in 1988. It supports a small-scale fishing industry that is very significant to the livelihood of the area, which is completely focused on fishing. The fishery is entirely artisanal and there are about 40 fish species in catches from Lake Malombe. Most abundant are Cichlid species, in particular about 20 benthic haplochromine species (*kambuzi*), of which three comprise 75% of the total catch (GOM/FAO/UNDP, 1993). The taxonomic status of most *kambuzi* species is not yet clear. They are quite small with a maximum size of 12 cm, mature at about 5 to 6 cm, and have a low fecundity. Recruitment is strongly related to the size of the adult stock, which makes it robust to withstand high fishing pressure (Tweddle, *et al.*, 1994).

The Cichlid of greatest economic value, the *chambo*, comprises three species (*Oreochromis* spp.), which are difficult to distinguish (GOM/FAO/UNDP, 1993). All are maternal mouth brooders, which mature at a size of about 15 cm. Other species in commercial catches are mainly "cat fishes" like *Clarias gariepinus* (*mlamba*) and *B. meridionalis* (*kampango*) and certain Cyprinids like *E. sardella* (*usipa*).

The agricultural potential of the land area around the lake is high, but carries a high risk of crop failure due to unimproved agricultural systems.

The principal potential for agriculture in the vicinity of Lake Malombe is on the better-drained sections of the alluvial rift valley floor. This area is severely limited by the surrounding escarpments and hills, and cultivation carries a high risk of erosion. Therefore, the area has a limited capacity to absorb dropouts from the fishing industry to the agricultural sector (Bell and Donda, 1993).

Lake Chiuta is a shallow lake shared by Malawi and Mozambique. It is located at an altitude of 620 m in the southern part of Malawi. The average depth of the lake is 5 m and it has a total surface area of about 200 km^2, of which 40 km^2 lie in Mozambique (FAO, 1994). The southern part is more or less permanently covered with emergent vegetation penetrable by canoes but not larger craft. The lake is fed by a number of affluent streams and is sometimes connected by a swampy channel to Lake Amaramba, from which flows the Lujenda River, a major tributary of the Ruvuma River in Mozambique. The main influent rivers of Lake Chiuta include Lifune, Chitundu and Mpili. Lake Chiuta's associated marshes are separated from those of Lake Chilwa by an extensive sand bar of about 20 m higher than present lake levels.

The Lake Chiuta fishery is procured by artisanal fishers who operate using either dug out or planked canoes and fish both for subsistence and commercial purposes. In terms of gears, fishers mostly use traps, gill nets, and long lines. The main fish species dominating catches include *chambo/ makumba (O. shiranus), chilunguni (Tilapia rendalli), matemba (Barbus paludinosus)*, and *mlamba (C. gariepinus).*

Lake Chilwa lies 624m above sea level in the centre of the low-lying Chilwa-Phalombe plain in the southern part of Malawi (Njaya, 2009). This is the second largest lake in Malawi. It is a shallow lake without any outlet and has a maximum depth of about 6m at peak water levels. It has a reed belt about 15km wide around the lake to the north and 1-2 km wide on the north-eastern side (Landes and Otte, 1983). Lake Chilwa goes through cyclic recessions due to the persistent droughts that usually last for three to four years. Records indicate that since 1879 the lake has dried up completely on six occasions, partially dried up eight times and flooded six times (Njaya 2009). When the lake recedes, the fishery collapses but recovers within three to four years after refilling.

The fishery, like in Malombe and Chiuta, is predominantly artisanal and on average contributes up to about 20% of the total fish landings in Malawi. The lake is shallow and highly productive. Lake Chilwa harbors 27 fish species but only three are of commercial importance: the small minnow (*matemba* (*B. Paludinosus*), the African catfish (*C. Gariepinus*), and tilapiine species of *O. shiranus chilwae* and *T. rendalli*.

Brief history of the genesis of co-management

Participatory fisheries management or commonly put, co-management, was introduced in Malawi on a pilot basis in Lake Malombe in 1993. This was a government initiated approach to improve fisheries management by enhancing fishing regulation compliance through the involvement of resource users. This approach was slowly adopted by fishing communities in the neighbouring lakes of Chilwa, Chiuta and Malawi. This section details experiences from Lakes Chiuta, Malombe and Malawi.

Lake Malombe

As previously stated, the co-management arrangement in Lake Malombe was introduced by the government. There were both internal and external factors that influenced the Department of Fisheries (DoF) to advocate participatory fisheries management in Lake Malombe. Two main internal influences were low capacity within the DoF to enforce fisheries regulations, and evidence of resource degradation based on scientific research.

Fishery regulations were generally based on biological research findings, and included the imposition of closed areas and closed seasons for fishing, and the use of specified mesh sizes for targeting various fish species. Under the "top-down" or centralized approach, these regulations were passed on to the communities without any explanations as to why they were formulated or why they should be followed. Since fishers did not appreciate their importance, high levels of non-compliance occurred. The situation required a very strong unit to effectively enforce the regulations, which the DoF was unable to accomplish, given its inadequate staff and insufficient funds.

Regarding the resource status, reports from the Traditional Fisheries Project (TFP) implemented towards the end of the 1980s, and the Chambo Fisheries Research Project, phased out in 1992, concluded that fish catches in Lake Malombe had been declining (Tweddle, Alimoso and Sodzapanja, 1994; GOM/FAO/UNDP, 1993). Recruitment overfishing was found in the case of *Chambo,* caused by excessive fishing of the parent stock. This was attributed to the use of fine mesh seine nets which captured immature fish and damaged fish habitats.

In addition, there were a number of external factors that forced the DoF to reconsider its way of managing the fishery resources, especially those in Lake Malombe. Following the decline of *chambo* catches, *chambo* beach seines were made redundant as the fishers decided to target the smaller species, *kambuzi*. Most of the *chambo* beach seines were converted into *kambuzi* beach seines, by removing the side wings of the net where mesh sizes were bigger than three inches, and introducing a smaller meshed bunt. This gear still proved inefficient. Therefore, a more efficient, offshore *kambuzi* seine net, called *nkacha,* was invented. The non-selectivity of *nkacha* nets, as well as their efficiency, was seen as the major cause of the decline in the fish stocks.

Apart from direct impact from fishing, degradation of terrestrial and aquatic habitats was also considered. Terrestrial environmental degradation, due to poor agricultural practices and increase in human population, resulted in heavy siltation and reduced water levels in the lake. Aquatic habitat degradation was alleged to have resulted from the use of *kambuzi* seine nets, either from the shore or offshore as *nkacha* nets. Habitat degradation had an adverse effect on the breeding grounds, which were either covered by silt or destroyed by fishing nets. Finally, as catches declined, and the demand for fish kept increasing, communities put pressure on the DoF to do something about the declining fish catches.

In summary, the driving force behind the DoFs decision to implement co-management was its discovery that there was overfishing and use of destructive fishing gears occurring in Lake Malombe. The problem was aggravated by human population growth, which resulted in increased fish demand and therefore increased fishing pressure. This was coupled with environmental degradation that ended in siltation and destruction of fish breeding grounds. However, the problem was exacerbated by the incapacity of the DoF to enforce fishing regulations.

Lake Chiuta

A different situation was observed in Lake Chiuta where co-management was introduced by the fishing communities themselves. In the mid-1980s, Lake Chiuta started to be invaded by seine netters who came to fish for *matemba.* According to village headman Kunawanga, these seine netters were from Thuto, a sand dune area at the southern tip of Lake Chiuta that separates it from Lake Chilwa. It is believed that these fishers could have been coming from Lake Chilwa, bringing with them the *matemba* beach seine nets.

By the early 1990s, another type of seine net, *nkacha,* was introduced from Lake Malombe. The *Nkacha* net is an offshore open water seine net, which targets small fish species, and may have a minimum mesh size less than one quarter of an inch (about 9mm) at the bunt. Both *matemba* and *nkacha* seine nets have similar bunt mesh sizes. These fishing gears are non-selective and are known to catch all sizes of different fish species.

The good catches realized from the seines initially drew fishers from Lake Malombe, but later also those from neighboring Lake Chilwa. The situation got out of control, with local fishers using gillnets, fish traps and longlines and the *nkacha* operators scrambling for fish. Conflicts arose between *nkacha* fishers and other fishers on the lake since *nkacha* nets constantly damaged other gears due to their nature of operation. Low catches were experienced by local fishers because their fishing gears were constantly being disturbed and their fishing efficiency was reduced. However, evidence indicates that the overall total catches for the lake did not decline (Donda, 2001).

In addition to these fisher to fisher conflicts, the rest of the village communities around the lake endured several socio-economic conflicts. During that time, Malawi, especially in the Chiuta area, was experiencing poor rainfalls, and all surrounding villages relied on the lake for domestic water supply. As a result of constant *nkacha* operations in the lake, the water was repeatedly being mixed and became too turbid for domestic use. Villagers who had nothing to do with fishing were also affected because the quality of water for domestic use was greatly reduced. As the problem with getting good drinking water escalated, most households around the lake became unhappy with the *nkacha* fishers.

Another problem associated with *nkacha* fishers – by local fishers and other members of the village communities – was stiff competition for the market. Fish landed by *nkacha* fishers was sold at a lower price than that landed by local fishers using small and cheaper gears. This resulted in low cash incomes for local fishers, as their landed quantities were far less than those of *nkacha* fishers. The other problem was that, with relatively higher cash incomes, the *nkacha* ancillary workers enticed school-age girls, which interrupted their educational prospects.

Fed up with the situation, about March 1995, one of the leading fishers, together with three fellow fishers from Misala beach, whose gillnets had just been destroyed by *nkacha* fishers, called a beach meeting to sensitize their fellow local fishers of the problem brought about by the *nkacha* fishers, and the dangers such nets posed to the fishery and the community as a whole. All the other fishers agreed and decided to report the issue to their village headman, on their intended action against the *nkacha* fishers, that they should no longer be allowed to fish in the lake, and that they should be chased away from the area. Having not been satisfied with the response from the village headman, they decided to approach their senior chief (Traditional Authority) in the area. Although the senior chief appreciated the problems, the other junior chiefs (Village Heads) disagreed with the decision to chase away the seine netters, because they were getting fish and money from these fishers as gifts or bribes. Seeing that things were not working out as the local fishers wanted, with the assistance of the chiefs, they decided to take the matter to the DoF for further assistance.

Towards the end of March 1995, the DoF representative based at Njelwa beach was approached by the leading four fishers, who by then had formed what they called a *"pressure group for change in fishing activities on the lake"* which they later called a fishers committee. They asked him whether it would be possible to completely suspend *nkacha* fishing in the lake. The failure of the DoF representative to respond in their favor, made the committee return to him on 28th April 1995 with a letter addressed to the District Fisheries Officer responsible for the Lake. The Committee asked the local DoF representative to escort their secretary to Zomba fisheries office to deliver the letter by hand and get its response. The main request in the letter was whether the DoF could give them an okay and help them evict the *nkacha* fishers from the lake.

While the committee was trying to get support from local leaders as well as the DoF, its members were busy conducting sensitization meetings and going from beach to beach forming fisher groups (initially called pressure groups), which were asked to make sure that the *nkacha* fishers did not use their nets in the lake. This move was supported by a lot of villagers due to the problems brought by the *nkacha* fishers.

Following the delivery of the letter in Zomba and the discussions that followed, a general meeting was held at Njelwa beach on 17th May 1995. In attendance were the Police Officer-in-Charge for Machinga District, the District Commissioner, Chief Kawinga, the Member of Parliament for the area, DoF staff, local leaders, the fisher pressure groups, local fishers and the *nkacha* fishers. It was resolved that effective that day (17 May 1995), *nkacha* fishers should stop operating in Lake Chiuta.

This peaceful resolution did not work; some *nkacha* fishers continued to fish in the lake. This made the fisher pressure groups resort to using force to evict the remaining *nkacha* fishers on the night of 19-20th May, 1995. Following this, an emergency meeting attended by all those who came to the 17th May meeting was held on 22nd May 1995. At this meeting, the resolution was the same; that *nkacha* fishers should stop fishing in the lake. Since then, no *nkacha* fishery has been allowed in the lake. This happened because, according to the Fisheries Act, there was no fishing regulation that applied to Lake Chiuta, let alone to ban the *nkacha* fishery anywhere in Malawi, and so the DoF had no jurisdiction. It was left to the local fishers to decide what action to take.

After the *nkacha* fishers left Lake Chiuta in May 1995, the DoF continued to work with the same fisher pressure groups, which by then had become more organized and were recognized as Beach Village Committees (BVCs). However, there were still no government regulations covering the Lake Chiuta fishing activities. This was the beginning of the change from locally-based fishery management to a co-management system. The DoF got more and more involved in the facilitation of capacity building of BVCs and their activities. For example, they provided some training to the BVCs in group leadership and dynamics. The provision of these and other services by DoF installed trust between the fishing communities and the DoF, and improved their relationship. The DoF then used these committees as the focal points of extension, message dissemination, and getting feedback from the communities.

Lake Malawi

The Lake Malawi case concentrates on the fishing activities that take place around one of the islands in the central part of the lake in Salima district, called Mbenje. Only about $3km^2$ in size, Mbenje Island is not a permanent home for anyone; it is only used during the fishing season. The fishers on the island come from the mainland districts of Salima, Nkhotakota and Likoma. The majority of fishers are from Salima, who also have management powers over the island.

The Mbenje fishery is dominated by pelagic haplochromine fish species (utaka) and *E. sardella* (usipa). The three main fishing gears are chilimira nets (open water seine nets), gillnets and long lines.

The management of the island fishery is not clearly marked whether it is locally based, centralized, or a mixture of the two. The island falls within the jurisdiction of the fishing regulation formulated by the DoF, which has the mandate to manage fisheries resources in the country. However, there is a long history of traditional fisheries management by people from the mainland. The Traditional Authority in the area runs a parallel management system for the island, to the extent that there are closed and open seasons for fishing, controlled by the Traditional Authority. This has at times brought conflicts between fishers from different areas over fishing powers on the island. However, the DoF recognizes the efforts of the Chief in controlling fishing on the island. It is believed that the Chief's management efforts date back to the 1950s.

Community benefits from participatory fisheries management
Malawi has no gold mine but fishers often refer to fishing activities as mining gold because of the returns they get from fishing. The following are accounts of fishers who have indicated the tangible benefits they are reaping from their fisheries.

Mbenje fishers openly declare that they are benefiting from the participatory fisheries management approach in Lake Malawi and have, on a number of occasions, referred to fishing activities at Mbenje as mining gold. Recent discussions with Mbenje fishers, and information gathered at the closure of the 2010 fishing season in November, bears testimony to this. Mr Feston, a fisher from Chikombe village in Traditional Authority Msosa who spent the fishing season at Mbenje Island, declared that his fishing activities at Mbenje in 2010 were very profitable:

> My fishing at Mbenje this year has been very good such that I have bought a new outboard motor engine worth K350,000,[2] have bought a new planked fishing boat, have built a new house, and I am looking for land in Lilongwe where I want to construct another house this coming fishing season. The fishery is my mine as I did not go to look for employment elsewhere or grow tobacco[3] this year.

Another fisher, Mr. Kalumbu, who also spent the fishing season at Mbenje Island, indicated that this fishing season (2010) was very good and profitable:

> I have made a lot of money this year from fishing such that I have built a new house with iron sheets on top with a cement floor. In addition to this, I have bought 137 cattle; had a chance to go to Dubai to buy a saloon car and a three-ton pick-up truck; and that is not all, I have even bought a new boat and an outboard engine. Fishing at Mbenje is like mining gold for me; I have never fished anywhere in all my life where I have this kind of fortune. Mbenje is the place to fish. However, for sustained fishing activities, we need to be part of the management group for the fishery.

A number of other fishers who made a fortune at Mbenje in the 2010 fishing season attributed the success of the fishery to the participatory management

2 Approximately US$2,300
3 Tobacco is often referred to as the green gold of Malawi because many people make a lot of money by growing and selling tobacco.

approach the fishing communities in the area are practicing, with traditional leadership and the DoF. Mr. Chikakuda, the chairman of Chikombe beach where fishers depart and land when going to and from Mbenje Island, hailed the good relationship between their traditional leader, Chief Msosa, the DoF and the fishers in the enhanced management of the fishery at Mbenje Island, and commented:

> The fish that we catch here has a livelihood impact not only to us here at Mbenje/Chikombe but to other communities around our inland satellite markets, for example at Msundwe in Lilongwe, Kasungu, Ntchisi, Dowa districts, and not forgetting the people that do fish processing and trading.

Speaking at the 2010 closing ceremony for fishing at Mbenje, the chairman of the Chikombe Beach Village Committee (BVC), Mr Banda, advised his fellow fishers that they should work together as a team and cooperate with the DoF for continued sustainable exploitation of the fishery at Mbenje:

> Let us follow regulations and seasonal events that affect fishing activities. For example, our observation of the time when fish are breeding coincides with the closing season regulations. These regulations are to our benefit and not the DoF, for we are the ones that benefit from the fishery. Fish is our gold here in Salima and Malawi in general as fishers also come from neighboring districts such as Mangochi in the south, Nkhotakota, Likoma and Nkhata Bay in the north. Even in times of poor agricultural production, we are at an advantage because the money we get from fishing is used to buy maize and other foodstuffs, making us food secure throughout the year, unlike our friends uphill who rely only on farming.

The chief for the area, Chief Msosa, who does not have fishing nets, is very supportive of the participatory fisheries management approach which is being implemented. He claims that his father, who was the chief before him, always asked fishers in his area to fish responsibly so that their children and grandchildren should also catch a lot of fish from the island. He said he always reminded the fishers that fish need time to breed, and therefore imposed a fishing ban during breeding season – which incidentally coincides with the rainy season – and urged the fishers to practice farming during that time of the year.

Discussions with Chiuta and Chilwa fishers, where co-management was initiated by the fishing communities themselves and is taken seriously, give similar pictures as those from Mbenje. A number of fishers claim to have seen the benefits of being involved in fisheries management. They indicated that the DoF should only play a facilitating role and give policy guidance to the management of the fisheries, and let most of the groundwork be done by the fishing communities because they are the ones that benefit materially from fishing. One of the fishers from Lake Chiuta, Mr. Chinyama of Limera village (Nafisi BVC) Traditional Chief Chikweo, commented:

> "It is for us fishers to make a choice, either to fish irresponsibly now and live miserable lives, or follow the fishing regulations and live happily thereafter".

However, he was quick to point out that for sustainable relationships between fishers, chiefs and the DoF, there is a need for trust and transparency in undertaking fishery management.

Conclusion

A number of lessons have been learned from implementing co-management in Malawi. The most notable is that if co-management is introduced by an outsider, like the DoF or an NGO, the sustainability of such an arrangement is less resilient than those initiated by fishing communities themselves. This is evidenced by the Lake Malombe case where co-management was introduced by the government, in contrast to the Lake Chiuta, Chilwa and Mbenje Island co-management cases where it was initiated by the fishing communities. Fishers in communities where the initiatives originated from them have openly declared the tangible benefits they are reaping from their fisheries, as opposed to communities where co-management was introduced by outsiders.

For sustainable participatory fisheries management, there is need for commitment, trust and a common goal shared by all management parties, as well as the need for transparency when working together. Fishers require those representing them to be accountable in all directions; that is, to the DoF, Traditional Chiefs and to them as fishers.

References

Bell, R. H. V. and Donda, S. J. (1993) Community fisheries management program: Lake Malombe and upper shire river consultancy report. Volumes 1 and 2. Mangochi: Government of Malawi, Department of Fisheries and the Malawi – Germany Fisheries and Aquaculture Development Project (MAGFAD).

Donda, S. J. (2001) Theoretical advancement and institutional analysis of fisheries co-management in Malawi: experiences from Lakes Malombe and Chiuta. PhD Thesis. Denmark: Aalborg University.

Europa (1993) The Europa World Yearbook. Volume 2. Europa Publications Limited.

FAO (1994) Fisheries characteristics of the shared lakes of the East African rift. CIFA Technical Paper 24. Rome: FAO.

GOM (1994) Government of Malawi, Fisheries Department, annual report 1993. Fisheries Bulletin 10.

GOM/FAO/UNDP (1993) Chambo fisheries research project: fisheries management in South-East Lake Malawi, the Upper Shire River and Lake Malombe, with particular reference to chambo (*Oreochromis spp*). CIFA Technical Paper 21.

Joyce, D. (2011) Biodiversity in Lake Malawi. University of Hull: Insight Magazine.

Landes, A. and Otte, G. (1983) Fisheries development in the Lake Chilwa and Lake Chiuta region in Malawi. Germany Agency for Technical Cooperation, MAGFAD. Malawi: Stufyfond.

National Archives, Malawi, 1931. District Commissioners Annual Reports, South Nyasa District, Fort Johnston, Item No. S1/60H/32

Njaya F.J. (2009) The Lake Chilwa Fishing household strategies in response to water level changes: Migration conflicts and co-management. PhD Thesis. South Africa: University of Western Cape

NSO (1994) Malawi Demographic and Health survey 1992. Zomba, Republic of Malawi: National Statistical Office.

Tweddle, D., Alimoso, S. and Sodzapanja, G. (1994) Analysis of catch and effort data for the fisheries of Lake Malombe, 1976 - 1989. Fisheries Bulletin, 11. Lilongwe: Fisheries Research Unit, Fisheries Department, Government of Malawi.

CHAPTER 15

Towards decentralized coastal management policy in Indonesia

Sapta Putra Ginting[1]

The Coastal and Small Island Management Act No. 27/2007, enacted by the Indonesian parliament on 17 July 2007, has triggered significant steps for local governments to establish a new system of integrated coastal management planning that will be formulated through cross-sectoral cooperation and public participation. A system of property rights (referred to as HP-3) in coastal waters is being established for individuals, legal entities and traditional communities. New authority has been given for the establishment of marine conservation areas by each level of government, subject to public consultations with local fishers. Additional criminal offences with heavy penalties have been instituted. At the regional level, a number of governments have embarked on coastal community empowerment programs to support the livelihoods of small-scale fishers. Each region is finding its own way on how to best coordinate the fisheries, forestry, industry, tourism, and mining sectors to sustainably use coastal and small island resources.

Introduction

Indonesian marine and coastal resources are found within Indonesia's maritime boundaries, which covers about 5.8 million km^2. This ocean space consists of approximately 0.8 million km^2 of territorial sea, 2.3 million km^2 of internal waters, and 2.7 million km^2 of Exclusive Economic Zone (EEZ). Home to over 17,500 islands, Indonesia's marine resources have been heavily utilized to generate economic growth, provide employment opportunities and enhance the welfare of coastal communities.

1 Corresponding author: Ministry of Marine Affairs and Fisheries, Republic of Indonesia, sapta@cbn.net.id

However, rapid economic development in Indonesia, based on the exploitation of natural resources, has been achieved at the expense of the coastal environment. Degradation of coastal resources is caused by many factors, primarily: i) socio-economic factors, such as poverty and lack of public awareness; ii) population growth; iii) bio-physical factors such as pollution and sedimentation; iv) coastal hazards such as earthquakes and tsunamis; v) user conflicts between various interest groups, and jurisdictional conflicts; vi) lack of law enforcement; and vii) ambiguity of property regimes (Ginting, 2003).

The coastal zone has continuously faced intense pressure from a wide range of development activities. Many of the fastest growing sectors of the economy like oil and gas, aquaculture, tourism, and trade need natural resources and environmental services from the coastal zone. Thus it was argued that in order to integrate perspectives and activities between the sectors, Indonesia needed to adopt an integrated coastal management (ICM) approach as endorsed by the Coastal Management Act (Idris *et al.*, 2007).

Until 2000, Indonesia did not have any specific legal policy in support of ICM. For this reason, in 2000, Indonesia started to draft a Coastal and Small Islands Management Bill to galvanize the implementation of ICM. The Bill's provisions emphasize that such implementation must include the primary stakeholders in coastal areas and promote integrated management between central, provincial and local government agencies, non-governmental organizations (NGOs), universities, and the private sector. On 17 July 2007, the Coastal and Small Island Management Act (CSIM Act) No. 27 was finally enacted. This Act is a new paradigm of coastal development demonstrating that a centralistic style of resource management has gradually changed to decentralized coastal and small island management. This essay describes the essence of the CSIM Act and reflects on some key aspects in the decentralization effort, three years after its enactment.

The legislative framework

The CSIM Act introduces a new system of coastal planning which includes 20-year strategic plans, 20-year zoning plans, five-year management plans and annual action plans that will be formulated through cross-sectoral cooperation and public participation. Obligations for collection and management of information on coastal management have been established alongside public rights to access this information. A system of property rights (*Hak Pengusahaan Perairan Pesisi*r or HP3) in coastal waters will be established for individuals, legal entities, as well as for traditional communities. Sectoral commercial licenses will be required to comply with the planning instruments in an effort to control activities that have traditionally been unchecked in exploiting coastal and marine resources without any consideration of sustainability. New authority has been given to each level of government for the establishment of conservation areas, which were formerly set up by central government authorities. Additional criminal offences with heavy penalties have been instituted; for example, minimum two years and maximum ten years in jail for those who destroy the coral reefs or mangroves.

However, the new law serves only as a national framework. A huge effort remains to draft operational regulations on subject matter ranging from the planning framework, management of data, geographic demarcation of jurisdictions with regional autonomy, establishment of conservation areas, licensing the commercial exploitation of coastal resources, management of most outer small islands, rehabilitation of coastal ecosystems, mitigation of coastal hazards, community involvement in management and control activities, and the establishment of accreditation programs. To date, eight ministerial decrees have been enacted, two government regulations have been enacted by the President, and three presidential decrees (executive orders) have been reviewed among concerned government agencies and submitted to the Presidential Office.

Regional autonomy and decentralized coastal management

Indonesia is made up of 33 provinces, each of which contains a number of districts/municipalities. As of 2010, 305 of 491 districts are coastal. The process of devolving authority for coastal resource management to the provinces and coastal districts/ municipalities started in earnest in 1999. Act No. 22 of 1999 which was amended to Act No. 32 of 2004 regarding Regional Governance, in conjunction with Act No. 27 of 2007, divided up the authority for management of marine resources as follows: (a) central government – sea between 0 and 200 nautical miles from the coastline; (b) provincial government – sea that lies between 0 and 12 nautical miles from the coastline; and (c) district/municipal government – the sea beyond the inter-tidal zone up to four nautical miles from the coastline.

The regions (provinces and districts/municipalities) share authority for: (a) exploration, exploitation, conservation and management; (b) administrative arrangements; (c) spatial planning; and (d) law enforcement regarding regulations issued by the regional government or referred by the central government. Central government authority in relation to marine affairs has been redefined. According to government regulations, it extends to the determination of policies on and regulation of the exploration, conservation, management and utilization of maritime natural resources in the maritime territories beyond 12 nautical miles, including local Indonesian waters and the seabed as well as Indonesia's EEZ and continental shelf. The central government also has responsibility for law enforcement throughout the whole maritime territory. It also has authority for the regulation of spatial planning from terrestrial to the Indonesian EEZ, and zoning for national interest areas, as well as for the regulation of environmental management in the use of marine resources for all waters. As mentioned above, the district government has responsibility for one-third of the provincial maritime zone. This means that most management responsibility in this area falls to district/municipal governments. This new arrangement puts a heavy burden on district governments.

At the regional level, a number of governments have embarked on law-making programs. This started in 2002, with the passing of regional government regulations in the province of North Sulawesi and the district of Minahasa. By mid-2010, 34 more regional government regulations endorsing the principles of ICM have been passed, and a further 26 are in draft form and/or being discussed in local parliaments. Each region is finding its own way on how to best coordinate the relevant sectors of fisheries, forestry, industry, tourism, mining and so on. Some have proposed a new management organization, while others have opted for adaptation of existing institutional arrangements. Recent regulations have given legal status to planning instruments as anticipated in the CSIM Act. An area that needs further attention is the control of sectoral licensing for marine and fisheries activities.

Initiatives and achievements

In view of the development of the decentralized coastal management policy, the government has undertaken a number of initiatives to address particularly pressing needs in coastal and marine resource management, most of which are relevant to coastal fishing communities, as detailed below.

Development of coastal management plans

The decentralized ICM policy has been tested through the implementation of the Marine and Coastal Resource Management Project (MCRMP), Coastal Resource Management Project (CRMP) and Coral Reef Rehabilitation and Management Project II (COREMAP II). These pilot projects applied a cross-sectoral decentralized approach. They integrated various sector development plans and regional development plans to collect, analyze and use data for ICM planning, and to develop business opportunities based on the characteristics of local coastal resources. The MCRMP is a decentralized model that focuses on strengthening the capacity of local governments to prepare and implement ICM plans in 15 provinces and 41 districts/municipalities. The COREMAP II is located in 8 provinces and 15 districts, and is being carried out over seven years from 2004/2005 to 2011. The CRMP is a model to empower the capacity of local government agencies and community organizations and to promote a community-based management approach.

Through lessons learned from the mentioned projects, the Ministry of Marine Affairs and Fisheries (MMAF) has formulated national guidelines for ICM, and seven other guidelines for: (i) marine conservation, (ii) small islands management, (iii) coastal community empowerment, (iv) monitoring-controlling and surveillance, (v) sea partnership, (vi) accreditation of coastal management plan and (vii) training and extension. These guidelines are legally binding under the authority of the CSIM Act and are endorsed by ministerial decrees, which give them legal significance. Under these guidelines, the provinces and districts are required to make coastal plans that are linked with licensing in all other resource sectors. It is expected that the regions will pass their regulations to implement provisions of the Act regarding such matters as access to information, community empowerment, and establishment of conservation areas. Through the MCRMP and local government initiatives, 222 coastal planning documents have been made, which consist of 56 strategic plans, 54 zoning plans, 56 management plans and 56 action plans.

Implementation of community-based management

Another step towards ICM policies has been to promote community-based management (CBM) or co-management, particularly in remote coastal areas where fisheries are important. Heavily developed and urban coastal areas are considered too complex for this approach. Through CBM, the local community is empowered to make their ICM plan, to generate alternative income activities, to reduce population pressure on coastal and fisheries resources, and to exclude intruders. The coastal villagers are expected to develop a permanent marine sanctuary of around 10 to 20 hectares for rehabilitation and protection of marine ecosystems. This model has been applied successfully in Blongko and Tumbak villages in North Sulawesi and several villages of Raja Ampat district through CRMP and COREMAP II, often with support from NGOs.

Major obstacles in implementing CBM are top-down decision-making from government, the interests of large developers, and illegal, unregulated, unauthorized fishing organizations. These external factors can have a very strong influence. For instance, agreements made at the community level are always not honored at the higher levels of government. A bureaucrat in Jakarta or in a provincial government may impose a sectoral project in a coastal region or provide a permit to a powerful developer against the wishes of the local community. As long as such external factors do not impinge on places where a CBM approach is being applied, it can succeed. The Directorate of Coastal Community Empowerment's role is to empower coastal communities so that they can effectively manage their own resources, and to facilitate them in accessing funding through block grants and/or credit schemes from government programs.

Marine conservation

A decision was made by the Government of Indonesia to set aside 13.5 million hectares of marine waters (5% of the total marine territory) as marine conservation zones by 2010. To date, the MMAF has declared the Savu Sea National Marine Park which covers of 3.5 million hectares of sea, micro islands and coasts. The MMAF and concerned local governments have also declared 35 local marine parks covering a total area of 4.6 million hectares. The remaining six national marine parks and marine sanctuaries comprising about 5.4 million hectares have been established under the jurisdiction of the Ministry of Forestry.

Small island management

The MMAF has identified about 13,466 small islands and, of those, 4,981 islands were deposited to the United Nations (UN) in 2008. The rest will be officially reported for the 2012 UN general assembly meeting. Prior to the implementation of the CSIM Act, more than 100 small islands have been utilized, especially for private resorts. However, most small islands were underdeveloped, isolated, and to some extent neglected. The MMAF has promoted investment in some small islands, subject to the provision of the CSIM Act. The investor may utilize the sea potency of small islands limited to conservation, research, education, eco-tourism, small-scale agriculture, mariculture, sustainable capture fisheries and livestock. New investments which comply with the provision of the CSIM Act have been started in Berhala Island, Nipa Island, Anak Sambu Island, Tabulu Island, Bawal Island, and Miang Besar Island, amongst others.

Coastal community empowerment

A provision of the CSIM Act is to empower coastal communities – including artisanal fishers, small-scale aquaculture operators, fishery middle women – and their participation in coastal management, starting from the planning stage and continuing on through implementation and control. In line with this, the MMAF has carried out a national program for empowering the economic conditions of coastal communities (*Pemberdayaan Ekonomi Masyarakat Pesisir* or PEMP) in more than 220 coastal districts/municipalities since 2001. This program covers micro finance, micro business development, applied technology adoption, fisher's regeneration, and coastal women's activities. To date the PEMP program has resulted in 259 micro finance units, 248 gasoline stations for fishers, and 197 mini coastal shops (Ginting, *et al.*, 2009). By 2009, most PEMP activities were incorporated into a national Community Empowerment Program (PNPM M-KP). During 2009 and 2010, the PNPM M-KP was implemented in 120 coastal districts/ municipalities. In 2011, the PNPM-MKP evolved into PUGAR (*Pemberdayaan Usaha Garam Rakyat* (community empowerment for salt production)), and is being implemented in 40 coastal districts. The government is currently designing a new Coastal Community Development (CCD) program, which will be implemented in late 2012 or 2013.

The Government has broadened coastal community empowerment programs through various activities, inter alia PEMP, PNPM M-KP, PUGAR and CCD. These programs cover a wide range of activities including poverty alleviation, alternative income generation, capacity building, and business development. The target groups are artisanal fishers, small-scale aquaculture operators, small-scale fishery middle women (*pengalas*), and other coastal communities.

Through COREMAP II, the government has promoted community-based management to empower artisanal fishers to protect the reefs and associated ecosystems from intruders or destructive fishing operators. The CSIM Act has empowered artisanal fishers to manage this area; and they are given exclusive rights to benefit from the reefs and associated ecosystems. The fishers are organized so that they engage in collective action, and they are also supported to develop alternative income generating activities.

Rehabilitation, mitigation and adaptation

Indonesia lies in the ring of fire; almost every month tectonic earthquakes occur under the sea. Some of these earthquakes have the potential to cause tsunamis. If a tsunami occurs, many coastal communities will be the victims. Under the CSIM Act provisions, central and local government agencies have an obligation to carry out mitigation plans and public awareness. In line with this, rehabilitation of degraded coastal ecosystems is also important to mitigate coastal communities from the impact of coastal hazards and the physical impact of coastal erosion. A coastal strip must be conserved to reduce the impact of coastal hazards, beach erosion and tsunamis. The coastal hazard area is being mapped out for coastal planning, and vulnerable coastal areas must be incorporated into local development plans. Furthermore, Indonesia also faces the impact of global warming. Hence, the MMAF promotes coastal community resilience to enhance the readiness of communities to adapt to the impact of sea level rise, cyclones or extreme winds.

Development of a marine property rights system

Under the provisions of the CSIM Act, a new marine property rights system, called Hak Pengusahaan Perairan Pesisir (or HP3), is being introduced for individuals and legal entities as well as traditional communities. HP3 for individuals aims to protect artisanal fishers and to provide property rights for local fishers who have lived on the coasts for at least a generation, and hold an area less than five hectares per household. HP3 for legal entities aims to promote economic growth of the sea and to ensure long-term investment of private and state companies as well as cooperative entities. HP3 for traditional communities aims to protect their marine tenure rights. If a traditional community system still exists, with a clear delineation of the marine management area, then the tenure rights will be recognized and will be granted HP3 protection. Government regulations for these matters are being formulated and discussed with the primary stakeholders and will be submitted to the Presidential Office to be enacted.

Conclusion

An inadequate legal and integrated coastal management framework coupled with inadequate coordination among government agencies has led to ineffective coastal management in Indonesia. Complex relationships and conflicting interests among sector agencies partly explain the failures in Indonesian marine and coastal management. Ineffective coastal management cannot regulate the utilization of coastal resources in a sustainable way. It triggers competition and conflict among resource users, which eventually lead to the "tragedy of the commons".

All these factors are known, but the real problem that we encounter is the absence of effective ICM regimes necessary to ensure the sustained use of coastal marine resources over time. The effectiveness of ICM depends on how the legislative framework can empower the integrated approach. Therefore, the CSIM Act coupled with national government regulations and presidential decrees as well as ICM guidelines is seen as an important step to crystallize decentralized ICM policy, as is the initiative taken by the regions to commence their own law-making programs on coastal management.

The CSIM Act will place MMAF in a better position to empower coastal communities in overcoming their management obstacles. Key decisions in policy formulation, planning and licensing will only be made after genuine public consultation, and sectoral government agencies will be bound by development plans for coastal management. The national or provincial governments will not be able to interfere and override decisions about activities in coastal areas. In this way, regional autonomy coupled with the CSIM Act will ensure true decentralization of coastal management policy.

To further facilitate decentralized ICM and to supplement a lack of institutional capacity, the government has promoted a Sea Partnership Program (Kemitraan Bahari), professional coastal management experts (HAPPI), and the Center for Coastal Community Empowerment (P3MP). The local marine affairs and fisheries agencies, universities and private enterprises form a consortium of Sea Partnerships in their region and share the resources among themselves. This consortium empowers coastal communities and local public institutions to implement the ICM, subject to local characteristics. These new organizations are partners of government agencies to oversee the ICM.

Three years after the enactment of the CSIM Act, several initiatives have been implemented. The government has broadened coastal community empowerment programs through various activities, inter alia MCRMP, COREMAP II, PEMP, CCD, PNPM M-KP, PUGAR. The programs cover a wide range of activities including poverty alleviation, alternative income generation, capacity building, and business development. The target groups are artisanal fishers, small-scale aquaculture operators, small-scale fishery middle women, and other coastal communities. It is still too early to fully gauge the success of this decentralized policy, especially in terms of its benefits to these target groups at the local level. Yet, it is fair to say that Indonesia is heading in the right direction with its decentralized and integrated initiatives.

References

Ginting, S. P. (2003) *Integrated coastal zone management in North Sulawesi, Indonesia*. Thesis. Townsville: James Cook University.

Ginting, S. P., Setyawati, A., Madeali, A., Ardarini, F., Arfiah, A., Fitrianto, I., Sutenti and Cahyo F. D. (2009) Buku profil kegiatan usaha mikro. Direktorat Pemberdayaan Masyarakat Pesisir, Direktorat Jenderal Kelautan, Pesisir dan Pulau-Pulau Kecil. Jakarta: Ministry of Marine Affairs and Fisheries.

Idris, I., Ginting, S. P. and Budiman (2007) Membangunkan raksasa ekonomi: sebuah kajian terhadap perundang-undangan pengelolaan wilayah pesisir dan pulau-pulau kecil. Jakarta: PT Sarana Komunikasi Utama.

Section 5:

Governance and principles

CHAPTER 16

The national fisheries department and small-scale fisheries management: 'fit for purpose?'

Robin Mahon[1] and Patrick McConney

In most countries, the national fisheries department (NFD) is responsible for managing fisheries resources on behalf of the public, who are the owners of the resource. The structure and function of NFDs has received very little attention in the fisheries development literature. In particular, there has been little examination of how a fisheries department should be structured to deal with small-scale fisheries. We explore several factors that might be relevant to determining an appropriate structure. These include the size of the country, the extent to which non-NFD expertise is available in other national or regional institutions, and the 'national fisheries portfolio'. In regard to these factors, we raise questions about the appropriate mix of technical and organizational skills in an NFD.

Introduction

This essay places national fisheries departments (NFDs) under a spotlight. It is not a detailed survey or analysis of the structure and function of NFDs. Rather, it reflects our work in the Wider Caribbean Region where there are 44 states, each of which has an NFD. The states range from large to small, rich to poor, and include 16 Small Island Developing States (SIDS). Between us, we draw on over 50 years of work with NFDs at the regional level, upon which to base our conclusions. Also, the fisheries of the Caribbean are mainly small-scale, so we consider our perspective to be particularly relevant to the structure and function of NFDs for small-scale fisheries. Indeed, our

1 Corresponding author: University of the West Indies, Barbados, robin.mahon@cavehill.uwi.edu

perspective is largely a Caribbean one, but conversations with colleagues from other parts of the world, and our own occasional activities outside the Caribbean, suggest that these issues and questions are widely relevant.

In most countries, the NFD is responsible for managing fisheries resources on behalf of the public, who are the owners of the resource. It may have jurisdiction over all of the fish chain/net (i.e., from hook to cook), or only part of it (for example, the harvest sector), with exports and other aspects of the fish chain being the responsibility of other departments or ministries. The NFD may often have other aquatic responsibilities (e.g., marine protected areas, aquaculture, coastal management) in addition to fisheries. A quick internet survey suggests that NFD structures in the Wider Caribbean are similar to those found globally. Typically, large to medium size NFDs are hierarchical with departments that may cover stock assessment, fishing technology, post-harvest technology, enforcement, and extension activities for community development. Smaller fisheries departments often try to mimic the structure of larger ones by collapsing these functions into fewer individual positions (Mahon and McConney, 2004).

The structure and function of NFDs has received very little attention in the fisheries development literature. In particular, there has been little examination of how a fisheries department should be structured to deal with small-scale fisheries (Berkes *et al.*, 2001). In this essay, we explore several factors that might be relevant to determining an appropriate structure. These include the size of the country, the extent to which non-NFD expertise is available in other national or regional institutions, and the 'national fisheries portfolio'. In particular, the idea of the national fisheries portfolio is considered important and, to our knowledge, the implications of the portfolio have not been explicitly explored.

In regard to the above factors, we raise questions about the appropriate mix of technical and 'enabling' organizational skills that should be included in a NFD. Would a portfolio that includes primarily large stocks, generates revenue and supports commercial fisheries require a different NFD structure than one which consists of primarily small stocks, generates little revenue and supports small-scale or artisanal fisheries? If fisheries management thinking based on the former portfolio is applied to NFD structure in countries with the latter portfolio, what are the implications for probability of successful management? In the latter case, should NFDs emphasize developing their own capacity for adapting to a variety of small-scale fishery situations by being self-organizing and resilient? Would this better enable them to develop these same qualities in the managing systems of the fisheries for which they are responsible?

Factors that might be relevant to determining the appropriate structure of an NFD

There is a diversity of factors that might be relevant to determining the appropriate structure of a NFD in any country. These include:

- The governance context or political arrangements within the country
- The size of the country
- The internal capacity of the country
- The capacity of any networks in which the country is a member
- The national fisheries portfolio which includes:
- The types of fisheries (small-, medium-, or large-scale) and their total values
- The sophistication of the fish chain
- The marketing arrangements

These factors are interlinked and interacting. To illustrate these interactions we use a typology of fisheries that is organized around three dimensions. These are (1) The capacity of the NFD; (2) The scale of the fisheries; and (3) The size of the fish stocks. We distinguish between fishery size and scale in the way described by Mahon (1997) and Berkes *et al.* (2001). The first dimension is largely related to the degree of development of the country, its size and, of course, the overall importance of fisheries, whether national or in coastal regions. The latter two dimensions define what we refer to as the 'fisheries portfolio'. In most cases the portfolio is a mixed one, so the country will lie somewhere along the continuum between two extremes: (1) a portfolio of primarily small stocks that generate little revenue and support small-scale or artisanal fisheries; and (2) a portfolio of primarily large stocks that generate considerable revenue and support commercial fisheries.

An important question is whether the NFDs of countries whose portfolios lie towards either extreme of this continuum would require different structures; and if so, what would these structures look like? Before we pursue this question, however, we will lay out the possible combinations of situations that arise from the typology (Table 1) and see how this might relate to the needs of small-scale fisheries.

As can be seen from Table 1, the implications for small-scale fisheries can vary considerably depending on the portfolio of stocks and also on the nature of the NFD. When there is a high capacity NFD because the portfolio includes large stocks that support large-scale, revenue-generating commercial fisheries, small-scale fisheries may get spillover attention from that capacity. When the portfolio is primarily small stocks fished by small-scale fisheries, especially in less developed countries, the attention received by the fisheries is inadequate and special thought is required as to how best to structure a low-cost NFD to deal with the prevailing management issues.

Emerging governance perspectives for fisheries, especially small-scale fisheries

To properly address the question of what NFD structure is appropriate in the various situations described, we must also look at emerging ideas regarding governance of social ecological systems (SESs) in general and small-scale fisheries in particular. There are a variety of such perspectives that are relevant to NFD structure. One might even argue that the world is in the midst of a conceptual paradigm shift with regard to the management of SESs. The pace of this shift has intensified considerably over the past decade, as thinking on several fronts converges towards a new approach to natural resources.

Much of the debate is around the implications of complexity and unpredictability of SESs. Complexity, diversity, and dynamics arising from many sources, not the least being the multi-scale nature of both ecosystems and governance systems in a globalizing world, are seen as resulting in SESs with high uncertainty and very low controllability. Most of the perspectives fuelling this paradigm shift relate to how to achieve good governance and sustainability in the face of these characteristics. In addition to the sustainability implications of SES complexity, there is increasing concern among stakeholders with a suite of principles that broadens the range of issues that must be taken into consideration in governance. These can be viewed primarily as human rights issues, including rights of access, the right to have the opportunity to rise above poverty, and the right to self-determination by participation in decisions that affect oneself (Kooiman *et al.*, 2005, Béné *et al.*, 2007). Thus the business of governance has become more complex.

Key emerging concepts relating to natural resource governance in response to the above circumstances include the capacity to adapt to changing conditions either by buffering against them – the idea of resilience (Berkes *et al.*, 2001) – or by changing with them in the most advantageous way possible – the idea of transformation (Olsson *et al.*, 2004). In both cases, characteristics that promote adaptive capacity such as the ability to detect changing situations, learn from past experience, and to innovate are increasingly recognized as valuable assets for complex, diverse and dynamic SESs (Folke *et al.*, 2002).

Enabling this type of capacity is seen as the way to deal with situations of high uncertainty and low controllability. A key complementary idea is that small-scale fisheries problems are 'wicked problems' that cannot be solved, only managed (Jentoft and Chuenpagdee, 2009). Hence there has been a growing emphasis on capacity building, processes and institutions across the entire suite of stakeholders, and attempts to define action frameworks that guide rather than prescribe.

Table 1 Typology of fisheries according to NFD development, fishery scale and stock size

		Large stock size	Small stock size
NFD well developed (usually developed country)	Large-scale	**Major resources upon which most conventional fishery management and assessment has focused** • Poor history of management, but nonetheless sufficiently valuable that managers continue to invest resources into assessing and managing	**Few large-scale fisheries develop on small stocks, except where many of them are in close proximity (e.g., off-shore salmon)** • These stocks usually get management attention due to the aggregate value and profile of the industry
	Small-scale	**Usually part of a system that includes large-scale fishing** • Tend to get management attention as part of the management package for the large-scale fisheries • Often take too small a proportion of stock to have impact	**Often by indigenous peoples and in rural areas.** • Tend to be less well managed than higher priority larger scale fisheries on large stocks • Get some spin-off attention from the capacity to deal with large-scale/large-stock fisheries; resources can be diverted to these fisheries
NFD small/poorly developed (usually developing country)	Large-scale	**Major resources upon which most conventional fishery management and assessment has focused; these fisheries often attract foreign investment and distant water fleets.** • The value of, and foreign interest in, these fisheries may lead to greater attention to assessment and management	**Few large-scale fisheries develop on small stocks, except where many are in close proximity; no known developing country examples** • If large-scale fisheries have developed on small stocks in countries with poor management, they usually attract outside attention, often because the fishery is based on outside investment
	Small-scale	**Usually part of a fishery system that includes large-scale fishing** • May take only a small proportion of the total catch and may be marginalized, even when there is substantial management attention on the stock for the large-scale fishery	**Very large number of such fisheries; often tropical, rural, coastal and inland** • Not sufficiently individually valuable to warrant conventional approaches • Receive little or no formal management attention • Vulnerable to depletion by small-scale, local fisheries

Frameworks such as the Sustainable Livelihoods Approach (SLA) have become more prominent in dealing with SES complexity. The SLA, for example, is a very people-oriented, practical framework that looks at the externalities that affect fisheries livelihoods and the various types of livelihood assets that fisher folk have at their disposal for coping with and adapting to change (Allison and Ellis, 2001). The Interactive Governance Approach, on the other hand, is a conceptual framework that takes a broad perspective on governance as an overarching issue, and breaks it out into component parts which treat processes and interactions on multiple levels, ranging from visions and principles through to action on the ground (Kooiman *et al.*, 2005). This is developed further into a governability framework that illustrates the complexity of the new expanded view of fisheries governance and the difficulty of developing a mechanistic assessment process (Chuenpagdee and Jentoft, 2009). Yet another framework constructed by the Earth Systems Governance Project is built around five analytical problems (Governance Architecture, Agents, Adaptiveness, Accountability and Legitimacy, Access and Allocation) and four cross cutting themes (Power, Knowledge, Norms, Scale). These words alone give a strong sense of the issues that are attracting increased attention in the emerging perspectives on governance of SESs (Bierman *et al.*, 2009).

These new lines of thinking are beginning to have a practical effect on the way that fisheries are assessed and managed. In Australia, a risk-assessment approach that is flexible enough to allow for the different conditions that might prevail at any point in the above typology is being applied to fisheries (Fletcher *et al.*, 2005). This approach allows the context to dictate the direction, technical complexity and cost of the proposed response. A scheme of fisheries governance benchmarking developed by Grafton *et al.* (2007) also shows how these ideas are being incorporated into fishery assessment. It includes aspects of transparency, accountability, adaptability and risk assessment/management. Within the FAO, the development of the Integrated Approach to Small-Scale Fishery Assessment includes a step in which fisheries are allocated into quadrants on two dimensions – complexity and value – as a preliminary guide to the best approach to managing them (Garcia *et al.*, 2008). At a regional level, the Wider Caribbean Region is pursuing ocean governance in the context of the Large Marine Ecosystem (LME) Governance Framework which emphasizes a network approach to governance and thus interactions among organizations at multiple levels (Fanning *et al.*, 2007).

What are NFD capacity needs in this governance environment?

It can be argued that, in the governance environment, an NFD with the following characteristics will be better able to cope with uncertainty and uncontrollability:

- Promotes shared visions and objectives
- Process-oriented
- People-centered
- Self-organization oriented
- Resilience and transformation oriented

This suite of characteristics will place the NFD in a better position to enable these same qualities in the fisheries systems for which they are responsible (Bavinck *et al.*, 2005, Mahon *et al.*, 2008). Contrary to conventional thinking, the emphasis here is not on more science and technology, although it in no way suggests that science and technology are abandoned. For example, the use of marine spatial planning and geographic information systems (GIS) is becoming commonplace. These tools can be wielded purely by experts and produce 'black box' results for stakeholders to consume. Alternatively, participatory GIS allows stakeholders to engage with the experts and the technology to produce a people-centred process and product. The Marine Information System (MarSIS) Project in the Grenadine Islands of the Caribbean is a case in which the authors expect that the participatory Geographical Information System (PGIS) (done as part of doctoral research) will become a valuable resource to fisheries, coastal and planning authorities as well as the many stakeholder groups in these small islands.

Shared visions and objectives are recognised as a powerful means of mobilizing actors towards cooperation and self-organization. This was the case in the formation of the Caribbean Network of Fisherfolk Organisations most recently, but before that the fisheries authorities of the English-speaking Caribbean undertook an exercise to agree upon the vision, mission and goals of both sub-regional fisheries bodies. The exercise was the first step in the formation of the Caribbean Regional Fisheries Mechanism as a regional fisheries body. Fisheries authorities have repeated this at the national level in the formulation of fisheries management plans guided, for example, by the Code of Conduct for Responsible Fisheries.

Attention to process in fisheries development and management promotes transparency, equity, accountability and participation, as it is clear to actors where and how they can become engaged. The fisheries management planning processes described above paid attention to these principles of good governance. As more attention turns to fisheries in the context of human rights, we can expect that there will need to be more emphasis on process, not just product. A people-centred perspective allows for more socially just and equitable outputs as the implications of decisions for various categories of actors are considered and weighed, rather than evaluating them only in terms of economic or food value.

A conscious orientation towards allowing room for actors to engage in self-organization is likely to promote solutions that are innovative and context appropriate. Allowing self-organization to occur requires a certain degree of elasticity (Mahon and Bavinck, 2005). Empowerment is not the same as enabling self-organization. Fisheries authorities may empower fisher folk groups only within a narrow range of options, such as prescribed by cooperative law, while not promoting the greater freedom of choice enjoyed by associations or other organizational structures. This may be done with good reason; such as to achieve economies of scale in the fishing industry. However, curbing the flexibility to choose organizational form is counter to enabling self-organization if alternatives to cooperatives are more adaptive.

Finally, a conscious orientation towards resilience and transformation will encourage actors to look at the system in a different way, and to ask different questions. Walker and Salt (2006) provide a useful primer and guide to promoting thinking about systems in this way. For some fisheries authorities, operationalizing these concepts will be the most challenging. Changing the ways in which uncertainty is perceived and addressed is one example of the possible constraints. In risk-averse bureaucracies, where stability or only incremental changes are highly valued, the notion of transformation may be problematic. Where fisheries systems are resilient in various undesirable domains, there may be no alternative but to push them past a threshold into another domain while appreciating that the outcomes or consequences cannot be accurately predicted.

Mahon *et al.* (2008) further develop the idea of the shift in the inputs required for management of fisheries from a complex adaptive systems perspective with reference to a complexity continuum, with predominantly small-scale fisheries at one end and predominantly large-scale fisheries at the other. The inputs needed at the SSF end of the continuum are argued to be primarily those that enable self-organization, as opposed to regulatory and directive inputs which maybe most appropriate at the other end of the continuum.

It follows that the mix of skills needed to develop the capacities described above, and to provide the enabling inputs needed for SSFs, will be much more people- and process-oriented than is often the case in conventional NFDs, especially those modelled after large developed country NFDs established to cope with large-scale commercial fisheries. The kinds of skills to be engendered in a small NFD would include the ability to: (a) Plan with stakeholders; (b) Develop and manage projects; (c) Mediate and manage conflict; and (d) Facilitate group processes and build consensus.

Adding process-oriented people skills could require radical rethinking in NFDs, especially small ones (Mahon and McConney, 2004). In small NFDs, it could mean that the predominately natural science trained staff would have to be replaced with a mainly social science trained one with an emphasis in the above skills.

Conclusions

The main points of this essay can be summarised as follows: (a) National fisheries departments (NFDs) are critical components of fisheries governance; (b) Their structure and function have received less attention than most fishing industry components; (c) Scale/size, especially of fish stocks, fishery and management capacity, are key issues; (d) There is a mix of features and skills that can help small NFDs cope with small and small-scale fisheries; and (e) Small or developing country NFDs are seldom structured to include these skills.

It is clear that there needs to be more attention to NFD structure and function. At the global level there are trends and changes suggesting that those countries seeking to reorient NFDs along the lines indicated above will find that there is significant support in the international community for these initiatives. The resources that are needed for these types of scale relevant and people-centered changes are becoming increasingly mainstream and available. These changes, while requiring considerable reorganization, appear likely to produce desirable results for small-scale fisheries.

Yet, despite the growing evidence that benefits outweigh costs, we can expect such change to be difficult and to require careful management to overcome administrative inertia, vested interests and repose in well-established comfort zones. In complex adaptive social-ecological systems we talk of transformation. In the interactive governance approach we appreciate that the governing system needs to reconfigure itself to interact with the system to be governed. Both perspectives recognize the role of self-organization in effecting change. Our final point is that the changes we advocate can most easily be made through collective action by authorities responsible for small-scale fisheries. Opportunities such as small-scale fisheries conferences and major projects foster information exchange and networking that facilitates the necessary learning and adaptation as a community of interest. They accelerate the process of proactive change required in small NFDs.

References

Allison, E. H. and Ellis, F. (2001) The livelihoods approach and management of small-scale fisheries. *Marine Policy*, 25, 377-388.

Bavinck, M., Chuenpagdee, R., Diallo, M., van der Heijden, P., Kooiman, K., Mahon, R. and Williams, S. (2005) *Interactive Governance for Fisheries: A Guide to Better Practice*. Delft, The Netherlands: Eburon.

Béné, C., Macfadyen, G. and Allison, E. H. (2007) Increasing the contribution of small-scale fisheries to poverty alleviation and food security. *FAO Fisheries Technical Paper 481.* Rome: FAO.

Berkes, F., Mahon, R., McConney, P., Pollnac, R. and Pomeroy, R. (2001) *Managing Small-Scale Fisheries: Alternative Directions and Methods.* Ottawa: IDRC.

Biermann, F., Betsill, M. M., Gupta, J., Kanie, N., Lebel, L., Liverman, D., Schroeder, H. and Siebenhüner, B. (2009) Earth system governance: people, places and the planet. Science and implementation plan of the earth system governance project. Earth System Governance Report 1, IHDP Report 20. Bonn: The Earth System Governance Project.

Chuenpagdee, R. and Jentoft. S. (2009) Governability assessment for fisheries and coastal systems: a reality check. *Human Ecology*, 37, 109-120.

Fanning, L., Mahon, R., McConney, P., Angulo, J., Burrows, F., Chakalall, B., Gil, D., Haughton, M., Heileman, S., Martinez, S., Ostine, L., Oviedo, A., Parsons, S., Phillips, T., Santizo Arroya, C., Simmons, B. and Toro, C. (2007) A large marine ecosystem governance framework. *Marine Policy*, 31, 434-443.

Fletcher, W. J., Chesson, J., Sainsbury, K. J., Fisher, M. and Hundloe, T. (2005) A flexible and practical framework for reporting on ecologically sustainable development for wild capture fisheries. *Fisheries Research*, 71, 175-183.

Folke, C., Carpenter, S., Elmqvist, T., Gunderson, L., Holling, C.S. and Walker, B. (2002) Resilience and sustainable development: building adaptive capacity in a world of transformations. *Ambio*, 31, 437-440.

Garcia, S. M., Allison, E. H., Andrew, N. J., Béné, C., Bianchi, G., de Graaf, G. J., Kalikoski, D., Mahon, R., Orensanz, J. M. (2008) Towards integrated assessment and advice in small-scale fisheries: principles and processes. *FAO Fisheries Technical Paper 515.* Rome: FAO.

Grafton, R. Q., Kompas, T., McLoughlin, R., Rayns, N. (2007) Benchmarking for fisheries governance. *Marine Policy*, 31, 470-479.

Jentoft, S. and Chuenpagdee, R. (2009) Fisheries and coastal governance as a wicked problem. *Marine Policy*, 33, 553-560.

Kooiman, J., Bavinck, M., Jentoft, S. and Pullin, R., eds. (2005) *Fish for Life: Interactive Governance for Fisheries.* Amsterdam: Amsterdam University Press.

Mahon, R. (1997) Does fisheries science serve the needs of managers of small stocks in developing countries? *Canadian Journal of Fisheries and Aquatic Sciences*, 54, 2207-2213.

Mahon, R. and McConney, P. (2004) Managing the managers: improving the structure and operation of small fisheries departments, especially in SIDS. *Ocean and Coastal Management*, 47, 529-535.

Mahon, R., McConney, P. and Roy, R. (2008) Governing fisheries as complex adaptive systems. *Marine Policy*, 32, 104-112.

Olsson, P., Folke, C. and Hahn, T. (2004) Social-ecological transformation for ecosystem management: the development of adaptive co-management of a wetland landscape in Southern Sweden. *Ecology and Society*, 9(4), 2.

Walker, B. and Salt, D. (2006) Resilience Thinking: Sustaining Ecosystems and People in a Changing World. Washington DC: Island Press.

CHAPTER 17

Fisheries networks in the Caribbean

Patrick McConney[1], Robin Mahon, Kemraj Parsram and Shelly Ann Cox

Emerging governance frameworks conceptualize, describe and analyze fisheries networks in the Caribbean from local to regional levels. They aim to better understand and manage such networks. How useful these frameworks will be for practical application in improving fisheries governance is a key issue. Creating enabling policy environments, facilitating self-organization, managing networks, using social science skills, partnering with the private sector and civil society, learning-by-doing, adaptive management, and grappling with uncertainty are all necessary changes in the institutional arrangements for fisheries. We present some ideas and approaches that may be useful in developing practical applications from the conceptual frameworks. We discuss how interdisciplinary research may advance this process in small-scale fisheries governance.

Introduction

These days almost everyone is aware of networks, at least as metaphors if not as analytical concepts. If you are not an active member of a social networking site on the internet, then chances are that your circle of friends makes you acutely aware of what you are missing. People around the globe are keenly interested in who they are, or can be, linked to. They crave information about the attributes of others with whom they share social ties. Barabasi (2002) and other authors have informed readers about the analytical and practical power of networks, at a deeper level, through their bestselling books. Academic and scientific articles on network research have recently increased dramatically in both number and breadth of coverage. Network analysis is as hot a topic in business studies and epidemiology as it is in mathematics, natural sciences

1 Corresponding author: University of the West Indies, Barbados, patrick.mcconney@cavehill.uwi.edu

and social sciences. What about interdisciplinary studies and the mobilization of network knowledge for practical application?

You would think that with all of this attention to networks there would be little new knowledge still to be researched and applied to problem-solving. On the contrary, network analysis continues to reveal the limits to our understanding of social and ecological structures. Consequently there are many exciting new fronts for interdisciplinary research and its practical application. One broad area of research that demonstrates this potential is the study of social-ecological systems, and the use of knowledge gained to improve the governance of living marine resources such as fisheries. The failure of conventional fisheries management to deliver expected results and benefits, as evidenced by the large number of severely depleted fisheries worldwide, has prompted a search for new and more successful alternative directions (Berkes *et al.*, 2001). Networks incorporated into current thinking about resilience, complexity, adaptation, capacity, self-organization and the like show much promise for contributing to new conceptual frameworks for practical application in fisheries management and governance.

Small-scale fisheries are of special significance in our search for understanding and improving the functioning of social-ecological systems (Mahon *et al.*, 2008). Unlike most large-scale industrial fisheries based on profit-driven corporate models and processes, small-scale commercial and recreational fisheries have conspicuous human dimensions. It is easy to see, even if not always fully understand, how most small-scale fisheries economies are deeply embedded in, and functionally inseparable from, culture and society. Resources, resource uses and resource users are networked in a multi-level matrix of ecosystems and governance. In the Caribbean, a very complex region of the world for marine governance (Mahon *et al.*, 2010), the necessity to think in terms of networks is prompted by the region's insular geography and diversity if nothing else (McConney *et al.*, 2007).

The majority of countries within and bordering the Caribbean Sea are small island developing states. Most of the living marine resources are, or may be, shared amongst multiple geo-political jurisdictions as they progress through stages of their life cycles. Connectivity is evident in the context of biological diversity and fisher mobility along areas of extensive reef and shelf, as both the resources and their human predators travel across ecological and anthropogenic boundaries. Social networks are evident at local, national and regional levels amongst government agencies, fisher groups, non-governmental organizations and various fisheries stakeholder communities of interest. If networks are ubiquitous, how can we draw on them to help us in Caribbean fisheries governance?

This essay explores some of the potential benefits of thinking analytically about fisheries networks in the Caribbean based on what we know, and some suggestions for how this knowledge can be put to immediate practical use. We also speculate a little about what we do not know and the ways in which interdisciplinary research applied to problem-solving and proactive interventions can assist further in the future. Going beyond just fisheries management, we take a much broader governance perspective on this topic. A common thread is the need to manage change for enhancing resilience. This includes changing our thinking and our interactions to improve outcomes for societal benefit.

What are we thinking now?

To examine our current thinking, let us work our way 'backwards' from the aim of societal benefits to ideas about how fisheries are structured. Although this may seem like moving in reverse, it is actually the appropriate direction in which to go. Only by starting with vision, goals or objectives can you get a true sense of where you are in relation to a meaningful context. For us, societal benefit is the aim of sustainable development. Many people do not like the term sustainable, but it is useful for communicating the notion of managing living resources while ensuring livelihoods, and seeking to meet the needs and satisfy the interests of diverse stakeholders at multiple levels in social-ecological systems. It is a complex concept reflecting complex realities. Terms such as sustainable development, sustainable use, sustainable management and sustainable livelihoods are included in the fisheries policy documents of the Caribbean region. They present images of desirable futures that resonate with the aspirations of the peoples of the Caribbean (Fanning *et al.*, 2009).

This is practically useful because shared images motivate stakeholders who have the capacity to take individual or collective action to self-organize, consciously or otherwise, to achieve a set of societal goals. These types of overarching and somewhat fuzzy notions are also ones that may fit most comfortably into our thoughts on culture, norms and societal values. When societies in nation-states share these ideals, you have a stronger foundation for transboundary (sub-regional and regional) linkages that encourage state agencies, civil society organizations and citizens to work together.

Relevance to the Caribbean is immediately apparent. For the purpose of illustration in this essay we focus primarily on the English-speaking Caribbean, which is a dispersed geo-political sub-region, but our arguments also apply to the much more diverse Wider Caribbean and other parts of the world. Current thinking, explained above, on how deriving societal benefits from fisheries networks rests on a foundation of compatible cultures and norms shared across national terrestrial and maritime boundaries is especially relevant to the insular Caribbean. Figure 1 shows schematically how the entities mentioned above may be connected across waters and boundaries to other entities in different island countries at multiple levels.

Figure 1 Networked islands and multi-level organizations

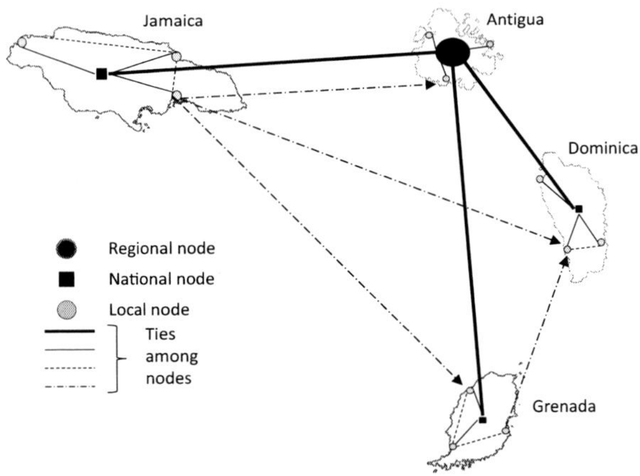

Later we go into more detail using concrete examples, but for now we can assume in this very simplified hypothetical illustration that nodes represent organizations (e.g., fisheries authorities, fisher folk groups or other fisheries stakeholders). In an ecosystem approach to fisheries we can expect the number of organizations (nodes) to be numerous and diverse. For our illustration they can simply be fisher folk bodies (e.g., cooperatives and associations). We show that each island in Figure 1 has a few local level (landing site) groups linked to each other and to a national level (secondary or umbrella) body by lines of different styles that are labeled as ties. The national organizations all have ties to a sub-regional node in one island, creating a three-tiered hierarchical network.

To add minimal complexity we show some of the local nodes linked across the sea to each other. Arrowheads on the ties show that the relationships are unidirectional. That is, there is a sending node and a receiving node. It could, for example, be the dissemination of information or issuance of directives without direct feedback. Where there are no arrowheads, we can assume that the ties are bidirectional or perhaps reciprocal (which can also be shown using double-headed arrows for clarity). Although we have only revealed a few of the possible ties (illustrated a network with low density), we see a picture emerging that would best be appreciated in a three-dimensional image (which can be done with some network analysis software) that is dynamic (but this poses greater challenges to modeling).

What have we left out? Even with all the simplifications above concerning a two-dimensional network structure we have not said anything about the network content. What do the ties contain? What is flowing between the nodes? In network analysis, content is key. Network analysis is less concerned about the attributes of the nodes than about the nature of the relationships or ties among nodes. This is what makes network analysis distinctive.

In our simplified illustration we can suggest that the ties show connections for communication, so the content is information. Of course we can further specify the type(s) of information of interest (e.g., ecological knowledge, management matters, planning data or strategizing political action). As we become more specific we will see differences in the structural patterns by type of information that relate to the roles of the various actors or nodes in the system. We may also want to measure the strength of ties. For example, how often is information of different types communicated? Or, how important is the information to particular outputs and outcomes? Numerous possible network analytical research questions exist. What gets asked and answered, the metrics used, and many other methodological details depend largely on the conceptual frameworks and the investigators. What we have described in summary is current, but cutting edge, thinking about fisheries networks.

We have argued that in fisheries, as in all other social-ecological systems, human and natural resources are inextricably linked through interactions at several levels. These linkages need to be taken into account in governance in order to reap societal benefits from fisheries, especially through an ecosystem approach. Networks are relevant to understanding and achieving the processes and products of fisheries governance. There are increasingly sophisticated methods available for analyzing networks, as well as a rapidly growing appreciation of the complexity of the conceptual constructs required for meaningful and practical interpretation, that stakeholders of all types can use to improve the success of agreed upon interventions. If, in our very brief and highly simplified overview, these are the main points made so far there will still be those who ask "so what?" In the real world, where investments have been sunk into building capacity for conventional fisheries management, it is legitimate and expected to have questions posed as to why we should shift thinking to new directions unless practical superiority, or a high likelihood of it, can be tangibly demonstrated.

We cannot fully answer "so what?", but we can identify some differences between conventional thinking and the emerging network governance perspective that we consider to be practically important. We suggest some changes that could be made to approach the question of "so what" through learning-by-doing or an adaptive design to fisheries governance. Finally, in the last section, we suggest how to manage the changes needed to transition from business as usual to a network governance perspective. Tackling this by a series of very short suggestions with examples is our means of providing a slate of emerging ideas that are awaiting opportunities for testing in order to construct a more robust conceptual framework for fisheries networks.

Practical application involves change

Conventional fisheries management focuses mainly on attempts at prediction (through quantitative models) and control (through regulatory management measures). The alternative directions that many fisheries now pursue tend to be more people-centered (Berkes *et al.*, 2001), and may introduce ecosystem approaches (Fanning *et al.*, 2009), but they do not go as far as fully incorporating network governance. In many cases it is not so much that the activities 'on the ground' are different, but the ways in which they are conceived, combined and fit into a coherent conceptual framework are different. Practical application involves translating the concepts of complex adaptive social-ecological systems into everyday thinking. We must think of fisheries systems in terms of resilience, adaptive capacity, self-organization and the like.

Tackling these in no particular order, let us start with creating an enabling policy environment. A classical conventional approach is to set policy top-down, expecting a fairly linear relationship between the guidelines of policy and the measurable planning objectives. Policy, a purview of analysts and experts in top-down approaches, reflects or even sets societal goals. A network perspective is more bottom-up and non-linear. While there is clearly a core policy domain or area of high and direct influence on policy-making, network analysis reveals that the actors in this domain are connected by several kinds of ties, influences and affiliations with diverse interest and stakeholder groups. These ties add uncertainty to policy, channel new knowledge, broker allegiances that facilitate action and otherwise shape policy.

Enabling policy is not intended to be rigidly prescriptive. Instead, it sets the stage for stakeholders and others to negotiate outcomes. It may rely more on incentives and compliance enhanced through legitimacy than penalties and regulation. A small example of networked policy is the way in which the Caribbean Network of Fisherfolk Organizations (CNFO), through connections to projects and influential actors, acquired a seat as an observer at the Caribbean Fisheries Forum and then obtained an audience with the Ministerial Council of the Caribbean Regional Fisheries Mechanism in 2009. It was the first clear case of fisher folk organizations, operating collectively as an informal network, getting direct input into regional fisheries policy. A major point in their policy brief was for fisheries authorities to facilitate, rather than control, the operations of fisher folk groups.

This leads us to self-organization. Contrary to conventional thinking in which stakeholders need to be strictly guided or constrained by rules, regulations, sanctions and the like in highly rigid and structured arrangements, complex adaptive systems incorporate notions of self-organization by constituent elements. With enabling policy in place, stakeholder organizations arrange their interactions to achieve societal goals through institutions that are more emergent than imposed. For example, the same Caribbean fisher folk organizations were able to form a multi-level network that interacts with the multi-level structure of the fisheries authorities. Both the governmental and non-governmental entities have other links with non-fishery nodes, but do not rely heavily upon them for inputs (e.g., funds, expertise and assistance) in order to keep the fishery system operational. This is very different from relations of external dependence in which one expects fisheries systems to collapse unless they are tuned and repaired from the outside. In this new view, nodes self-organize into different configurations depending upon changing circumstances. If, despite perturbations and changes, their arrangement performs essentially the same functions, then the fishery system can be judged, at least in some respects, to be resilient.

What we just described as self-organization requires adaptive capacity. This is also a key ingredient in resilience. The enabling policy environment provides the space in which to adapt, but this may make no difference unless the actors in the fishery system also have the capacity to adapt. Adaptive capacity is a multi-dimensional concept in which networks play a large part. This ranges from the options entities see as their scope for self-organizing (their world vision) to the hard inputs required (financial and physical assets). Network ties may limit or expand knowledge mobilization as well as other resources or capital. Ties with other entities, such as fisher folk groups or fisheries authorities, may be needed to effectively make use of resources already acquired for adaptation, such as when one group has the funds and the other has the expertise to build new capacity, and they realize mutual benefits from collaboration.

These differ from the typical view of capacity building as simply more training. Another difference is that the capacity and structures developed using networks are expected to be dynamic. For some fisheries authorities, this will go against the grain of trying to build solid monolithic administrative and technical structures where one size or set of skills is supposed to fit all situations. In network thinking, allowance is made for flexibility and fluidity to adapt to changing circumstances, linking to other nodes (private sector, non-governmental, academic etc.) to achieve the level of adaptation necessary. Again the example of the CNFO is useful. This entity has very limited internal capacity, yet by forming ties with other bodies it has been able to adapt to the requirements for holding technical workshops, formulating policy advice, identifying potential leadership and more. In doing so there has been transfer of skills and views that build internal capacity.

The need to use more social science skills is another consequence of fisheries network governance. Social science provides frameworks with which to address the human dimensions of social-ecological systems. In conventional fisheries management, biology has been joined by economics to become the two main disciplinary platforms. Additional disciplines are peripheral even in developed countries. Most Caribbean fisheries authorities are still focused on fisheries biology with slowly growing attention to economics but little on sociology and anthropology, except for those who incorporate these into their interdisciplinary studies. Since social scientists in fisheries authorities are still rare everywhere, interdisciplinary studies provide an avenue to integrate the social and ecological aspects of ecosystems. However, there remains a need to incorporate more people skills such as facilitation, negotiation, stakeholder analysis and conflict management, and to match fisheries authorities to meet situational demands rather than seeking to mimic the large agencies of more developed countries (Mahon and McConney 2004). It will take some time for Caribbean fisheries authorities to change their in-house capacity for a better fit. Meanwhile it will be necessary for them to bring in such expertise through partnering and contracting suitable short-term agents such as consultants and institutional allies.

Uncertainty can be as much a part of conventional fisheries management as a network perspective, but the approach to dealing with it is different. Conventional management often seeks to minimize and control uncertainty, whereas network approaches tend to readily accept and work around uncertainty. Controlling for certainty fits well with the aversion to risk typical of most fisheries authorities and policymakers. Taking an experimental, adaptive, learning-by-doing approach can be considered reckless. For example, politicians anywhere may be reluctant to admit that they approve of using funds from the public purse to experiment with fisheries policy, planning and adaptive management, even though this may be a better use of funds with less wastage than business-as-usual. Politics is based a lot on promises, even if we know they are not always kept. To admit that outcomes are uncertain is politically risky. High prices are often paid for 'failure' even though, in an adaptive context, negative or unexpected outcomes are part of the learning process. Since policymakers respond to their constituents, it is reasonable to consider the extent to which ordinary citizens and civil society organizations do or would endorse network governance perspectives. In the Caribbean this poses a problem. Conceptual frameworks for fisheries management and governance are much less in the public domain than in other parts of the world. Fisheries governance is not a political issue, despite depleted nearshore resources and increasingly exploited offshore resources.

The examples given above do not exhaust the range of relevant topics. However, they provide a good cross-section for illustrating the potential benefits of network perspectives in Caribbean fisheries as a means of breaking free from conventional fisheries management and opening new vistas for learning. In some examples we were able to show projects through which we are already testing the practicality of the concepts for application to fisheries. As we progress towards the end, however, we also admit that some aspects of making the change remain challenging even before one can consider whether they are superior or more useful and successful alternatives to current practice. Change has to be managed.

How do we manage change?

Change management has become an occupational field specialization in the wider arena of organizational management. We will not give it a detailed treatment here, but we emphasize how networks can play a role in managing the change in direction towards a fisheries network perspective. Demonstration is a powerful incentive for change. There are currently several projects underway in the Caribbean that demonstrate, at linked levels from local to regional, how a network perspective can help to improve the actual processes and products of fisheries governance. The most elaborate are under the

Caribbean Large Marine Ecosystems (CLME) project. These pilot projects can assist the development of new practical applications by field-testing the conceptual research frameworks for marine resource governance (Mahon *et al.*, 2009). They shorten the path to adoption or adaptation since they incorporate participatory monitoring and evaluation.

Successful change also depends heavily on having effective leaders, champions and other change agents. These individuals and organizations need to understand and buy into network concepts before they can lead, facilitate or convince others. This is proving challenging. However, the project on Marine Resource Governance in the Eastern Caribbean (MarGov) is paying much attention to communication (McConney and Haynes 2011). This will pave the way for advocacy and policy influence as well as develop a demand for governance research. Small grants disbursed under this project allow fisheries authorities, NGOs and community-based organizations to try out some of the concepts on their own, with guidance from researchers. There has been capacity development on governance, resilience and network topics (Deane *et al.*, 2010), and the demand for further capacity-building remains high.

Another way in which fisheries network proponents are managing and directing change is by doing more interdisciplinary research to advance knowledge on the topics. The approach is to use participatory action research to engage target audiences in the research process itself. Thus, through research networking, a small but influential community of interest and practice is being built. This approach also takes the perspective out of the purely academic realm and firmly embeds it in practice and problem-solving (McConney and Charles 2010). An example of this is the design of the network structure of the CNFO previously mentioned. Several meetings were held between fisherfolk leaders, network governance researchers and the regional fisheries authority in order to discuss and decide on the best structural model for the organization and to test it in practice (McIntosh *et al.*, 2010).

Conclusion

We believe that there are many potential benefits to introducing a network perspective on fisheries governance in the Caribbean, and probably elsewhere (Parsram and McConney 2011). This is needed to shift away from the failed prescriptions of conventional fisheries management even as it slowly evolves to become more people-centered. We believe that, rather than this slow evolution, a radical transformation is urgently needed. However, the transformation is more in the way we think about problems and problem-solving than what is done on the ground. Many of the tools and skills will remain the same. Yet, shifting to this new perspective is easier in some respects than others. There will be fast and slow arenas of change, and change will be different at multiple levels on different scales. These complex changes have to be managed. We, and others around the world, have made a start. However, the more supporters that join the collective action, the closer we will be to a tipping point that significantly transforms fisheries to better achieve societal goals.

References

Barabasi, A. L. (2002) *Linked: The new science of networks*. Cambridge, MA: Perseus.

Berkes, F., Mahon, R., McConney, P., Pollnac, R. and Pomeroy, R. (2001) Managing small-scale fisheries: alternative directions and methods. Ottawa, Canada: International Development Research Centre.

Deane, L., Pena, M. and McConney, P. (2010) Report of the Workshop on the application of resilience thinking to fisheries governance in the Eastern Caribbean. CERMES Technical Report No. 32., 1-2 September 2010. Barbados: UWI Cave Hill Campus.

Fanning, L., Mahon, R. and McConney, P. (2009) Marine ecosystem-based management in the Caribbean: an essential component of principled ocean governance. Report of Caribbean Regional Symposium, CERMES Technical Report No. 17., December 10-12, 2008. Barbados: UWI Cave Hill Campus.

Mahon, R. and McConney, P. (2004) Managing the managers: improving the structure and operation of fisheries departments in SIDS. *Ocean and Coastal Management*, 47, 529-535.

Mahon, R., McConney, P. and Roy, R. (2008) Governing fisheries as complex adaptive systems. *Marine Policy*, 32, 104-112.

Mahon, R., Fanning, L. and McConney, P. (2009) A governance perspective on the large marine ecosystem approach. *Marine Policy*, 33, 317-321.

Mahon, R., Fanning, L., McConney, P. and Pollnac, R. (2010) Governance characteristics of large marine ecosystems. *Marine Policy*, 34, 919-927.

McConney, P. and Charles, A. T. (2010) Managing small-scale fisheries: moving toward people-centered perspectives. In Grafton, R. Q., Hilborn, R., Squires, D., Tait, M. and Williams, M., eds., *Handbook of Marine*

Fisheries Conservation and Management. New York: Oxford University Press, pp. 532-545.

McConney, P. and Haynes, C. (2011) Communication. In Pomeroy, R. S. and Andrew, N., eds., *Small-scale Fisheries Management: Frameworks and Approaches for the Developing World*. Oxfordshire: CABI Publishing.

McConney, P., Oxenford, H. A. and Haughton, M. (2007) Management in the Gulf and Caribbean: mosaic or melting pot? *Gulf and Caribbean Research*, 19, 103-112.

McIntosh, S., Lay, M., McConney, P. and Phillips, T. (2010) The development of a Caribbean regional network of fisherfolk organisations and its role in influencing fisheries policy. *Proceedings of the Gulf and Caribbean Fisheries Institute*, 62, 298-305.

Parsram, K. and McConney, P. (2011) A network approach to understanding coastal management and governance of small scale fisheries in the eastern Caribbean. In Ommer, R., Perry, R. I., Cury, P. and Cochrane, K., eds., *World Fisheries: A Social-Ecological Analysis*. Oxford, UK: Wiley-Blackwell.

CHAPTER 18

Good practices for governance of small-scale fisheries

Anthony Charles[1]

Governance of small-scale fisheries is a challenge, given the complexity of these social-ecological systems. While there are no one-size-fits-all solutions in small-scale fisheries, there is a need for guidance on 'good practices' in fishery governance – priorities and approaches that seem, from theory and practice, to be most important for success. This paper identifies eight 'good practices' for policy and governance of small-scale fisheries, with particular attention to practices relating to rights-based management (viewing rights broadly as incorporating fishery rights and human, social and economic rights). Other 'good practices' include integrated systems thinking, a balanced sustainable development approach, attention to community well-being and food sovereignty, links to beyond-fishery policy, and effective information flows.

Introduction

Small-scale fisheries, wherever they are found globally, are highly complex systems, involving close interactions with coastal communities, harvesting of a wide range of multi-species fish resources, use of traditional management and knowledge systems, and pursuit of a multitude of objectives (Berkes *et al.*, 2001). As such, governance is a challenge. There are no one-size-fits-all solutions; context-sensitive perspectives are crucial. Still, there is a need to seek out what might constitute 'good practices' in fishery governance.

Accordingly, this paper proposes eight specific 'good practices' for policy and governance of small-scale fisheries. These were originally presented in an earlier version of the paper, prepared as a concept note for a series of FAO Regional Workshops held in 2010 on *Securing Sustainable Small-Scale*

1 Corresponding author: Saint Mary's University, Canada, tony.charles@smu.ca

Fisheries: bringing together responsible fisheries and social development (FAO, 2010). The present paper also draws on previous work reported in three book chapters (Charles, 2009, 2011; McConney and Charles, 2009), and several key policy documents, including the Code of Conduct for Responsible Fisheries (FAO, 1995), related technical guidelines, the inputs and outputs of the 2008 Global Conference on 'Securing Sustainable Small-Scale Fisheries' (FAO, 2008), and the 'Bangkok Statement' produced for that conference by the Civil Society Preparatory Workshop (2008).

Good practices

In this paper, eight elements are identified and proposed as a key set of 'good practices' in the governance of small-scale fisheries. It should be noted that these are not intended to represent an exhaustive list of such practices, but reflect aspects that seem especially important based on both theory and practice. The list of 'good practices' is as follows:

1. Adopt a 'rights-based' approach that links fishery rights and human rights
2. Ensure secure access rights to a fair share of fishery resources
3. Enhance legal space, empowerment and organizational capacity
4. Adopt an integrated, system-oriented, community-focused approach
5. Adopt an appropriate sustainable development approach
6. Strive for food sovereignty and household/community well-being
7. Include beyond-fishery policy measures and livelihood diversification options
8. Ensure effective, transparent and inclusive information

These are discussed consecutively in the following sections. In the first three elements, particular attention is given to rights-based 'good practices' in the governance of small-scale fisheries (Charles, 2008). This builds on the understanding that a key aspect of success in fisheries is the recognition and support for effective and well-accepted rights, and runs parallel to the focus of the Code of Conduct (FAO, 1995) on the responsibilities of fishers (e.g., Article 6.1: *'The right to fish carries with it the obligation to do so in a responsible manner...'*).

Adopt a 'rights-based' approach that links fishery rights and human rights

The Code of Conduct for Responsible Fisheries, while containing few explicit references to small-scale fisheries, does contain the statement:

> Recognizing the important contributions of artisanal and small-scale fisheries to employment, income and food security, States should appropriately protect the rights of fishers and fishworkers, particularly those engaged in subsistence, small-scale and artisanal fisheries, to a secure and just livelihood, as well as preferential access, where appropriate, to traditional fishing grounds and resources in the waters under their national jurisdiction (FAO, 1995, article 6.18).

This statement refers to two specific kinds of rights: the right 'to a secure and just livelihood' and the right to 'preferential access, where appropriate, to traditional fishing grounds and resources'. This highlights the two ideas arising when the word 'rights' is used in fisheries discussions. On the one hand, 'fishery rights' are held by those entitled to access the fishery and/or to be involved in managing the fishery. On the other hand, human rights, as well as social and economic rights, are held by fishers, households and communities, and can interact with management decisions taken in the fishery.

Until recently, these two categories of rights have been treated quite separately. One would have been hard pressed to find any mention of human rights in fishery policy debates. While the latter has been increasingly focused on so-called 'rights-based management', this tended to focus solely on 'fishing rights'. Human rights were largely left in the domain of those dealing with development concerns and human well-being.

However, there is now an emerging emphasis on good practices that link human rights and fishery rights. Attention has been drawn to this through the efforts of the World Forum of Fisher People (WFFP), the International Collective in Support of Fishworkers (ICSF), and other civil society organizations, to define a human rights approach for fisheries, in the 'Bangkok Statement' (Civil Society Preparatory Workshop, 2008). The WFFP and ICSF (ICSF-WFFP, 2009, p. 3) subsequently highlighted that a human rights approach in fisheries:

> 'recognizes that development efforts in fisheries should contribute to securing the freedom, well-being and dignity of all fisher people everywhere' and specifically recognizes the rights of fishing communities:

a) 'to their cultural identities, dignity and traditional rights, and to recognition of their traditional and indigenous knowledge systems',
b) 'to access 'territories, lands and waters on which they have traditionally depended for their life and livelihoods',
c) 'to use, restore, protect and manage local aquatic and coastal ecosystems',
d) 'to participate in fisheries and coastal management decision-making',
e) 'to basic services such as safe drinking water, education, sanitation, health and HIV/AIDS prevention and treatment services',
f) of all fish workers 'to social security and safe and decent working and living conditions'.

ICSF-WFFP (2009, p. 3) also specifically notes the rights of women to *'participate fully in all aspects of small-scale fisheries'*, *to have access to 'fish resources for processing, trading, and food'*, and to utilize *'fish markets, particularly through provision of credit, appropriate technology and infrastructure at landing sites and markets'*.

Fishery rights – which include use rights (over access to fishery resources) and management rights (over involvement in decision-making) – are typically discussed in the context of achieving more effective management, both by specifying who is involved in the fishery (and potentially how much), and by bringing fishers and others more actively and supportively into the management process. However, in a small-scale fishery setting, such fishery rights may also have impacts on the well-being and security of fishers and fishing communities; the effects may be positive (given suitable recognition, design and implementation of rights) but negative impacts are also possible (Charles, 2001; Béné *et al.*, 2010). Notably, fishery rights can affect human rights and, conversely, the pursuit of human rights can alter how fishery rights are designed and implemented, such as decisions concerning who should hold those rights, and how they should be managed (Charles, 2009).

Such linkages are important in addressing the challenge of poverty in fishing communities. Indeed, attention to livelihoods and poverty reduction in the context of small-scale fisheries is directly related to fishery rights. Allison and Horemans (2006, p. 760) noted that *'Addressing governance... remains the key challenge for both poverty reduction and responsible fisheries'*. Furthermore, the connection of artisanal and subsistence fishing to food security and livelihoods is an important element in considering human rights and fishery rights (Schumann and Macinko, 2007).

Ensure secure access rights to a fair share of fishery resources

Perhaps the most common concern of small-scale fishing communities is that of access to the resources they need for their livelihood and food security. Such 'use rights' to access and 'use' the resources, are referenced in the Code of Conduct (FAO, 1995: Article 6.18). Thus, *'when designing management measures, it might be appropriate to consider those which provide exclusive or preferential access for small-scale fisheries. Zoning, for instance, could favour and protect access to the resource by small-scale fishers, amongst whom the poorest are likely to be found'* (FAO, 2005, paragraph 2.7.6).

The Bangkok Statement (Civil Society Preparatory Workshop, 2008) focuses specifically on access rights in calling for a number of policy measures. For example, the statement notes the need to *'Guarantee access rights of small-scale and indigenous fishing communities to territories, lands and waters on which they have traditionally depended for their life and livelihoods' (#1), to 'Reverse and prevent the privatization of fisheries resources, as through individual transferable quotas (ITQs) and similar systems...' (#6), and to 'Protect access of women of fishing communities to fish resources for processing, trading and food, particularly through protecting the diversified and decentralized nature of small-scale and indigenous fisheries' (#15).*

Secure access rights are clearly beneficial to those holding them, whether an individual fisher, a fishers' organization, or a fishing community. On the other hand, caution is needed since restricting fishery access can have equity considerations and impacts on poverty and vulnerability of households and communities (Béné *et al.*, 2010). Overall, if access rights are managed well, they can serve social, cultural, economic and ecological goals, such as reducing conflict, enhancing food security and livelihoods, and helping protect local ecosystems (by inducing conservation actions). In addition, as the FAO (2005) notes, *'Promoting the small-scale over the industrial sub-sector may bring efficiency gains for the fisheries as a whole in addition to social benefits for the small-scale sub-sector.'* The perceived benefits of access and use rights to fishers and communities has motivated the emergence of such rights arrangements in a wide range of locations around the world (e.g., Johannes, 2002; Ruddle, 1989).

There are many choices to be made about the type of use rights to be utilized, and the allocation of these rights. An important option in small-scale fisheries is having fishing rights held collectively within a particular community (Charles, 2006), which has the potential to better reflect community values and objectives, to empower the community, and to allow for local social and ecological conditions (Willmann, 2000; Charles, 2001). As

Panayotou (1982, p. 44) suggested some time ago, '*The revival and rejuvenation of traditional community rights over coastal resources offer, perhaps, the best possible management option for scattered, remote and fluid, small-scale fisheries.*' However, one must ensure that possible imbalances in power within the community do not lead to inequitable results in the allocation of rights.

If a use rights system is well accepted, suitably effective, and meeting current objectives in the fishery, as well as criteria of equity and sustainability, then the key may well be to accept and reinforce these traditional rights. However, if the status quo use rights are seen as inappropriate or ineffective, some change may be needed. If there happens to be no use rights in place, a new system will need to be designed and implemented. Various use rights options exist, and no single approach will be applicable everywhere – each option has inherent advantages and limitations, so that what is 'best' will depend on the specific situation. Particular aspects to consider include the community's and society's objectives, the social, cultural and economic features of relevance, the key features of the fish and the ecosystem, and the governance capacity of the fishery system (Charles, 2002, 2009). The stability of access rights is a crucial factor, which relates to rights transferability, i.e., whether use rights can be permanently transferred to others. This may be considered reasonable if carried out within households or families but not through the use of market mechanisms (buying and selling rights), which tend to concentrate control over rights and shift them out of small communities (Copes and Charles, 2004).

Enhance legal space, empowerment and organizational capacity

Experience in fisheries throughout the world has demonstrated that while the secure access rights discussed above are undoubtedly crucial for the well-being of small-scale fishers, this must be accompanied as well by 'management rights'. This right to be involved in fishery management decision-making is referred to in the Code of Conduct's (FAO, 1995, paragraph 6.13) call to '*facilitate consultation and the effective participation of industry, fishworkers, environmental and other interested organizations in decision-making with respect to the development of laws and policies related to fisheries management, development, international lending and aid*'.

The Bangkok Statement (Civil Society Preparatory Workshop, 2008) calls on policymakers to 'Recognize and implement the rights of fishing communities to restore, protect and manage local aquatic and coastal eco-systems' (#2), and 'Guarantee the equal participation of small-scale and indigenous fishing communities in fisheries and coastal management decision-making, ensuring their free, prior and informed consent to all management decisions' (#10).

This reflects a shift away from a mode of operation in which management was done largely (within a commercial fishery context) by government officials. As McConney and Charles (2009, p. 538) note, '*A key trend in institutional design lies in acknowledging the value of participation by all fisheries stakeholders, and some level of empowerment of those stakeholders. This move, one that is crucial in supporting self-organization, leads to forms of co-management...*'. Such co-management involves the joint development of management measures by fishers, government, and possibly local communities, which the FAO (2005) notes, '*is also expected to promote improvements in public accountability and to foster empowerment of poor and vulnerable groups*'. In implementing management rights, and co-management in particular, a set of 'good practices' have emerged from the analysis of abundant examples worldwide – see, for example, FAO (2005) and McConney and Charles (2009) for lists of success factors in co-management.

The particular management rights option of community-based fisheries takes a 'place-based' approach to involving fisherfolk, and potentially others within a coastal community or region, in fishery management decision-making. This is essentially a matter of assigning management rights specifically on a community basis (whether to the set of fishers in a community, or to the community itself). Its suitability and effectiveness will depend on a range of factors (Charles, 2011); in the absence of suitable conditions, community rights may not be feasible but, given suitable conditions, the benefits of such a system of rights can be considerable (Graham *et al.*, 2006).

Real participation also requires accompanying efforts at empowerment and, fundamentally, the 'legal space' (notably through legislation and supportive policy) for fishers' organizations and fishing communities to be able to take on management responsibilities – with policy based on '*a subsidiarity approach that places decision making at the lowest, most local level feasible*' (McConney and Charles, 2009; Bavinck and Jentoft, this volume). If the focus is on 'pro-poor' approaches, empowerment must go beyond policy and legislation to look at how participatory processes are implemented at the community level, including local initiatives that go beyond the fishery per se.

While good governance of small-scale fisheries requires the effective *participation* of fisherfolk and fishing communities, there is often a need to support and build the *capacity* of organizations/associations of fishers, as well as community institutions, in order to accomplish this. Allison *et al.* (2010) argue the need to support fishing communities '*both through their social inclusion and building their capabilities*'. Capacity development is often needed in governments as well, so that governmental staff and institutional arrangements understand the needs and rights of small-scale fishers and communities. The Bangkok Statement expresses the need to support the capacity building of fishing and indigenous communities to participate in governance of coastal and fisheries resources.

Adopt an integrated, system-oriented, community-focused approach

While the community perspective in a small-scale fishery is likely to be 'integrated' (in that the community necessarily deals simultaneously with multiple economic sectors, social issues and environmental realities) and 'system-oriented' (in that interactions between people, among organizations and between society and nature are inherent within a community), conventional fishery management and policy has been constrained in 'silos' that neglected interactions with other productive sectors, with local environments, and local institutions. There is a need to reinforce the more integrated, systems-oriented approach to small-scale fisheries, an approach that simply reflects the reality of these fisheries.

One avenue to achieving this lies in finding an appropriate form of the broadly popular 'ecosystem approach to fisheries' (EAF) – a 'small-scale fishery EAF' with inherently strong human dimensions (De Young *et al.*, 2008; FAO, 2009). There is also a need to link small-scale fishery thinking more explicitly with integrated management of coastal and ocean areas. As McConney and Charles (2009, p. 540) note, '*This is a natural connection, especially in the context of SSF, since integrated approaches are characterized by a multiplicity of resources and habitats, a range of environmental variables, and a balancing of attention to both natural and human systems and dynamics...*'.

Adopt an appropriate sustainable development approach

A suitable sustainable development perspective involves the pursuit of development goals (notably the Millennium Development Goals) in a manner compatible with environmental sustainability. Since small-scale fisheries, like others, are susceptible to over-exploitation, environmental sustainability must be recognized as a prerequisite for food security, poverty alleviation and other development objectives.

Fortunately, a range of studies from around the world show that effective local self-management has often evolved within small-scale fisheries. There are 'benefits of the commons', in potentially producing strong local stewardship (Kurien, 2007). However, small-scale fisheries may nevertheless suffer from serious over-exploitation, whether driven from within the fishery, or as a result of external influences, including those from governments (e.g., through 'modernization' drives), that harm local stewardship practices. Thus, the need for rebuilding depleted resources and addressing the roots of over-exploitation are crucial in small-scale fisheries. Key elements in achieving this include avoiding open access conditions, and finding the appropriate form of rights over access and use – to ensure that small-scale fisherfolk have secure access and incentives to support and be involved in stewardship of fishery resources.

Strive for food sovereignty and household/community well-being

The need to strive for food sovereignty and household/community well-being as priorities in small-scale fisheries has been increasingly recognized in international policy, with the realization of the major role such fisheries play, or could play, in enhancing food security and poverty alleviation (Béné, 2003; Béné *et al.*, 2007). This counteracts the '*tendency globally to focus on industrialized, larger scale fisheries involved in commodity production and trade*' (McConney and Charles, 2009, p. 537).

The importance of small-scale fisheries to food sovereignty and community well-being is also reflected in the Code of Conduct (FAO, 1995). Article 11.2.15, in addressing international fish trade and export production, notes that '*States, aid agencies, multilateral development banks and other relevant international organizations should ensure that their policies and practices... do not result in environmental degradation or adversely impact the nutritional rights and needs of people for whom fish is critical to their health and well-being...*'.

In connecting small-scale fisheries with goals of food sovereignty and local well-being, Allison *et al.* (2010) recommend that policymakers '*Integrate responsible fisheries policies with wider poverty reduction policies in countries where fisheries are economically important*'. They suggest that this is important not only for a more integrated approach, but also to ensure '*that fisheries agencies receive an appropriate allocation of central and local government budgets, and that small-scale fishing communities are included equitably in national economic development and social development planning.*' This leads into the seventh 'good practices' element below.

Include beyond-fishery policy measures and livelihood diversification options

A 'bigger picture' perspective recognizes the strong connections between small-scale fishery systems and the well-being of coastal communities, as well as connections with policy and legal frameworks in coastal and marine regions, and in society beyond. For example, ensuring access rights to subsistence fishing in coastal communities may serve food security goals, and incorporating post-harvest aspects in rights discussions may help reinforce the rights of women involved in marketing fish. Incorporating this broader perspective is needed in order to holistically address the many issues facing small-scale fisheries. It will also involve a better understanding of the linkages within the fishery system itself, and in a multi-sectoral context, so as to produce more comprehensive approaches to managing small-scale fisheries; ones that are better able to improve well-being and safeguard livelihoods.

With respect to multi-sectoral considerations, the Code of Conduct technical guidelines (FAO, 2005), in considering governance of fisheries, highlight that '*Those wishing to support the contribution of small-scale fisheries to poverty alleviation and food security should thus strive to engage in policy processes in other sectors.*' In addition, the Code of Conduct's (FAO, 1995) Article 10.1.2 specifically refers to fisher participation in broader decision-making for coastal areas: '*In view of the multiple uses of the coastal area, States should ensure that representatives of the fisheries sector and fishing communities are consulted in the decision-making processes and involved in other activities related to coastal area management planning and development.*'

Finally, attention to livelihood diversification, both within the fishery system and beyond, is well recognized as an important ingredient of governance, albeit one that faces barriers in terms of governmental and agency mandates. Diversification does, however, closely link with moves toward integrated coastal management.

Ensure effective, transparent and inclusive information

If small-scale fisheries are to maintain their role as a key provider of sustainable livelihoods and food, achieve their potential for poverty alleviation, and shift toward a good governance framework that is more participatory, suitable information generation and use, as well as appropriate communications, are essential. As FAO (2005) highlights, '*The current policy shift towards decentralization of management responsibilities and the importance of small-scale fisheries in poverty alleviation and food security requires a reconsideration of the type of data and information necessary, and the way data are collected, used and disseminated*'.

This perspective on good practices with respect to information and communication is reinforced with a link to governance (McConney and

Charles, 2009, p. 541): '*The new approach in [small-scale fisheries] is to pay more attention to converting data from several sources into information for decision making and communicating the information to all stakeholders – to facilitate the transparency and accountability needed for good governance.*'

A related aspect concerns the source of information, in particular the common existence of 'local knowledge' held by fishers and communities in a certain location (and also known as traditional ecological knowledge or fisher knowledge). The FAO (1995), in Article 12.12, notes that '*States should investigate and document traditional fisheries knowledge and technologies, in particular those applied to small-scale fisheries...*'.

Finally, good practices in information acquisition involve the participation of small-scale fishers in determining knowledge needs and in the resulting research itself. This reflects an increasing recognition that the participation of fishers in management, as part of good governance, is more effective if they are also involved in dealing with the underlying information base. The result has been a trend toward participatory research involving fishers and fishing organizations in many locations. In the best situations, this becomes institutionalized (for example, within community-based fishery management) as a built-in participatory research component.

Conclusion

This paper has described a set of 'good practices' for policy and governance of small-scale fisheries. Particular attention has been paid to practices involving rights-based management, focusing on the linkage of fishery rights and human, social and economic rights. Ensuring secure access rights to a fair share of fishery resources, and enhancing legal space, empowerment and organizational capacity of fishers are important in this regard. Other 'good practices' include integrated systems thinking, a balanced sustainable development approach, attention to community well-being and food sovereignty, links to beyond-fishery policy, and effective information flows.

An assessment and understanding of key policy needs in small-scale fisheries, and the 'good practices' required to move in the direction of these goals, is a first step. However, it is crucial to take the next steps to utilize this understanding in a concrete manner, to achieve real progress in meeting the potential of the world's small-scale fisheries – such as an international instrument on small-scale fisheries, whether through an addition to the Code of Conduct or a Plan of Action, and a global program of assistance to these fisheries. While moving from words to actions will require considerable care in formulation and implementation, meeting the great potential of small-scale fisheries globally to help achieve Millennium Development Goals, and particularly to help alleviate poverty and contribute to food security, will undoubtedly make the efforts worthwhile.

References

Allison, E. H. and Horemans, B. (2006) Putting the principles of the Sustainable Livelihoods Approach into fisheries development policy and practice. *Marine Policy*, 30, 757-766.

Allison, E. H., Kurien, J., Pomeroy, R., Ratner, B. D., Willmann, R. and Asgard, B. (2010) Rights-based fishing: should it include human rights? Penang, Malaysia: The WorldFish Center.

Béné, C. (2003) When fishery rhymes with poverty: a first step beyond the old paradigm on poverty in small-scale fisheries. *World Development*, 31, 949-975.

Béné, C., Macfadyen, G. and Allison, E. H. (2007) *Increasing the Contribution of Small-Scale Fisheries to Poverty Alleviation and Food Security*. FAO Fisheries Technical Paper 481. Rome: FAO.

Béné, C., Hersoug, B. and Allison, E. H. (2010) Not by rent alone: analysing the pro-poor functions of small-scale fisheries in developing countries. *Development Policy Review*, 28, 325-358.

Berkes, F., Mahon, R., and McConney, P., Pollnac, R. and Pomeroy, R. (2001) *Managing Small-scale Fisheries: Alternative Directions and Methods*. Ottawa: International Development Research Centre.

Charles, A. (2001) *Sustainable Fishery Systems*. Oxford: Blackwell Science.

Charles, A. (2002) Use rights and responsible fisheries: limiting access and harvesting through rights-based management. In: Cochrane, K.L., ed., *A Fishery Manager's Guidebook. Management Measures and their Application*. FAO Fisheries Technical Paper 424. Rome: FAO, pp. 131-158.

Charles, A. (2006) Community fishery rights: issues, approaches and Atlantic Canadian case studies. In Shriver, A.L., ed., *Proceedings of the Thirteenth Biennial Conference of the International Institute of Fisheries Economics & Trade*, Portsmouth, July 11-14, 2006. Corvallis, Oregon: International Institute of Fisheries Economics and Trade (IIFET).

Charles, A. (2008) The right form of rights. *SAMUDRA Report, International Collective in Support of Fishworkers (ICSF)*, 51, 4-6.

Charles, A. (2009) Rights-based fisheries management: the role of use rights in managing access and harvesting. In Cochrane, K.L. and Garcia, S.M., eds., *A Fishery Manager's Guidebook*. Oxford: Wiley-Blackwell, pp. 253-282.

Charles, A. (2011) Human rights and fishery rights in small-scale fisheries management. In Pomeroy, R.S. and Andrew, N., eds., *Small-scale Fisheries Management*. Frameworks and approaches for the developing world. Wallingford, UK: CABI Publishing, pp. 59-74.

Civil Society Preparatory Workshop (2008) Statement from the civil society preparatory workshop to the global conference on small-scale fisheries (4ssf), Bangkok, Thailand. *SAMUDRA Report, International Collective in Support of Fishworkers (ICSF)*, 51, 7-9. Retrieved from http://sites.google.com/site/smallscalefisheries/statement

Copes, P. and Charles, A. (2004) Socioeconomics of individual transferable quotas and community-based fishery management. *Agricultural and Resource Economics Review*, 33, 171-181.

De Young, C., Charles, A. and Hjort, A. (2008) *Human Dimensions of the Ecosystem Approach to Fisheries: An Overview of Context, Concepts, Tools and Methods*. Fisheries Technical Paper 489. Rome: FAO.

FAO (1995) Code of Conduct for Responsible Fisheries. Rome: FAO.

FAO (2005) Increasing the contribution of small-scale fisheries to poverty alleviation and food security. FAO Technical Guidelines for Responsible Fisheries 10. Rome: FAO.

FAO (2008) Global conference on small-scale fisheries. Securing sustainable small-scale fisheries: bringing together responsible fisheries and social development, Bangkok, October 13-17, 2008. Retrieved from http://www.4ssf.org

FAO (2009) Fisheries management. 2. The ecosystem approach to fisheries. 2.2 The human dimensions of the ecosystem approach to fisheries. FAO Technical Guidelines for Responsible Fisheries, No. 4, Suppl. 2, Add. 2. Rome: FAO.

FAO (2010) Securing sustainable small-scale fisheries: bringing together responsible fisheries and social development. Concept note for regional workshops. Retrieved from ftp://ftp.fao.org/FI/DOCUMENT/small_scale/2010/concept_note.pdf

Graham, J., Charles, A. and Bull, A. (2006) Community fisheries management handbook. Halifax: Gorsebrook Research Institute, Saint Mary's University. Retrieved from http://www.coastalcura.ca/resources.html

ICSF-WFFP (International Collective in Support of Fishworkers and World Forum of Fisher People) (2009) Briefing note for delegates to the 28th session of FAO's Committee on Fisheries (COFI), Rome, March 2-6, 2009. Retrieved from http://sites.google.com/site/smallscalefisheries

Johannes, R. E. (2002) The renaissance of community-based marine resource management in Oceania. *Annual Review of Ecology and Systematics*, 33, 317-340.

Kurien, J. (2007) The blessing of the commons: small-scale fisheries, community property rights, and coastal natural assets. In Boyce, J. K., Narain, S. and Stanton, E. A., eds., *Reclaiming Nature: Environmental Justice and Ecological Restoration*. London: Anthem Press, pp. 23-54.

McConney, P., and Charles, A. (2009) Managing small-scale fisheries: moving towards people-centred perspectives. In Grafton, R.Q., Hilborn, R., Squires, D., Tait, M. and Williams, M., eds., *Handbook of Marine Fisheries Conservation and Management*. Oxford: Oxford University Press, pp. 532-545.

Panayotou, T. (1982) *Management Concepts for Small-scale Fisheries: Economic and Social Aspects*. FAO Fisheries Technical Paper 228. Rome: FAO.

Ruddle, K. (1989) The organization of traditional inshore fishery management systems in the Pacific. In Neher, P., Arnason, R. and Mollett, N., eds., *Rights Based Fishing*. Dordrecht, The Netherlands: Kluwer Academic Publishers, pp. 73-85.

Schumann, S. and Macinko, S. (2007) Subsistence in coastal fisheries policy: what's in a word? *Marine Policy*, 31, 706-718.

Willmann, R. (2000) Group and community-based fishing rights. In Shotton, R., ed., *Use of Property Rights in Fisheries Management. Proceedings of the FishRights 99 Conference*, Fremantle, Western Australia, November 11-19, 1999. FAO Fisheries Technical Paper 404/1. Rome: FAO, pp. 51-57.

CHAPTER 19

Towards a sufficiency economy for small-scale fisheries

Kungwan Juntarashote and Ratana Chuenpagdee[1]

Fisheries sustainability is a goal that remains largely elusive despite various initiatives proposed and implemented around the world. While this goal appears to be universally shared, it is not necessarily based on common values. The image of sustainable fisheries, and thus plausible pathways toward it, may indeed differ between groups of stakeholders. For several small-scale fishing communities in the south of Thailand, sustainability is less about economic efficiency and more related to the concept of 'sufficiency economy.' In this essay, we explain how this thinking has contributed to protection of aquatic ecosystems, improvement of fishers' livelihoods, and empowerment in resource management.

"There is sufficiency in the world for man's need but not for man's greed."
Mahatma Gandhi

Introduction

Sustainability is a common goal for small-scale fisheries development, with its prospect of maintaining fishing livelihoods and improving well-being of fishing communities. Several visions of approaches and pathways towards sustainability have been projected by planners, policymakers, developers and scientists from natural and social science disciplines. While the sustainability goal may be shared, these visions are not. Different perspectives, such as biodiversity conservation and economic development, are not necessarily compatible, and can cause difficulties and raise doubts in attempts to achieve local and global sustainability.

1 Corresponding author: Memorial University of Newfoundland, Canada and Coastal Development Centre, Thailand, ratanac@mun.ca

This was the situation in Thailand, where increased revenue to fishing households was the main aim of small-scale fisheries development for more than half a century. Fisheries policies were generally formulated to promote economic growth and development, as well as diversification of livelihood activities including aquaculture, fish processing and other supplementary employment. This development direction and livelihood options were influenced by two discourses: fisheries technology and economics. Large investments from private and public sources went to support research and development in fishing gear and post-harvest technology, hatcheries and aquaculture, and feed and fish meal production. A lot less funding was provided for conservation, restoration, and sustainability. While these development programs may have led to increased income to fishing households, small-scale fishers started to raise questions about their longevity and the sustained economic benefits. They doubted, for instance, that they would be able to catch fish at the same rate for generations to come, and to receive the same level of profit. High competition among small-scale fishers, as well as with larger scale operators, and resource decline were observed. Small-scale fishers were also concerned about the increasing cost of their operations (due, for instance, to high fuel prices). On a whole, small-scale fisheries livelihood was believed to be at risk, especially if capitalism continued to influence development decisions. These concerns led small-scale fishers in some parts of Thailand to explore other options that better correspond with long-term perspectives and their way of life. They turned to the 'sufficiency economy' philosophy that had been used to address problems faced by farmers in Northern Thailand.

This essay is drawn from research experience in small-scale fisheries in Thailand, as well as the latest study conducted for the PovFish project,[2] to understand what this philosophy means and how it has contributed to fishers' livelihoods and improved their well-being. We begin with an overview of the Thai small-scale fisheries and conclude with lessons about living a satisfying and sufficient lifestyle.

2 The research was conducted in 2008-2009 for the "Unravelling the Vicious Circle: Poverty Alleviation and Sustainable Livelihoods in Small-scale Fisheries" project, funded by the Norwegian Research Council.

Small-scale fisheries of Thailand

Marine fisheries in Thailand consist of small-scale, artisanal and large-scale, commercial sectors, with an intermediate, middle-scale sector that shares some characteristics of both. Small-scale fisheries can involve fishing without a boat, with a non-powered boat, or with an outboard engine-powered boat. Furthermore, in order to be considered small-scale, inboard engine boats must be less than five gross registered tonnage (GRT) in hold volume. According to the latest fishery census in 2000, about 57,000 fishing households in Thailand, or almost 90% of the total, were small-scale (DOF 2010). The majority of small-scale fishers (about 82%) use outboard engine-powered boats, followed by those who fish without boats or with non-powered boats (12%), and inboard engine boats. Without official statistics or proper studies to illustrate what their contribution is to the national economy and society, small-scale fishing people are seen as poor, with low status and capital, and with little bargaining power. They are perceived as lacking options, due partly to their low education, and are physically and politically marginalized (Juntarashote and Dausukho, 1986).

Our research shows that this perspective is simply a misconception among non-fishers, including government officials and policymakers. Small-scale fishers are greatly diverse and, unless this multiplicity is understood, we risk creating more problems in fishing communities rather than resolving them. Based on the research that we have conducted about small-scale fisheries in Thailand in the past 30 years, particularly the latest study for the PovFish project (see details in Chuenpagdee and Juntarashote, 2011), we propose three main categories to describe small-scale fishers. We refer to the first group as the "impoverished." These are small-scale fishers who live on the remote part of the coast where facilities and infrastructure, such as roads, electricity and sanitation, are generally poor. They fish without boats or with a small boat that has no engine. They use basic gears, such as simple hand lines and small traps, made of natural materials that they can find in their surroundings. Their daily catch is comprised of a small amount of fish and other animals, used to feed their families; and they only sell to the markets when they have surplus or when they get high-valued fish in their catches. This group of fishers is smallest in number, and they usually live near mangrove forest areas in the southern part of Thailand, which enable them to collect non-timber products as food and income supplements.

The second category is a "modest" group, to which the majority of small-scale fishers all across the country belong. They go out in long-tailed boats, about 8m in length, fishing with a mixture of gears, mainly gill nets and traps, depending on season. They sell almost all their catches to local fishmongers, and reserve a small portion for household consumption and/or distribution in the communities. Some fishers do primary processing of their catches, such as boiling shrimp and crabs, for sun-drying or picking, respectively. Comparatively, these 'modest' small-scale fishers earn higher average annual incomes than small-scale rice farmers. We will talk more about this group, to demonstrate clearly how the sufficiency economy philosophy has benefited them.

The final group is the "well-heeled" fishers who are able to take advantage of modern technology in advancing their fishing operations. Few fishers with entrepreneurial skills and access to capital, including their own savings, belong to this group. Their fishing boats are bigger, with inboard engines, but are still under five GRT in capacity. This enables them to fish further offshore but, similar to the other two groups, fishing is still a daily activity (e.g., going out in the morning and returning in the evening of the same day, or going out in the evening and returning the next morning). Common gears used are gill nets and traps, which are larger in size and higher in quantity compared to those used by the modest fisher group. Unlike others, well-heeled fishers normally hire 3-4 local men as crew members. Because of the high volume of catches, they can sell to fishmongers in the main ports, and receive better prices. Under favorable conditions, it is likely that this group of fishers can develop full-scale, commercial fishing enterprises.

Sufficiency principle

Sufficiency as a principle has long been in existence, as espoused, for instance, by Mahatma Gandhi. It aligns well with the contemporary ecological philosophy of 'deep ecology,' which emphasizes the interdependency between humans and nature, and the need to ask deeper questions, such as those related to environmental ethics. In the modern context, it is often used to discuss resource sustainability, particularly in contrasting it with the popular concept of 'efficiency.' As contended by Sachs (1999), the two concepts must go hand-in-hand since 'efficiency without sufficiency is counterproductive' (p. 41). Similarly, Princen (2005) argues that environmental degradation can be reversed to a sustainable path if efficiency is used complementarily with sufficiency.

But what is sufficiency? Drawing again from Princen:

> sufficiency as an idea is straightforward, indeed simple and intuitive, arguably 'rational'. It is the sense that, as one does more and more of an activity, there can be enough and there can be too much (p.6).

Questions remain about how much is enough, what is too much and how to know when it is too much. If it is as intuitive as Princen suggested, should fisheries resources be in better shape than they are today? Is it possible that the transition from economic efficiency to sufficiency must initially take place before we can expect the two concepts to operate together? We submit that such transition requires a deep reflection about meta-order elements, e.g., individual and societal values, norms, principles and ethics, as well as a major shift in people's mindset (Kooiman and Jentoft, 2009).

In Thailand, the revered King Phumibol incorporated the sufficiency principle as a guiding philosophy for the country in 1974. He explained that sufficiency is about living in a modest but reasonable way, with security and protection against uncertainty and threats, and suggested that taking the 'middle path' in pursuing economic development is more in keeping with globalization (TRF, 2006). The middle path concept aligns well with the teaching of Buddhism, the main religion of the Thai people, but also with the Muslim people who dominate the south. The government of Thailand has since embraced the concept and promoted numerous 'sufficiency economy' programs to help improve the quality of life of rural people. It is used, for instance, as the guiding principle in drafting the current 9th National Economic and Social Development Plan. It was first promoted in the agricultural sector, after repeated failure in small-scale farming that relies on a single crop, with heavy use of fertilizer and pesticides. Farmers ended up in debt, and their health was also deteriorating because of persistent toxic absorption. The sufficiency economy philosophy was introduced to enable small-scale farmers to plant multiple crops, to make their own organic fertilizers, and to use natural herbicides only when necessary. Importantly, agricultural products are used, first and foremost, to satisfy the dietary needs of their own households, and only the excess supply is sold locally. Local marketing is emphasized to minimize high transportation costs and to reduce dependency on outside buyers. The success of the program for rural farmers prompted small-scale fishers in the south of Thailand to adopt the philosophy and make it work in their settings, as further described.

Drawing from our research and recent interviews with several small-scale fishing communities in southern Thailand, we posit that the sufficiency economy concept consists of five domains. The first is directly related to fishing practices. It begins with the recognition that fisheries resources are not inexhaustible, and that some conservation measures are necessary to maintain their productivity. For the sufficiency economy concept to work, small-scale fishers need to adjust fishing effort and gear efficiency such that the amount of fish caught and the income from fisheries match the needs of their households. Further, they have to refrain from using fishing gears or practices that cause severe damage to the resources and the aquatic environment. The second component of the sufficiency economy concerns rational decision-making that is not driven by profit maximization. Instead, other considerations such as ecosystem health, fair resource allocation, and opportunities to educate their children are important in fishing decisions. Next, in the globalized and uncertain world, small-scale fishing communities are constantly exposed to natural, social and political changes which may have both short- and long-term effects on their livelihoods. The sufficiency economy principle is about being aware of one's own vulnerability, which is the third domain. Once factors increasing or reducing vulnerability are understood, it is possible to promote activities that strengthen fishers' capacity to cope and adapt to the changing conditions. Mangrove reforestation, protection of critical habitats such as spawning grounds and seagrass, and artificial reef installation are examples of initiatives that small-scale fishers in the south of Thailand have embarked on and benefited from.

Knowledge is the fourth element in support of the above and is essential for the practice of a sufficiency economy. The role of scientific knowledge in cooperation with local or traditional knowledge for resource management is well recognized. The sufficiency economy calls for ongoing knowledge acquisition in order to keep up with the changing world. Decisions about fisheries can thus be informed by current understandings and holistic knowledge about the entire fisheries system. The final tenet, and perhaps the most important, in the sufficiency economy philosophy is ethics. As previously stated, ethics is one of the meta-level elements that underlies what goes on in fisheries governance, starting with a daily management routine and progressing to institutional design and implementation. According to fishers we interviewed, some of the basic ethical values for small-scale fishing communities in achieving a sufficiency economy are related to honesty, hard work, perseverance and compassion. It also requires that fishers participate in rule- and principle-setting that conforms to their tradition and culture so that they can work in harmony by complying with the agreements and rules.

Factors and conditions for sufficiency economy

Several factors enable small-scale fishing communities to follow the sufficiency economy philosophy. However, because basic needs must first be met, some communities may not be in a position to follow the principle, at least not without initial support from governments. In other words, governments need to institute infrastructure for health, education, transportation and communication, appropriate landing and post-harvest facilities, and rules and regulations that protect and preserve fisheries resources for small-scale fishing people. In the case of Thailand, the majority of small-scale fisheries take place mostly in semi-developed areas, with reasonable roads, water and electricity. Small-scale fishers mostly live in relatively well-built houses, with proper roofing and basic amenities. Many of them own refrigerators, televisions and motorcycles. Mobile phone service is available throughout Thailand, and many fishing people use it for marketing purposes, in addition to communicating and sharing news.

Thai small-scale fishers are largely motivated to explore the sufficiency economy philosophy because of concerns with resource and habitat degradation due to overfishing, and increasing conflict with large-scale fisheries. Once the basic conditions are met, other enabling factors for small-scale fishers to pursue the sufficiency economy principle are possible. Despite modernity, Thai people place high value on families and, unlike in urban areas, rural communities are mostly closely-knit, with a relatively strong social network. Elders and others in need are taken care of. Respect for seniors is an important community value that is still practiced today. The wisdom of the elders and the energy of the youngsters is a good combination in moving forward, especially when treading new water, so to speak, which is what they need to move toward a sufficiency economy.

In a supportive society such as in many small-scale fishing communities, fishers share food and give it out to those who cannot fish. They help each other with boat repairs and net mending. Women also participate in net mending and preparing gears, as well as in post-harvest activities. Their relationship is even closer than that among men. They depend on each other, particularly on elderly women, for childrearing. In a typical small fishing village, it is common to see a group of women sitting in a circle with young children nearby, sorting shrimp or small fish into different sizes, after they are boiled and dried, while chatting about their lives. Everyone knows each other, and many of them are related through family lineage and marriage. They also have a similar level of well-being, except perhaps fishmongers, thus reducing social tension and ambition to obtain higher status. Notably, fishmongers are an important part of the social network, having come from the community or lived there for a long time. Rather than taking advantage of the people, they serve as safety nets for the whole community. These characteristics facilitate coordination and harmony that need to take place to implement actions aligning with the sufficiency principle.

A remarkable aspect about Thai fishing communities in the south is how Muslim and Buddhist fishers cohabit, despite the majority of the former. Local leaders draw from both religions, which have a similar orientation towards a sufficiency economy, in working with community members. Together they explore ideas about fishing gears that can improve catches without destroying habitats or overharvesting; post-harvest technology that minimizes waste; and livelihood activities to provide supplementary income to households. Education of children, skills training for women and men, and knowledge and information sharing are key aspects that support the transition from economic efficiency to sufficiency economy among small-scale fisheries in Thailand. In essence, these are social indicators for sustainability that need to be recognized and supported.

Sufficiency economy and sustainability

In the following, we present three local sustainability initiatives which originated in the Thai small-scale fisheries, and are founded on the sufficiency economy philosophy. The first is a story about a voluntary conservation effort referred to as a 'crab bank.' After witnessing decline in the crab population, and through consultation with fisheries scientists, small-scale crab fishers in many coastal villages in the south entered an agreement that requires them to separate female berried crabs from the catches, and release them into holding pens (or cages). This practice is not only inconvenient, it also implies less income from the current catch. Fishers understand, however, that this sacrifice is quite small for the long-term future of their crab fisheries. Even when the female crabs are sold after laying eggs in the cage, the money from the proceeds is donated to schools or used to support other public programs. In some villages, another step in conservation is taken by designating certain areas of the coast that is rich with seagrass beds as marine protected areas. Female berried crabs are released in these areas, allowing them to continue to reproduce. The ultimate goal of the crab bank project is to encourage fishers not to retain female berried crabs in their catches. The sufficiency economy philosophy makes it possible for crab fishers to forego current benefits, and to feel assured that everyone will enjoy the fruitful outcome of this conservation effort in the future.

The second example is related to the establishment of a locally-based central market for a group of neighboring small-scale fishing communities. Normally, fishers sell their catches at prices determined by fishmongers. This local trading works well when fishmongers are from within the village and have been supportive of the fishing communities. Still, fishers have little say about the transaction, and are not in a position to bargain for better prices when they are perceived to be unfair. This is because fishmongers can always buy from other fishers in nearby areas. The pressure is then on fishers who often feel that they need to increase their catches in order to maintain their income. To bring the situation in line with the sufficiency economy, fishers consulted with scientists and local leaders about alternative market models. The locally-based central market program emerged as a result, and has been implemented in several pilot communities. Instead of selling individually to fishmongers, fishers bring their catches to the central market for collective sale at minimum prices set by the local coordinator. Fishmongers have to auction for the fish in an open bid process. The aims of this initiative are multi-fold. First, it enables fishers to receive high prices for their products, consequently earning the same amount of income for less amount of catch. It also reduces the power of buyers and distributors in price setting, and lessens fishers' heavy debts with moneylenders. Finally, organizing buy and sell in the communities allows on-site monitoring of landed products, which can eventually lead to minimizing undersized fish. The success of the program has attracted other communities to join the central markets in their areas.

The third, and final, example is about the roles that fishers play in local resource management. Thailand went through a major governance reform to a decentralized system in 1997, which resulted in formulation of a sub-district administration office (known locally as 'Au-Bor-Tor'). This governance arrangement, consisting of elected members of the communities, is tasked with daily local management, as well as overseeing how resources, including fisheries, are governed. When Au-Bor-Tor leaders are well-respected members of the communities, goals such as fisheries sustainability through a sufficiency economy can be achieved. This is the case with several coastal communities in the south of Thailand where fishers and the Au-Bor-Tor actively participate in research, monitoring and surveillance, and rebuilding of the fisheries resources. One example is a co-management initiative in Bang Saphan Bay, where fishers and Au-Bor-Tor joined the fisheries department, the Food and Agriculture Organization of the United Nations, and the Southeast Asian Fisheries Development Centre, in creating 'shared rights' over fishing areas adjacent to the nine local fishing communities around the bay. The pilot project began in 1999 and initially ran for eight years, during which time fishers, community leaders and government officials met regularly to discuss issues and ways to address them. They agreed, for instance, to designate fishing areas for small-scale fishers, participate in patrolling the areas, install artificial reef, share information, and acquire new knowledge. Although the initial phase of the project ended in 2007 when the funding stopped, the communities have been able to keep the project going, thus turning it into a real community-based management program. According to the people we interviewed, this success has been attributed to strong leadership and good vision for community well-being and, to some extent, to the sufficiency economy philosophy, which provides key guidance on how small-scale fishers connect fishing, fisheries and their environmental surroundings to their livelihoods.

Conclusion

In this essay, we illustrate that, at least in the context of Thai small-scale fisheries, a sufficiency economy is a possible alternative to the economic efficiency model. It supports the image that coastal communities have of their fishing livelihoods and the sustainability of fisheries resources. What may seem like a sacrifice to some people, like when fishers willingly release female berried crab in holding pens, is a sensible pathway for many communities. Several factors and conditions enable fishing communities in Thailand to embrace the sufficiency economy philosophy, such as basic amenities, social capital and networks, and leadership and vision. There are tangible and observable outcomes, such as increased income and improved protection of the aquatic environment, that appeal to many communities, not only in the fisheries but also in small-scale farming, small business enterprises and other sectors. While it may seem that the sufficiency economy philosophy works only in a small-scale, rural, developing context, there is value in careful deliberation about the applicability of this principle in large-scale operations in the industrialized world. This is very important in the case of fisheries because sustainability requires that small- and large-scale fisheries work together to care for the resources and the aquatic ecosystem that they share and depend on. It also provides the foundation needed for small-scale fisheries to avoid being annihilated by the large-scale fishing sector. The sufficiency economy principle, however, is not a prescription for fisheries sustainability. It is simply one of the possible pathways, which has been shown to work in many small-scale fishing communities in Thailand, and it may also work in others in similar situations.

Acknowledgments

Funding for this research was provided by the Norwegian Research Council to the "Unravelling the Vicious Circle: Poverty Alleviation and Sustainable Livelihoods in Small-scale Fisheries" project. We acknowledge the critical inputs and feedback from Svein Jentoft, the project leader. We also thank the small-scale fishers of Thailand who shared with us their stories and philosophy.

References

Chuenpagdee, R. and Juntarashote, K. (2011) Learning from the experts: Attaining sufficiency in small-scale fishing communities in Thailand. In Jentoft, S. and Eide, A., eds., *Poverty Mosaics: Realities and Prospects in Small-Scale Fisheries*. Dordrecht, The Netherlands: Springer.

DOF (2010) National Status and Trend of Fisheries of Thailand: Marine Fisheries. Department of Fisheries. Bangkok.

Juntarashote, K. and Dausukho, S. (1986) Handbook for Fishery Extension Workers in Southeast Asia. Southeast Asian Fisheries Development Center. Bangkok.

Kooiman, J. and Jentoft, S. (2009). Meta-Governance: values, norms and principles, and the making of hard choices. *Public Administration*, 87(4), 818-836.

Princen, T. (2005) The logic of sufficiency. Cambridge: The MIT Press.

Sachs, W. (1999) Planet dialectics: explorations in environment and development. London: Zed Books.

TRF (2006) *Economic Sufficiency*. Bangkok: The Office of Thailand Research Fund, Government of Thailand.

CHAPTER 20

Subsidiarity as a guiding principle for small-scale fisheries

Maarten Bavinck[1] and Svein Jentoft

Imagination is a key ingredient for governing fisheries, and the subsidiarity principle is one of its more interesting fruits. Born in the realm of political theory, it emphasizes the rights of lower units vis-à-vis higher ones and the need to 'build bias' in decision-making. In line with others, such as Mathew (2005), we suggest applying subsidiarity to issues of scale and technology. The argument is that when small-scale fishers can do the job just as well (or better), they are given priority over large-scale fishers.

Everyone who is less important has to get out of the way
of everyone who is more important. (Lewis 1954)
Subsidiarity is the principle that each social and political group
should help smaller or more local ones accomplish
their respective ends.
Carozza (2003)

Introduction[2]

Small-scale fishers are used to being pushed around. As C.S. Lewis points out, describing traffic rules in the fictive country of Tashbaan, the fate of those who are 'less important' is frequently 'to get out of the way' and make the best of the situation. The images available for the future of small-scale fisheries are not always very favorable – neither among fishers, nor among scientists and policymakers.

1 Corresponding author: University of Amsterdam, The Netherlands, j.m.bavinck@uva.nl

2 A shorter version of this paper was published in Samudra (2008, volume 51) and in Common Voices (2011, issue 5).

Governance requires the capacity to think beyond the confines of sectoral interests and immediate needs. For this, imagination is a key ingredient. Images of how society might look are critical to efforts for solving problems and opening opportunities (Kooiman *et al.*, 2005; Jentoft *et al.*, 2010). The very definition of what constitutes a problem or opportunity depends on the way the future is imagined. To take this discussion to the field of capture fisheries, do we imagine the world's 25 million or so small-scale fishers to happily leave their occupations and blend into the industrial workforce? This is, after all, what has happened to countless other professional groups in history. Or, to present a contrary view, do we imagine a world in which small-scale fishing communities are given historical rights to the resources which they have always relied on, and will hopefully live happily ever after? Regardless of how appealing either image may be, they do have potential shadowsides: If fishers move out they also leave behind a cultural heritage, a community, and a potential source of food supply. Similarly, historical rights for fishers not only keep others out, they can also lock people in. While it is therefore important to possess images, one needs to investigate their possible consequences too.

Governing by principle

Governance principles are derived from images. Where images paint pictures, formulate ideas, and sometimes also express hopes, principles constitute measuring rods to distinguish the wanted from the unwanted, and the good from the bad. Images on the future of fisheries not only articulate ideas about problems and opportunities as they exist in real circumstances, they also contain values and preferences as to what avenues to take. There are many governance principles floating around, which influence the way we think about fisheries. In some instances, such as with human rights and biodiversity conservation, they are written into law. As for human rights, which are now increasingly being referred to in academic literature and debates on small-scale fisheries (Allison *et al.*, 2010), principles rely at the outset on notions of what it means to be human. Similarly, the precautionary principle which is incorporated into environmental law in many countries could hardly have been formulated without the natural environment imagined as an ecosystem, rather than as made up of individual species and organisms, and the idea that humans have a moral responsibility to protect and preserve.

In many instances principles are unspoken. What these principles are, and how they function therefore requires in-depth empirical inspection. This is also the perspective of "legal pluralism", which we believe holds promise for understanding small-scale fishing communities and their governance regimes (Bavinck, 2005; Jentoft *et al.*, 2009).

The subsidiarity principle

The subsidiarity principle is one of the more powerful ideas suggested for re-imagining and hence (re-)structuring societal governance. It finds adherence, as we shall see below, in a strange collection of bedfellows, such as the Catholic Church, the European Union, and the anti-globalization movement. We shall add the fisheries sector to this list.

The notion of 'subsidiarity' originated in the realm of political and legal thought, and refers to the relationship between higher and lower political units and how the responsibility for societal functions should be distributed hierarchically. Certain societal functions are obviously better handled by higher than by lower authorities, and vice versa. For example, state authorities are better able to organize a country's defense or foreign policy, whereas most would agree that the upbringing of children is better left to the family. The Catholic Church has fiercely debated issues of subsidiarity over the ages. In his encyclical *Quadragesimo Anno* (1931) Pope Pius XI formulated its meaning as:

> a fundamental principle of social philosophy, fixed and unchangeable, that one should not withdraw from individuals and commit to the community what they can accomplish by their own enterprise and industry (79).

Although the encyclical emphasizes the responsibility of individuals and family, it does not leave the community without responsibility. In fact, the subsidiarity principle also leads to an ethical imperative, as formulated by Carozza (2003), quoted in the motto to this chapter. In his view, social and political groups have a moral obligation to assist those that are smaller and local to fulfill their particular responsibilities "without, however, arrogating those tasks to itself" (p. 38, note 1).

Marked out in this fashion, the subsidiarity principle has clear relevance for poverty alleviation in small-scale fisheries (Fourage, 2004; Jentoft *et al.*, 2011). The principle suggests that, although there is a need for macro-economic policies that only the state can implement, fine-tuned policies at the local level are also necessary. Such policies would be better left with the local community because they require local knowledge and a sense for detail that a distant authority typically does not have. But central government still has a supportive role to play; for instance by installing legislation or providing services that enable local parties to perform their tasks. This is also why Kooiman *et al.* (2005) stress the need for thinking about governance as opportunity creation.

Other authors in the field – much in line with Pope Pius XI, although not with his religious interpretation – highlight the *rights* of lower administrative, political or social units vis-à-vis higher ones, and the notion that whatever *can* be decided at a lower level *should* be done there. The subsidiarity principle is thereby a potentially potent force in protecting inferior entities from interference by superiors: it is only if the task or issue cannot be effectively addressed by the inferior unit that the higher level unit is allowed to step in. A significant detail is that the burden of proof regarding who is better capable to handle societal functions (read for example fisheries management functions) always rests with the higher and not the lower unit. The higher authority cannot just disprove the capability of the lower unit by not fulfilling its responsibility for support. What the lower unit can do in principle is something the higher unit should help it do in practice.

The anti-globalization movement too is in favor of the subsidiarity principle. In an interview, Jerry Mander, founder of the International Forum on Globalization, explains:

> Subsidiarity is [...] the effort to bring real economic and political power as close to the local as is practical and feasible. Where the currently dominant globalization system tries to diminish the power of the local, the regional, even the national, and move all power to centralized global mega-institutions far away from affected communities, "subsidiarity" tries to stimulate the opposite. It seeks to build a bias in all governing systems, operating systems and rules to favor the local – or as close to the local as possible (Multinational Monitor, 2009).

Mander's view of 'building a bias' in governing systems most clearly highlights the political dimension of governance and the imperative of crafting an appropriate institutional framework. His emphasis on the role of power coincides with what we have elsewhere identified as the understated aspect of fisheries and coastal management (Jentoft, 2007). Governing from the subsidiarity principle would help to level the playing field, curb whatever ambition a higher authority may have in dominating the less powerful, and thus help create a more authentic democracy and a governing system based on social justice.

In the political realm, subsidiarity has been part of the thinking on the meaning of federalism, in the United States and in the European Union. The European Union has accepted it as one of its constitutional principles. The Edinburgh European Council of December 1992 issued a declaration on the principle of subsidiarity, which was subsequently developed into a protocol by the Treaty of Amsterdam. Hereby, subsidiarity came to play an important role in structuring the relationship and the distribution of competences between European and national-level authorities, also with regard to economic policy (Gelauff *et al.*, 2008). In this case, the subsidiarity principle is adopted to ensure that decisions are taken as closely as possible to the citizens of the Union, with the implication "that matters ought to be dealt with by the smallest, lowest or least centralized competent unit" (http://www.aer.eu/main-issues/subsidiarity.html).

Subsidiarity in fisheries governance

The European Union has long been criticized for not applying the subsidiarity principle to its fisheries sector. With the regionalization of this policy in recent years, and with the new common fisheries policy which is underway, efforts are being made to accommodate this critique, as the following citation from the website of the European Commission – which does not actually busy the term but approaches the substance of subsidiarity – suggests:

> It is critical to the success of reform that industry should understand the need for it, support it and have a genuine stake in its successful outcome. In a mostly top-down approach, which has been the case under the CFP[3] so far, the fishing industry has been given few incentives to behave as a responsible actor accountable for the sustainable use of a public resource. Co-management arrangements could be developed to reverse this situation (http://ec.europa.eu/fisheries/reform/green_paper/top-down_approach/index_en.htm) (Accessed April 7, 2011).

In the fisheries field, authors have often referred to subsidiarity to describe the relationship between government and user groups, and the role of participation therein (McCay and Jentoft 1998). Following the 2004 tsunami

3 Common Fisheries Policy

in Asia, Kurien (2005) also used the term to address the responsibilities of various parties with regard to disaster relief. More specific to small-scale fisheries, the Brussels Workshop on Common Fisheries Policy Reform in the European Union and Small-Scale Fisheries, that took place on September 28, 2009, included as recommendation number 13 to "apply the principle of subsidiarity to the management of small-scale fisheries, where management systems incorporate and are otherwise guided by local knowledge, experience and proven good practice." In all these instances, including the statement from the European Commission above, resonates the concerns of co-management, and the most appropriate way to distribute rights and responsibilities between the parties involved.

Mathew (2005) added a significantly different ingredient to the interpretation of subsidiarity. In his contribution to the Sixth Meeting of the United Nations Open-Ended Informal Consultative Process on Oceans and the Law of the Sea (2005), he suggested the implementation of what he calls 'scale subsidiarity'. By this he refers to the process "whereby larger fishing units are considered in a fishery only after exhausting the possibility of employing smaller fishing units in the same fishery." Small is hereby given priority over big.

Matthew's suggestion has been taken up in a proposal (2009) of the National Fishworkers' Forum (NFF) of India to revise the Marine Fisheries Regulation and Management Act, which is being drafted by the Central Government. NFF formulates seven principles of resource allocation, the second of which is to apply the principle of subsidiarity, which means that resources that can be harvested by smaller and less powerful units should be reserved for them and larger and more powerful units should be given access only to catch what cannot be caught by smaller and less powerful units.

It is too early to know whether or not the Indian legislators in question will actually incorporate a subsidiarity principle. The suggestion fits in closely, however, with other proposals for the support of small-scale fishers. Articles 1 and 2 of the Statement from the Civil Society Preparatory Workshop to the Global Conference on Small-Scale Fisheries in Bangkok (October 2008) thus request access and management rights over local or traditional sea territories. Article 3 lends priority to small-scale fisheries in Exclusive Economic Zones, and Article 4 strives to prohibit industrial fishing in inshore waters. In all these cases, small-scale fishers are given territorial rights. These are motivated and anchored in a human rights discourse that provides small-scale and indigenous fishing communities a preferential position for protecting and sustaining their livelihoods and cultures.

Although an application of the subsidiarity principle to technologies has similar consequences, it is deliberately rooted in a discussion of effectiveness. The argument is that when small-scale fishers can do the job just as well (or better), they are given priority; when they are not – yet – up to the task, however, other parties have a role to play. But one may ask: effectiveness with regard to what? In our view, four criteria suggest themselves (in this order):

1. Prevention of harm to the marine environment, which nurtures the fishery;
2. Ability to catch what the ocean allows, taking account of environmental limitations, thereby contributing to the well-being of human society;
3. Generation of a maximum of livelihood opportunities, in accordance with the need thereto;
4. Providing high quality protein for consumers in local, national and international markets

The advantages of small-scale versus industrial fishing are proven quite easily for points 1, 3 and 4 above (although there will always be exceptions). This is not to deny that small-scale fishing may also have negative environmental consequences, and that improvements are required. But point 2 is more difficult to establish. Can small-scale fishers indeed replace industrial fishers in capturing maximum sustainable yield? Are there not many instances where this would be dismissed as wishful thinking? Some fishing grounds are distant, and some target species might not be within reach of small-scale fishing technology.

It would be a misrepresentation of the subsidiarity principle to say that small is always beautiful. For instance, a smaller vessel might expose the crew to danger to a degree that a larger vessel does not. The sea is, after all, not a safe working space. As Mathew recognizes, applying the subsidiarity principle technologically would therefore always need careful consideration of the particular ecological and social context because, at the end of the day, it is that context, and the manner in which it is evaluated by fishers themselves, that determines what technology is or is not appropriate. Then we would also need a finer gradient than big versus small; the technology most appropriate to the situation at hand may well be of intermediate scale.

It is easy to see that the scaling up or down of fishing technology that is already in place is challenging. It would need a governance mechanism with sticks and carrots, and a design that allows decision-makers to know and understand the particularities of the social and ecological system within which the technology will operate. Thus, to use Mander's notion of 'building a bias', organizational subsidiarity accompanies technological subsidiarity. Decisions about appropriate fishing technology are among those that lend themselves for devolvement to the local level – as compared with, for instance, quota management where scientific knowledge is needed.

Conclusion

In this essay we have tried to imagine the consequences of applying a subsidiarity principle to the governance of small-scale fisheries. Our image rests upon specific interpretations of governance. First of all, it assumes a governance approach that pays attention to social justice and the rights of small-scale fishers to live in a healthy environment where their livelihoods are secure. It also assumes a decision-making process where small-scale fishers are empowered to participate effectively, and that protects them from the abuse of – in C.S. Lewis' (1954) terminology – 'everyone who is more important'. Small-scale fishers certainly would benefit from a more principled form of governance, where social values, norms and rules are respected and implemented in ways that are transparent, accountable and consistent both with local realities and internationally accepted standards.

Given their enormous global diversity, the condition of universal rules presents a challenge for small-scale fisheries. It is also for this reason, however, that the subsidiarity principle has merit. After all, by locating decision-making at the lowest possible level, diversity can be taken into account. But of course there are always issues of a more general nature too, such as those that relate to human rights. These call for policy development at higher scale levels. The Code of Conduct for Responsible Fisheries, which was drafted by the Food and Agriculture Organization, is a good example of such international policy.

Like other principles generally acclaimed for their moral superiority, subsidiarity risks ending up as an unfulfilled promise and as window dressing. It obviously requires more than a declaration to move it forward, and it is primarily up to small-scale fishers and their organizations to do so. In conclusion, however, we also refer back to Carozza (2003) and his interpretation of subsidiarity, which is that social and political groups should 'help' smaller or more local ones to accomplish their respective ends. Translated to the empowerment of small-scale fisheries, it suggests that large-scale fishers – who are more fortunate and capable – should assist small-scale fishers in doing their work, before defining their own work field. A start would be for small- and large-scale fishers to get together and negotiate a deal on how to share resources and territories. This needs to take place at many scale levels. It goes without saying that small-scale fishers, as the weaker party, have limited capacity to implement agreements, and that intervention of a higher authority, like government, is therefore required.

We argue, however, that a deal that is developed from the bottom up is likely to be more sustainable than one that is imposed on fishers (whether small- or large-scale) from above. Facilitating such encounters is among the responsibilities that government agencies – aspiring to meet the conditions set by the subsidiarity principle – might assume. Thus, subsidiarity is an image – an image for joint action.

References

Bavinck, M. (2005) Understanding fisheries conflicts in the South – a legal pluralist perspective. *Society and Natural Resources* 18 (9): 805-20.

Carozza, P. G. (2005) Subsidiarity as a structural principle of international human rights law. *The American Journal of International Law*, 97, 38-79.

Fourage, D. (2004) *Poverty and Subsidiarity in Europe. Minimum Protection from an Economic Perspective*. Chelterham: Edward Elgar Publishing Limited.

Gelauff, G., Grillo, I. and Lejour, A. (2008) *Subsidiarity and economic reform in Europe*. Berlin/Heidelberg: Springer-Verlag.

Jentoft, S. (2007) In the power of power: the understated aspect of fisheries and coastal management. *Human Organization*, 66(4), 426-437.

Jentoft, S., Bavinck, M., Johnson, D. and Kaleekal T. (2009) Co-management and legal pluralism: how an analytical problem becomes an institutional one. *Human Organization*, 68(1), 27-38.

Jentoft, S., Chuenpagdee, R., Bundy, A. and Mahon R. (2010) Pyramids and roses: alternative images for the governance of fisheries systems. *Marine Policy*, 34, 1315-1321.

Jentoft, S., Eide, A., Bavinck, M., Chuenpagdee, R., and Raakjær, J. (2011) A better future?: prospects for small-scale fishing people. In Jentoft, S. and Eide, A., eds., *Poverty Mosaics. Realities and Prospects in Small-Scale Fisheries*. Dordrecht, The Netherlands: Springer.

Kooiman, J., Bavinck, M., Jentoft, S. and Pullin, R., eds. (2005) *Fish for life – interactive governance for fisheries*. Amsterdam: Amsterdam University Press.

Krishna, A., ed. (2008) *Poverty, Participation, and Democracy. A Global Perspective*. Cambridge: Cambridge University Press.

Kurien, J. (2005) Tsunamis and a secure future for fishing communities. *Ecological Economics* 55(1), 1-4.

Lewis, C. S. (1954) *The Horse and His Boy*. New York: Harper Trophy.

Mathew, S. (2005) *Fisheries and their contribution to sustainable development.* Contribution to Discussion Panel A, 6th Meeting of the United Nations Open-Ended Informal Consultative Process on Oceans and the Law of the Sea, 6-10 June 2005.

McCay, B. J. and Jentoft S. (1996) From the bottom up: participatory issues in fisheries management. *Society and Natural Resources*, 9(3), 237-250.

Multinational Monitor (2009) A bias to the local: the subsidiarity principle. http://findarticles.com/p/articles/mi_hb6664/is_200903/ai_n32327188/?tag=content;col1 (accessed March 24, 2011).

National Fishworkers' Forum (India) (2009) *Let's protect fish and fishers. Towards revising the draft Marine Fisheries Regulation and Management Act, 2009, A position paper.* Kolkatta: NFF.

Section 6:

Reflections and visions

CHAPTER 21

Governance arrangement for small-scale fisheries: some reflections from negotiating small-scale fisheries governance arrangements in real life

John Kurien[1]

Much has been written and reflected upon about fisheries governance arrangements -- both in the industrial and small-scale fishery sectors. However, understanding the theory of governance and putting in place governance arrangements in real life are two very different processes. In this essay, I am using my own personal experiences of direct involvement in fisheries development and management work in three locations in Asia, to reflect on real-time and real-life issues which surfaced when negotiating governance arrangements with small-scale fishing communities.

Introduction

Much has been written and reflected upon about fisheries governance arrangements in both the industrial and small-scale fishery sectors. However, understanding the theory of governance and putting in place governance arrangements in real life are two very different processes. In real-life situations, one is confronted with an evolving reality which is influenced by a large number of social forces. One is therefore compelled to carefully negotiate – give and take – the way forward rather than move along a fixed trajectory of steps. Sometimes, totally unexpected events, larger social changes, new alignment of circumstances, or altered interactions between participants prompt the need to change track. Earlier path projections get

1 Corresponding author: International Collective in Support of Fishworkers (ICSF), India, kurien.john@gmail.com

derailed; new approaches must be created. In this essay, I use my own personal experiences of direct involvement in fisheries development and management work in three locations in Asia to reflect on real-time and real-life issues which surfaced when negotiating governance arrangements with small-scale fishing communities. The locations are:

- Kerala, India (1973-1978). Here the effort was on creation of economic enterprises and administrative governance arrangements for socially disadvantaged fishers to assist them in obtaining greater control over the price of fish and inputs required for fishing.
- Tonle Sap Lake Region, Cambodia (2005-2006). Here the effort was to make a top-down, state constituted "community fishery" decree acceptable to riparian communities and convert their new freedom to access resources into a means of achieving genuine development.
- Aceh Province, Indonesia (2007-2010). Here the effort was to help negotiate creation of participatory fisheries co-management governance arrangements in coastal areas physically and demographically devastated by the 2004 tsunami.

I take one issue from each of the locations. From Kerala I discuss the issue of "identity" and its impact on governance. From Cambodia I examine "freedom" and its implications on people's ability to take control over the governance of resources. Lastly, from Aceh I talk about the role of "custom" and legal pluralism in governance arrangements.

Kerala, India

> *In much of mainstream economic theory, humans are assumed to be purely self-interested, egoistic individuals who make no efforts to identify with others in deciding what objectives to pursue or choices to make. Reality is (thankfully) very different – especially when we are thinking beyond economic exchange. Social identity is one important factor which influences human decisions and actions in all realms – including exchange. Social identities are also multiple and overlapping. They evolve and change.*

In Kerala, my initial work of organizing fish marketing cooperatives with socially disadvantaged small-scale fishing communities was consciously focused on stressing their identity as members of a "laboring or working class" standing in opposition to the merchant class and the capitalist class which were exploiting them. The organization of fish marketing cooperatives required a clear class distinction to ensure that the spirit of the organization and the enthusiasm for self-governance over the exchange process be maintained. Fishery cooperatives in Kerala (and elsewhere) were a gross failure primarily because of the control merchants had over their functioning. Keeping merchants out and ensuring that the cooperatives were wholly fisher-operated democratic organizations – of the fishers, for the fishers, by the fishers – was an end in itself. It soon became apparent that though the intention of the fishers to make such committed action was undoubtedly resolute, they were not adequately literate or trained to take this responsibility. The solution was to appoint a paid managerial class to take care of day to day affairs and be answerable to the fishers. This arrangement worked well for over a decade. Organizations flourished and were recognized for their committed and innovative actions. Despite some mismatch between the knowledge levels and *modus operandi* of the fisher governors and the managerial class, the basic goals of increased incomes, control over credit, technology and fishery inputs was largely achieved. All seemed to be going well until a major shift occurred in the national discourse on social-cultural identity and development.

Reflecting changes in India from the late 1980s, identity discourse in Kerala shifted considerably from class to caste. Positive discrimination measures, as a means of empowering socio-economically backward communities, became a powerful instrument in the hands of the state. Organizing along the lines of one's caste – overlooking class contradictions – became the preferred mode of adversarial collective action. For a capitalist state, such actions were easier to co-opt and neutralize. Among the fishers with whom I was closely associated, the move toward redefining their identity in terms of caste was spearheaded by the middle-class youth among the fishing castes, with the actual worker fishers following their lead. The immediate consequence of this was to require all persons associated with the governance of economic organizations to be exclusively from the fishing castes – often, irrespective of their qualifications or competence.

The fishers' initial identity as members of a larger group of persons who labor hard to earn a livelihood linked them immediately with people of similar status in other occupations. It also helped them to seek the best possible value during the exchange of the products of their labor. But the caste identity created an exclusivity that drew fishers away from the mainstream. It created a strong within-group affinity that often went hand in hand with between-group discord and suspicion. This new assertion of a unique caste identity, which had an important political role in India, came at the expense of the plurality of identity – a reality which faces all of us.

The shift from an inclusive to a more exclusive identity had important implications for the governance priorities and practices of the fisher organizations. It blunted their focus on getting the best value for their products by negating the contradictions between workers and merchant-capitalists of the same caste. It affected relationships within fisher organizations as well as between them and other organizations.

The issue of identity – how one defines oneself, how one sees oneself – has a strong impact on the making and working of governance arrangements. In a multi-stakeholder context, identities can trump other, more direct considerations related to the fishery issues at hand. And, if identities change in the course of a working relationship, the terms of the discourse and actions will also change.

Identities such as race, ethnicity, religion, caste and creed are increasingly entering into community mindsets in developing countries, as part of larger socio-cultural and political machinations by powerful forces. While these tendencies are clearly 'far away' from fishery concerns, their appearance quickly 'spoil' our neat and tidy schematic flow charts of what fishery governance should be about.

Discussions on fishery governance need to be rooted in the political economy of the country/region in question. This is primarily because the fishery is usually a small or insignificant economic activity in most countries, and unlikely to 'call the shots' in relation to what the mainstream, or a dominant section of it, considers as the key identity. Well-functioning governance arrangements can get rude shocks, particularly if we assume that identities are fixed or uniquely defined. The issue of identities needs to be even more seriously considered in multi-stakeholder governance initiatives, such as fisheries co-management arrangements.

Tonle Sap, Cambodia

> *Amartya Sen's treatise "Development as Freedom" has changed the way we look at development and freedom. However, freedom is a state of affairs which many people have been deprived of for long periods of time. So, when they are 'given' freedom in order to achieve development, it can create a situation which they are unable to handle.*

Aquarian reforms in Cambodia were pronounced in 2000 when Prime Minister Hun Sen decreed the removal of large tracts of the Tonle Sap Lake, which were organized as fishing lots, from individual control and conferred them to many hundreds of rural communities living around the Lake. The decree granted the communities new rights of access to the fishery, flooded land, flooded forests and other resources. The communities were mandated to organize 'community fishery organizations' through which they were to manage the resources collectively and sustainably. There was a major initiative to set up appropriate administrative structures within the Fisheries Administration to support the community initiatives. International agencies such as the Food and Agriculture Organization of the United Nations (FAO) and financial institutions like the Asian Development Bank were closely involved with the efforts.

I was personally involved, initially with the government's fishery research and development institute, in helping to formulate governance policies related to the new community fisheries institutions. Later, I was also involved with efforts at the community level in ensuring that these institutions actually functioned as co-management initiatives.

The dynamics of implementing a top-down, hierarchy-driven, participatory community approach to aquatic resource management fuelled its own suspicions among the households in the *phum* (village), which were now grouped together into one community fishery organization. Because these households had often become part of the village as a consequence of a forced settlement plan, there was little 'sense of belonging' to the village. There was a huge 'trust deficit' among the inhabitants. Households pursued their own multiple sources of livelihood. Unfettered access to the vast and rich resources of the Lake was never part of this scheme. Consequently, when they were granted access by decree, the main barriers to negotiating use of the resources were thought to be lack of investment and the appropriate technology.

At a very early stage in this process, during extensive field visits to the communities, I was confronted with the revealing observation that, for many communities, the main challenge with respect to the new community fishery management regime was how they were to make full use of the 'freedom' conferred on them, as individuals, to relate to and access a resource from which they were previously banned.

We need to place this new freedom in historical context if we are to understand the anxieties of the communities about its utility and genuineness. Cambodia is a country with recent mass killing of persons claiming occupations other than agriculture; uprooting of communities leading to the deliberate creation of distrust; torture of people of non-Khmer ethnicity; persecution of religious leaders, and so forth. Against the background of this violent history, it was an immense challenge to convert this newly conferred individual freedom into collaborative collective action at the village level. The governance structures for the community fishery organizations were rather rigidly determined by rules and regulations in the sub-decree. The sub-decree did not give much scope for genuine expression of this newfound freedom in the governance initiatives of the community fishery organizations.

It became apparent to us that the huge trust deficit in the community was at the root of this reluctance to work together in this new context of freedom. It was also the reason for the 'wait and watch' approach to participation in governance matters. But trust does not develop in a social vacuum. Trust develops best in the slow process of working together. Thus there was a vicious circle which had to be broken.

The Fisheries Administration, with the assistance of the FAO, systematically organized community fishery organizations and established legally sanctioned governance arrangements through open, democratic procedures. This laid the foundation for converting the freedom granted to individuals and communities to relate to the resources into trust building exercises. The process was slow. Suspicions and doubts always surfaced about the real intentions of government, the responses of neighbors, the possible counteractions from former lot owners and so forth. The sub-decree was a legal instrument and became "the book" for the community fishery organizations.

But putting its provisions into action in each community's specific context required more than the occasional visits of government officers to the villages. Good moral leadership from the Buddhist monks; the presence of resourceful women in the village; earlier experiences of working together; and the unwavering support of good commune council officers were some of the important catalytic elements that helped individual freedoms to coalesce into collective action for genuine development in some of the villages.

In many developing countries which have experienced repressive political regimes, there is a legitimate 'fear of freedom' which tends to inhibit people's public participation even in changed circumstances. Governance initiatives in fisheries must provide the incentives and space to help individuals to translate the freedom granted to them into actions which create a sense of community and fellowship in the collective enterprise. There is always a strong influence of the community and of the people with whom we identify and associate in shaping our beliefs, our knowledge, our comprehension, as well as our ethics and norms. Facilitating this must be a deliberate act of governance within new organizational forms.

Aceh, Indonesia

> *Social institutions are important bonding arrangements which hold communities together in times of turmoil and devastation. Custom is an important base of social institutions. However, only when custom responds to the needs of changing socio-cultural circumstances and times does it remain vibrant and relevant.*

Aceh came into global prominence following the December 2004 tsunami which totally devastated the coastal areas of this socio-politically troubled province of Indonesia. The loss of life in the tsunami played havoc with the demographic profile of the coastal communities and ruined Aceh's coastal fishery. It was a province with a long and separate political history and rich natural resources. Aceh also evolved its own village and natural resource governance institutions based on customary (*adat*) principles. Having been repressed and isolated from the rest of Indonesian society for over four decades, these customary institutional arrangements were pursued with a special socio-cultural vengeance by the communities, while they were being forced to adopt homogenized institutional arrangements which were enforced during the New Order period (1966-1998) by the national government in Jakarta.

I was involved with an initiative of the FAO to introduce new fisheries co-management arrangements into the four districts of Aceh most affected by the tsunami. The earlier neglect of Aceh, the devastation of the tsunami, and the weak organizational structure of the fisheries administration provided fertile ground for introducing new ideas for fishery resource management approaches. Since co-management was a new concept, we first spent considerable time to create awareness about it and also to build the necessary capacity among the three lead stakeholders – the fishers, the coastal community and the state. During this process we were introduced to the customary fishery institution called Panglima Laot, and made aware of its diminishing role, largely because of the death of many of its leaders in the tsunami. The Panglima Laot played an important role in all aspects of fishery governance. Its most active role was in conflict resolution.

The existence of a certain legal pluralism became well apparent. However, the decision which needed to be factored into the evolution of co-management arrangements was the role and relevance of customary practices in the coastal fishery. These customary practices, which had a history of over 400 years, survived because they had wide social acceptance which was in turn a result of the dynamic nature of fishery norms and practices. It was living law which would lose its relevance the moment it was codified. The challenge was to bring custom into co-management governance and thereby foster legal pluralism in a meaningful and fruitful manner. If custom was to be revitalized to meet the governance challenges in the fishery, then several enabling conditions had to be met. Custom had to be revived and revitalized within the coastal community. It had to be formally acknowledged by positive law as an important component of the identity of the coastal community and, hence, a very useful tool for resource management. Custom also had to be recognized within the larger society.

One of the important methods used to achieve this objective was the production of a participatory film along commercial lines and in Bollywood style. The film was intended to provide a ready reference guide for those who wished to follow customary law for conflict resolution in the fishery. It was scripted to create a general awareness about the historical and current role of the Panglima Laot in Acehnese society. It was to provide wholesome entertainment to the general Acehnese audience. The commercial sales and the 'impact assessment' made among the viewers of the film vouches for its success in meeting its objectives. The film was also selected for an international festival of fishery films.

The centrality of customary law and the institutional arrangements needed to ensure its place in Acehnese legal discourse has been assured through provincial legislation (qanun). The role of the Panglima Laot in the development and management of Acehnese fisheries is incorporated in framework fishery legislation now (2011) before the provincial parliament in Aceh. The extent to which this will lead to genuine recognition of custom as living law, rather than its co-option by the state, remains to be seen. Only the future can tell.

In many developing countries, traditional regulations and customary practices are commonplace in fishing communities. This is particularly strong among island communities. The hoary tradition and the continued community recognition of many of these rules, norms and practices is testimony to their ability to 'keep up with the times'. Perhaps it is also evidence of customary practices being able to achieve inclusiveness, fairness and equity – facets which ensure community sustainability.

New governance arrangements in fisheries need to scan the socio-cultural environment to identify and recognize custom in fishing. When establishing new governance arrangements in fisheries, fostering legal pluralism without falling into the trap of the two extremes – romanticizing custom or allowing its total co-option by the state – is a desirable scenario to work towards.

Conclusion

I have tried to highlight that putting in place governance arrangements with real-life participants in the small-scale fishery sector requires tackling socio-cultural, economic, technological and political specificities. These are 'real time' processes and cannot be delineated totally according to elegant theories and flow charts. There is an important need to negotiate – involving give and take – which creates its own complex social and political dynamics. Both the initial and evolving conditions (social, cultural, economic etc) influence this process significantly. While each of the groups of participants may have their own specificities, when they agree to negotiate governance arrangements, their interactions result in transformations which are hard to predict. This creates dilemmas which can be intractable and difficult to solve, and greatly affect the planned governance arrangements. Finding solutions to these dilemmas are some of the most important challenges in negotiating governance arrangements for small-scale fishing communities in developing countries in the future.

CHAPTER 22

Views from below: student reflections on fisheries research

Andrew Song[1] and Ahmed Khan

This essay is based on our observation of several inconsistencies in fisheries research. They include the high degree of polarization characterizing the many scholarly debates concerning the status of the world's fisheries. As two graduate students conducting fisheries research in developed and developing countries, we reflect on the discourse of marginalized small-scale fisheries and the imperative of interdisciplinary fisheries research. These trends, though at times baffling, have also been a source of enlightenment and motivation. We recognize, for instance, a need for a more genuine engagement in human-nature interactions, one that also draws on the potential merits of fisheries governance theory.

Introduction

This essay offers our perspectives, as two fisheries students whose enthusiasm is somewhat stymied by what we observe to be certain inconsistencies in the realm of fisheries research. We list several key observations that have puzzled us as we continue with our training. We also try to link these observations to small-scale fisheries, a loose assemblage of fishery type that we hold much faith in. We hope that by bringing forward these reflections, we will be better able to make sense of fisheries realities around us. It must be pointed out that, despite our regular interactions with other fisheries students, we do not purport the views contained in this essay to be a ubiquitous representation of the general sentiment shared by our generation of fisheries scholars, who likely formally started studying fisheries sometime in the new millennium. We are institutionally based in North America with some familiarity with

1 Corresponding author: Memorial University of Newfoundland, Canada, amsong@mun.ca

African and Asian fisheries, and are academically grounded in geography, whose sharp attention on human-nature interactions helps foster an interdisciplinary mindset.

We first observe that we have been trained in a sustainability era where global fisheries are deemed to be under crisis, and research efforts are sought to reverse this trend. The debates concerning the health and status of fish stocks have been featured prominently in academic circles, and have extended to decision-making and policy arenas in national and international forums, and also to popular media. Concurrently, we have also seen troubling concerns about the future of small-scale fisheries and the negative impacts its relegation could have on national economies and coastal livelihoods. In addition, amidst a general call for interdisciplinary research and an explicit promotion of social science-based contributions, we observe certain contradictions that still prevail. These trends inevitably influence our research agendas, but we may also bring in alternative perspectives, especially those rooted in human-nature interactions and governance. We share these reflections in the hope that they add to the dialogue on the visions for fisheries research, as well as create an opportunity for further learning and refinement in our ongoing training.

Are fisheries under threat?

In the 1970s and 1980s, the development aspiration of fisheries held sway in an effort to modernize the fishing sector and communities. This was partly led by the development of the United Nations Convention on the Law of the Sea that placed much emphasis on ownership and national benefits, especially in the exclusive economic zones leading to a large increase in fisheries landings. Since then, there has been a shift towards curbing, or at least balancing, the development agenda with safeguarding the health of fisheries resources. Flourishing in the late 1990s and continuing to date, concerns with the status of the world's fisheries have received considerable attention among fisheries researchers. Consequently, we represent a generation immersed in the accounts of fisheries decline and crisis. Whether one believes that fisheries have degenerated into a serious and even irrevocable state, or that the situation is less alarming, the very intensity of this debate itself is a sign that fisheries crisis is one of the most significant driving forces that shapes the focus of much of contemporary fisheries inquiry. A simple reflection on the number of fisheries research articles that presuppose the premise of fisheries decline or crisis serves a crude, yet potent, indication of its dominant status (for example, see Garcia and Grainger, 2005). Thus, to many researchers, including us, depletion of fisheries resources and ensuing crisis naturally have become the ultimate concern, and is taken for granted as the chief problem that should essentially be addressed, regardless of the ecological, social, economic or cultural angle one may take. It works as a paradigmatic foundation upon which our current research is based – the main raison d'être of the fisheries training within which we are ingrained.

As an extension of the fisheries crisis ethos that encapsulates our academic mindset, we are also a generation embedded in the context of small-scale fisheries as being increasingly under threat. Despite the important contribution of small-scale fisheries to national and regional economies in terms of income, livelihoods and food security of millions of people worldwide, they are said to be consistently under-valued in socio-economic analyses and under-appreciated in the political realm (see Pauly, 1997; Berkes *et al.*, 2001; Béné *et al.*, 2006; Chuenpagdee and Pauly, 2006; Andrew *et al.*, 2007). Thus, anything less than the full appreciation of this sector has become a major concern to many fisheries researchers. Earlier accounts of its marginalization in the face of modernization and development agendas of many states around the world, through the introduction of mechanization and production-oriented strategies, provide historical precedence for the small-scale sector's weakening role (for example, see Berleant-Schiller, 1981; Bailey, 1988; Chauveau and Samba, 1989; David and Cillaurren, 1992; Flaherty and Karnjanakesorn, 1993). Its weaker standing vis-à-vis its large-scale counterpart is shown to arise in part due to its prevalently remote setting, the dynamic and plural nature of its operation, and a more fragile integration into the formal economic trade. Furthermore, the associated difficulty in obtaining adequate and reliable data does not bode well with quantitative analyses needed for conventional management purposes and national policy formulation.

This 'small-scale fisheries under-threat' mentality can perhaps be adequately illustrated via widespread episodes of conflicts generated by the frequent overlaps of coastal fishing grounds between large-scale and small-scale sectors in the offshore and inshore zones respectively. In Newfoundland, Sinclair (1985) documented how changes in gear use from traps to draggers, buoyed by government subsidies, generated tension and conflict in social relations between the two gear user groups, which subsequently put small-scale inshore fishers at a great disadvantage. On the contrary, in the Philippines, the national government enacted a regulation that created municipal fishing waters designated for the use of small-scale fishers, within which larger-scale commercial fishing is strictly prohibited.

Several West African countries, such as Sierra Leone, have also set a similar rule, where small-boat fishers have exclusive access rights to inshore areas of 12 nm while large-scale boats are restricted within the rest of the 200 nm. Though clearly favoring small-scale fisheries, such spatial restriction can be interpreted as a conscious attempt to *protect* and *defend* small-scale fishing rights against the intrusion of the more (mechanistically and spatially) powerful large-scale sector. Even with these legislative mandates in place, persistent problems of invasion and gear destruction by large boats further instill a sense of precariousness in the small-scale sector. In addition, the recent trend in the globalization of seafood markets and the political economy of eco-certification poses risks of additional marginalization in terms market access, quality control, and the loss of affordable protein to a large segment of fishing dependent coastal populations (Kurien, 2004; Jacquet and Pauly, 2007). What all this means is that small-scale fisheries have been put in a position where it is always 'playing defense'. To us, it seems as though the fisheries research community is firmly entrenched in the idea that small-scale fisheries, especially when juxtaposed with large-scale fisheries, is fading away and thus requires delicate protection and insistent revival efforts.

But what exactly are small-scale fisheries? Though we acknowledge the futility of attempting to define a label that is inherently relativistic and context-based, we believe a general characterization will help to guide our observations:

> Small-scale fisheries can be broadly characterized as a dynamic and evolving sector employing labour intensive harvesting, processing and distribution technologies to exploit marine and inland water fishery resources. The activities of this subsector, conducted full-time or part-time, or just seasonally, are often targeted on supplying fish and fishery products to local and domestic markets, and for subsistence consumption. Export-oriented production, however, has increased in many small-scale fisheries during the last one to two decades because of greater market integration and globalization... Small-scale fisheries operate at widely differing organizational levels ranging from self-employed single operators through informal micro-enterprises to formal sector businesses. This subsector, therefore, is not homogeneous within and across countries and regions... (FAO, 2005, p. 4)

The striking point about small-scale fisheries is the expansive breadth and depth of fishing operations and orientations it encompasses. Whether considering the commitment to fishing (i.e., both full-time and part-time), marketing mechanism (i.e., subsistence to global trade), formality of fishing enterprise (i.e., self-employed to incorporated), attached meaning (i.e., a source of cultural identity and community cohesion to a source of wage income and business opportunities), and spatial range of occurrence (i.e., marine and inland, and both in the South and the North), small-scale fisheries covers a wide spectrum and vast majority of fishing types and people, such that we wonder whether it might well be easier to define and isolate what is *not* small-scale fisheries. Given the prevalence of small-scale fisheries and the major contribution it makes to livelihoods and economies, we cannot help but feel curious as to why small-scale fisheries has to necessarily assume the defensive position. Is it possible to imagine and approach small-scale fisheries in a way that celebrates its universality, instead of regarding it as a 'weak sibling' who requires constant protection? Shedding the marginalized image of small-scale fisheries and imbuing it with a higher sense of faith and assertiveness seems consistent with the high merit attached to this sector. At the same time, adopting this vision may not change the fact that offshore, large-scale, industrial fishing fleets (one type of fishing that is unambiguously not small-scale) typically hold much political clout and wield substantial lobbying power. We figure there is a still clear and urgent need to promote small-scale fisheries and amplify the benefits it provides to communities, nations and the world. This is a conundrum that underlies our fundamental understanding of small-scale fisheries.

Polarized debates in fisheries

Related to the crisis rationale and 'under-threat' mentality is the polarizing views increasingly observed in fisheries. The scientific debate on the health status of global fisheries is perhaps the most prominently featured one. Propelled by the 'fishing down the marine food webs' thesis in the late 1990s (Pauly *et al.*, 1998), a series of scientific studies and arguments have been brought forward that warn us of the dire state of the world's fisheries (for example, see Worm *et al.*, 2006). On the other hand, alternate interpretations that claim to invalidate or trivialize such 'alarmist' views have argued that the state of current and future fisheries is not as bleak as the picture painted by Pauly and his colleagues, and instead point to a number of well-managed fisheries in various parts of the world (see Hilborn *et al.*, 2005; Essington *et al.*, 2006). This debate is often characterized as stemming from disciplinary differences between 'marine ecologists' and 'fisheries scientists' (Hilborn, 2007; Pauly, 2009; Salomon *et al.*, 2011). The very latest exchange of such polarized views, published in the blog of the Nature Conservancy (Cool Green Science) by the two leading figures of each 'camp', Pauly and Hilborn, shows no sign of reconciliation. So, to students like us, who are trained to embrace interdisciplinary ideals, it is not obvious why the debate remains so fierce. Even more perplexing is that these various opposing streams of scholars all seem to submit to the fundamental preamble that fisheries decline is real and consequential.

We observe that scientific debates are nothing new in fisheries, especially on issues relating to causes of fish stock collapse, farmed and wild salmon interactions, marine protected areas, just to name a few. In fact, often fascinating to follow, they offer not only the means through which knowledge gets contested and rejected to instigate paradigm shifts, but also how it gets clarified and accepted as a 'scientific truth'. The expanding global reach of many fisheries researchers, enabled by technological sophistication (e.g., in monitoring, mapping, and statistical computation), seem to have spun off elevated debates on numerous other issues. Fishing gear impact on ecosystems has created a heated discussion between conservation scientists and the fishing industry over towed fishing methods such as trawling and dredging (Malakoff, 1998; Watling and Norse, 1998; Collie *et al.*, 2000; Chuenpagdee *et al.*, 2003). Also, on the socioeconomic side, the merits of rights-based management and Individual Transferrable Quotas (ITQs) have been fiercely contested within and across the discipline of economics and the social sciences, respectively (Copes, 1986; Grafton *et al.*, 2006; Costello *et al.*, 2008; Bromley, 2009). Unfortunately, we find ourselves a bit disoriented amidst the many polarized debates in recent decades. These extremes put us in a delicate position where the range of middle ground is narrowed, while some of us may even feel compelled to take a side in order to align our research within the prevailing discourses. When many of the viewpoints in the debate seem to converge on the same overall vision for fisheries (e.g., abundant fish in the sea), it can be additionally confusing to students who are still trying to ascertain our own set of research positions. Perhaps this is why collaboration (Worm *et al.*, 2009) that attempted to reconcile the contrasting ideas on the status of the world's fisheries offers us a certain degree of 'comfort'.

We certainly do not doubt the scientific virtue or the societal implications of studies that attempt to accurately advance technical understanding of an issue. Students with a strong disciplinary aptitude may also welcome deeper findings on particular issues and enjoy the continuing debates. However, our understanding is that fisheries is an inherently complex field of study that entails multiple emphases and meanings, which draw upon marine biology, oceanography, economics, geography, sociology, political science, environmental ethics, anthropology, history, law, to name but a few. Much advantage can thus present itself if the various dimensions are linked to each other and approached as a whole (also see Ludwig *et al.*, 2001). We believe that the many polarized debates deeply-seated in disciplinary differences do not necessarily assist in our endeavor to see things in a more harmonious and synergistic manner. In this regard, we sometimes even wonder whether they are detracting our attention from engaging with the other central concerns in fisheries research, such as finding ways for small-scale fishers to better cope with ecological and societal changes, and to sustain and rebuild their economies and livelihoods. This is another area of concern for us as we try to crystallize our interdisciplinary research agendas and contribute to a better future for small-scale fisheries.

Social sciences and interdisplinary research

Another observation relates to the emphasis on the human aspect and the role of social sciences in fisheries research and management. While we are by and large comfortable with the increasing attention these topics receive in the mainstream fisheries literature, we note with confusion several aspects that do not quite align well with our trained thinking. The late Peter Larkin, a renowned fisheries ecologist, commented that the future of fisheries management lies in managing fishers, rather than in habitat protection or enhancement of fisheries production (Larkin, 1988). His argument is that the preferences and behavior of fishers is constant across space and time, thereby implying that people are easier to manage. Though innovative at the time, the idea that regulating fishers is a straightforward task, on the basis of universal and predictable preferences, seems somewhat out of touch with the present thinking. Much emerging evidence demonstrates the complexity of fishers' socioeconomic, cultural, aesthetic, and institutional motivations (see Pollnac and Poggie, 1978; Ostrom, 1990; Charles, 2001). Larkin's assumptions for highlighting the commonality of fishers are particularly problematic when viewed from the social science perspective, which sees fishing places and practices as socially and culturally constituted and embedded in local contexts. Recently, Hilborn (2007) reinforces the claim that fisheries is about managing fishers in his argument for incentive-driven fisheries management. There are many other observed cases in which studies of overfishing culminate in general recommendations that frame the issue as ultimately a human problem, rather than one of fish stocks or ecosystem dynamics. Furthermore, shortcomings in scientific research, either through high levels of uncertainty, resource and time constraints, institutional rigidities, and path dependency, have led to calls for the precautionary approach and ecosystem-based management with close consideration of human factors (FAO, 1995; Charles, 2002).

We feel this sentiment of managing humans instead of fish is now so widespread that this is no longer an epiphany, no longer an argument to be asserted or reaffirmed. This emphasis of the human aspect makes us feel a little perturbed. When we imagine fisheries, we are attracted to its diversity, the various parts that make up the whole fisheries system (i.e., 'oceans to plate'), as well as an array of diverse disciplines, each of which adds to the holistic picture of fisheries perspectives and logic. Being trained in interdisciplinary research allows us to see both humans and other ecosystem features as uniquely important. In fact, what makes more sense is the emphasis on *interactions* in wider nature-society settings, by which we mean mutually reinforcing relations among fish and fishers, various other actors, as well as the agency of nature and institutions. The focus is neither the action and motivation of fishers alone nor the dynamics of fish populations, but the exchanges between them that together influence how fisheries are shaped and carried out. We, thus, feel that such reorientation towards interactivity may help erase the redundancy conveyed by any one-sided claim that 'essentializes' fisheries management research.

A similar observation can be drawn with the role of social sciences in fisheries. Historically, fisheries management has mainly been dominated by the biological and economic sciences, as often reflected in the use of the Gordon-Schaefer bioeconomic model in the study of commercial large-scale fisheries (see Clark, 2006). In recent decades, however, it is increasingly recognized that fisheries research would benefit from a greater integration of the social sciences. This heightened interest in the relevance of social science research centers around wide-ranging topics such as institutions, social capital, vulnerability, gender, power relations, attitudes and beliefs, knowledge systems, culture, social justice, and well-being. This contemporary surge builds upon the rich set of anthropological and sociological studies including the works of Acheson, McCay, McGoodwin, Jentoft, and Ommer, to name a few. What sets the current trend apart from the earlier period, however, is not the rise in the degree of insightfulness in the research findings *per se*, as social science fisheries research has always had much to offer. Rather, in line with the widespread call for an interdisciplinary research platform, a concerted emphasis on the social dimension of fisheries research is being explicitly promoted and engaged in discussion by both social and natural science researchers (for example, see Clay and McGoodwin, 1995; Christie *et al.*, 2003; Mascia *et al.*, 2003; Pauly, 2006; Ommer *et al.*, 2008; Symes and Hoefnagel, 2010; Urquhart *et al.*, 2011). This tendency was also evident at the World Small-Scale Fisheries Congress[2] convened in Bangkok in October 2010, which featured a significant representation of social science research and viewpoints, achieving a good balance with ecosystem-based natural science studies.

To us, the importance of social science research is not a moot point. Throughout our training, we feel that we have been well doused in the merits of interdisciplinary work, and seeing a genuine cooperation and blending between divergent sets of research expertise as a norm. What troubles us, however, is that despite explicit concerns to promote social science studies as part of interdisciplinary work, we still observe much discipline-specific research that does not breach into new arenas. Ontological differences between the social sciences, and natural sciences and economics, appear too big of a hurdle for constructive engagement that bridges deductive, positivistic and quantitative approaches with those that are more qualitative and inductive (for details, see Jentoft, 2006; Urquhart *et al.*, 2011). For example, the focus of most social science studies still very much lie in locally-based inshore fisheries with poor attention to the mechanisms of industrial fleets, which remain largely in the domain of economists and biologists. Also, social scientists giving due consideration to natural scientists' suggestions (for example, Pauly, 2006) have been more of an exception than a regular practice. Similarly, natural scientists' analysis of socio-cultural constructs of leadership and social capital is often carried out without rethinking their ontological basis, but by simply following a positivistic and reductionist methodology that fails to fully appreciate the complex and subjective meanings attached to such concepts (for example, see Gutierrez *et al.*, 2011). Likewise, when interdisciplinary research is said to be practiced, we notice that it is about working on an issue of an interdisciplinary nature and merely recognizing the goodness of it, rather than fully embracing the disciplines outside one's comfort zone and genuinely making an amalgamating effort. We wonder whether this is simply a part of the time lag that occurs before interdisciplinary work takes root, or whether much of this will stay as rhetoric or 'fishful' thinking (Pitcher and Lam, 2010). It remains to be seen. In the meantime, we may find ourselves having to try harder to communicate to the wider audience (e.g., journal editors and the mainstream research community) about the merits of our work that relies on interdisciplinary, mixed-method, and loosely established ontological ground.

2 www.seafdec.or.th/wsfc2010

The results of a journal article scan on small-scale fisheries research provide a set of augmenting insights into the status of interdisciplinary cooperation. Using "ISI Web of Knowledge" and "Google Scholar" as search engines, we collected a total of 1717 peer-reviewed journal articles gathered from 350 English-based journals published in the period 1917 to 2009. Focusing on small-scale fisheries, the scan was based on the identification of search words ('small-scale', 'artisanal', 'traditional', 'subsistence' and 'inshore' with the combination of 'fishing' and 'fisher*') contained in the article title, abstract or keywords. The top 20 journals that featured the greatest number of small-scale fisheries articles according to these search criteria are listed in Table 1. Assuming that the journal in which an article is published reasonably reflects the disciplinary nature of that article (which is often part of the guidelines to authors), we find that the list is dominated by journals that exhibit a stronger receptiveness towards natural science research. Only five journals whose stated aims explicitly invite social sciences or interdisciplinary contributions appear in the list: Marine Policy, Ocean & Coastal Management, Environmental Conservation, Human Ecology, and Human Organization, while the other fifteen journals mainly hinge on biology, ecology, oceanography and other applied natural sciences. We regard this as another reminder that natural science work (and economics-based, in some cases) has occupied a predominant place in small-scale fisheries research, and that practical and structural hurdles remain in facilitating more robust disciplinary interactions.

Table 1 *Journals that have most frequently published small-scale fisheries articles and the number of published articles as identified in the scan*

Journal	**Number of articles**	**Journal**	**Number of articles**
Fisheries Research	176	Environmental Conservation*	24
Marine Policy*	87	Marine Ecology-Progress Series	24
Ocean & Coastal Management*	61	Biological Conservation	23
Fisheries Management and Ecology	38	Human Ecology*	23
Aquatic Living Resources	34	Journal of Applied Ichthyology	23
ICES Journal of Marine Science	31	Conservation Biology	22
Bulletin of Marine Science	26	Revista de Biologia Tropical	22
Canadian Journal of Fisheries Aquatic Science	26	Human Organization*	21
Scientia Marina	26	Reviews in Fish Biology and Fisheries	20
South African Journal of Marine Science	26	Journal of Fish Biology	19

* Journals that explicitly invite social science based or interdisciplinary contributions in their stated aims and scope

There are, however, several encouraging findings from the article scan. For instance, as part of this exercise, the first and the second authors of the articles were recorded.[3] This allowed identification of those who have contributed most to small-scale fisheries research in terms of the number of journal articles on which they are primary or secondary author. Table 2 lists 15 authors who were most frequently featured in the scan result. The list includes many prominent researchers such as Berkes, Beggosi, Béné, Ruddle, Johannes, and Charles, who are well known for their influential work; generally well-respected and referred to widely by varied audiences (natural, social and interdisciplinary scholars and practitioners) in the fisheries community. They have consistently crossed disciplinary, methodological and even in some cases ontological barriers by publishing in a broad range of journals of different scholarly bent, such as Fisheries Research, Human Ecology, Marine Policy, Ecology and Society, and Ocean & Coastal Management, among others. This demonstrates their flexibility and openness to engage in the type of research that can speak to wider readership and perhaps create a greater impact. We see this as an enormously admirable quality that motives us – a quality we also strive to sharpen as we look ahead in our academic career.

Table 2 Top contributing authors in the small-scale fisheries article scan with the number of authored articles

Author	Number of authored articles	Author	Number of authored articles
Fikret Berkes*	12	Christope Béné*	7
Alpina Beggosi*	10	Francisco Arreguín-Sánchez	6
Joshua Cinner	10	Robert Pomeroy	6
Omar Defeo	10	Kenneth Ruddle*	6
Tim McClanahan	9	Robert Johannes*	5
Joanna Burger	8	Karim Erzini	5
Juan Carlos Castilla	8	Anthony Charles*	4
Konstantinos Stergiou	8		

* Authors who have shown considerable capacity to cross disciplinary boundaries in small-scale fisheries research based on their authorship in a wide spectrum of journals

3 If more than three authors were listed, only the first author was recorded

Fisheries governance

In our graduate theses, we have attempted to highlight human-nature interactions, employing tools and techniques that go beyond the traditional disciplines in fisheries. Yet, our motivation for better understanding fisheries from holistic and interdisciplinary perspectives creates many challenges, as well as research opportunities. Our search for negotiating these challenges and also the observed inconsistencies has led us to explore the concept of governance, which may present an intriguing entry point to reorient the fisheries research agenda. According to Jessop (2003), interest in governance is a response to paradigm crises in the social sciences. His argument is that traditional disciplines that reflect the organization of modern industrial societies, such as economics focused on markets, political science on sovereign national states, and sociology on civil society, have become less relevant in dealing with contemporary issues. Thus, he says "governance is being introduced to bridge disciplines and to provide alternative ways of understanding" (p. 102). This viewpoint is especially pertinent as environmental issues now widely spill over to political and institutional domains. Similarly, another perspective of governance, being referred to as adaptive governance of social-ecological systems, also exhibits the integrating tendency by incorporating the broader social dimension into a coupled ecological framework of ecosystem-based management (Folke *et al.*, 2005). Generally attached with optimistic and synergistic connotations, such as partnership, network, problem-solving and co-operation (for example, see Kooiman *et al.*, 2005), however, governance is not without its own set of controversies. It comes with a multitude of possible meanings and usage; as Offe (2009, p. 554) contends, "when one refers to something as an instance of 'governance' one has not yet expressed much". Skepticism towards the theory and practice of governance is also linked with potential blindness for the question of power, its distribution and conflicts of interest veiled by its harmonizing rhetoric, and the neglect of social norms and value (and moral) commitments, which de-substantiates the *politics* claim of governance. Such requests to reconsider and refine the concept of governance are a critical, but welcoming gesture.

Our perspectives are particularly rooted in 'interactive governance' (Kooiman *et al.*, 2005), which conceptualizes the fisheries reality to be composed of interactions among various inherent and constructed components. We feel that it offers one promising roadmap towards fisheries research and policy that gives due consideration to the coherent synthesis of natural and social sciences. For example, species recovery and fisheries rebuilding are arguably best approached from a 'fish chain' framework supported by governance thinking (Khan and Neis, 2010). It necessitates an interdisciplinary endeavor that draws from fish biology, ecosystem dynamics, socioeconomics of fishing activities and livelihoods, policy and legislative requirements, and an ethical dimension concerning food security and the rights of future generations. Hence, what mechanisms prevent fish species from recovering and where in the 'chain' the blockage occurs are not questions that one angle of inspection can adequately address. It could be due to climate factors, loopholes in the policy, certain cultural habits of fishers, or a combination of all of the above. Governance-informed research, therefore, appears beneficial in explaining the intricate and complex causal effects and confounding factors that are latent in the rebuilding process.

The potential of governance research for small-scale fisheries is further validated by the findings of the journal article scan of small-scale fisheries research topics, a component of the same scan described earlier. The research topics identified in each journal article were framed in two ways – as 'issues' or 'solutions' – depending on the way the article approached the topics. Issues were defined as research gaps or specific data needs that an article intends to address, while solutions were regarded as central topics that are elaborated and justified throughout the article in response to certain issues, rather than mere recommendations or suggestions tacked on at the end of the article. Eight broad research themes surfaced in the scan, one of which concerns governance-related topics. This theme focuses on representation and participation in governance through various arrangements such as community-based management, co-management, traditional institutions, spatial planning and other integrated efforts. The theme also seeks to draw attention to the diversity of interests within small-scale fisheries (including women and future generations), organizational aspects such as cooperatives, multiple geographical and governing scales and linkages, and the diverse governance perceptions, visions and values associated with the sector. All eight themes are sequentially listed in Table 3 based on the tally of how many times they were framed as 'solutions' in the entire scan. The governance theme was prominently featured in small-scale fisheries literature both as 'issues' to be addressed and as 'solutions' to be explored. We feel encouraged to discover the strong interest in governance research among many fisheries researchers over time. This affirmation represents a positive outcome for us in utilizing governance research to study and promote small-scale fisheries.

*Table 3 Research themes identified in the small-scale fisheries research scan**

Research themes	Frequency of being framed as 'solutions'	Frequency of being framed as 'issues'
Governance	*87*	*410*
Fisheries information, knowledge and research	85	586
Fisheries development, livelihoods, well-being, and integration	59	405
Access, rights and responsibility	14	79
Ecosystem complexity, dynamics and interactions	4	552
Poverty, marginalization, vulnerability	3	69
Post-harvest processes and globalized markets	3	57
Overexploitation, gear conflict, resource/space competition and rule compliance	2	216

* The frequency number is out of 1717 articles, and identification of multiple themes in one article was allowed

Conclusion

We learn a great deal each day in our graduate training. Reading articles, mentorship by supervisors, attending conferences, listening to talks and discussing with peers all add to our understanding of fisheries realities by increments. However, it also highlights elements of uncertainty, as we witness the inconsistencies in fisheries research. What we are taught is sometimes at odds with what we observe, and what we perceive to be dominant trends in fisheries research. As students who are yet to be equipped with a robust and refined sense of academic standpoints, philosophy or identity, we can be easily swayed by various interpretations and views, and slight contradictions can seem bewildering. This essay was a small attempt to flesh out, reflect on and make sense of these inconsistencies, with particular regard to small-scale fisheries and interdisciplinary thinking. A writing process certainly presents one way of streamlining one's ideas and a chance to further learn from them. We feel fortunate to be given the chance to do so in this volume, and to engage in this dialogue.

Borrowing the Aristotelian language of episteme to predominantly represent fisheries natural science, phronesis to refer to fisheries social science, and techne to signify fisheries management or practical management solutions, Jentoft (2006, p.678) comments that "it would help a great deal if fisheries education were to become more multidisciplinary and interactive across disciplines, with students inducted into the episteme, phronesis and techne of fisheries research and management." We are motivated by such sentiment to further engage in and promote interdisciplinary interactions.

Though recognizing that it will need to be accompanied by better facilitating institutional frameworks and a culture of genuine encouragement, there is a high sense of optimism. On our part, exploring this possibility through governance research, we look forward to an insightful academic journey as we hope to make meaningful contributions down the road towards creating healthier fisheries.

Acknowledgment

We thank our supervisor, Ratana Chuenpagdee, for encouraging our interest and resolve in writing this essay. Both of us receive graduate student fellowships from the Social Sciences and Humanities Research Council of Canada.

References

Andrew, N. L., Béné, C., Hall, S. J., Allison, E. H., Heck, S. and Ratner, B. D. (2007) Diagnosis and management of small-scale fisheries in developing countries. *Fish and Fisheries*, 8, 227-240.

Bailey, C. (1988) The political economy of fisheries development in the third world. *Agriculture and Human Values*, 5(1-2), 35-48.

Béné, C., Macfadyen, G. and Allison, E. H. (2006) *Increasing the Contribution of Small-Scale Fisheries to Poverty Alleviation and Food Security*. FAO Fisheries Technical Paper 481. Rome: FAO.

Berkes, F., Mahon, R., and McConney, P., Pollnac, R. and Pomeroy, R. (2001) *Managing Small-scale Fisheries: Alternative Directions and Methods*. Ottawa: International Development Research Centre.

Berleant-Schiller, R. (1981) Development proposals and small-scale fishing in the Caribbean. *Human Organization*, 40(3), 221-230.

Bromley, D. W. (2009) Abdicating responsibility: the deceits of fisheries policy. *Fisheries*, 34(6), 280-290.

Charles, A. T. (2001) *Sustainable Fishery Systems*. Oxford: Blackwell Science.

Charles, A. T. (2002) The Precautionary Approach and 'Burden of Proof' challenges in fishery management. *Bulletin of Marine Science*, 70(2), 683-694.

Chauveau, J. P. and Samba, A. (1989) Market development, government interventions and the dynamics of the small-scale fishing sector – an historical-perspective of the Senegalese case. *Development and Change*, 20(4), 599-620.

Christie, P., McCay, B. J., Miller, M. L., Lowe, C., White, A. T., Stoffle, R., Fluharty, D. L., McManus, L. T., Chuenpagdee, R., Pomeroy, C., Suman, D. O., Blount, B. G., Huppert, D., Eisma, R. V., Oracion, E., Lowry, K. and Pollnac, R. B. (2003) Toward developing a complete understanding: a social science research agenda for marine protected areas. *Fisheries*, 28(12), 22-26.

Chuenpagdee, R., Morgan, L. E., Maxwell, S. M., Norse, E. A. and Pauly, D. (2003) Shifting gears: assessing collateral impacts of fishing methods in US waters. *Frontiers in Ecology and the Environment*, 1, 517-524.

Chuenpagdee, R. and Pauly, D. (2006) Small is beautiful? A database approach for global assessment of small-scale fisheries: preliminary results and hypotheses. In Nielsen, J., Dodson, J. J., Friedland, K., Hamon, T. R., Musick, J. and Verspoor, E., eds., *Reconciling Fisheries with Conservation. Proceedings of the 4th World Fisheries Congress*, Vancouver, May 3-6, 2004. Bethesda: American Fisheries Society, pp. 587-595.

Clark, C. W. (2006) *The Worldwide Crisis in Fisheries: Economic Models and Human Behaviour*. Cambridge: Cambridge University Press.

Clay, P. M. and McGoodwin, J. R. (1995) Utilizing social sciences in fisheries management. *Aquatic Living Resources*, 8, 203-207.

Collie, J. S., Hall, S. J., Kaiser, M. J. and Poiner, I. R. (2000) A quantitative analysis of fishing impacts on shelf-sea benthos. *Journal of Animal Ecology*, 69(5), 785-798.

Copes, P. (1986) A critical review of the individual quota as a device in fisheries management. *Land Economics*, 62, 278-291.

Costello, C., Gaines, S. D. and Lynham, J. (2008) Can catch shares prevent fisheries collapse? *Science*, 321,1678-1681.

David, G. and Cillaurren, E. (1992) National fisheries development policy for coastal waters, small-scale village fishing, and food self-reliance in Vanuatu. *Man and Culture in Oceania*, 8, 35-58.

Essington, T. E., Beaudreau, A. H. and Wiedenmann, J. (2006) Fishing through marine food webs. *Proceedings of the National Academy of Sciences of the USA*, 103, 3171-3175.

FAO (1995) Code of Conduct for Responsible Fisheries. Rome: FAO.

FAO (2005) Increasing the contribution of small-scale fisheries to poverty alleviation and food security. FAO Technical Guidelines for Responsible Fisheries 10. Rome: FAO.

Flaherty, M. S. and Karnjanakesorn, C. (1993) Commercial and subsistence fisheries conflicts in the Gulf of Thailand – the case of squid trap fishers. *Applied Geography*, 13(3), 243-258.

Folke, C., Hahn, T., Olsson, P. and Norberg, J. (2005) Adaptive governance of social-ecological systems. *Annual Review of Environment and Resources*, 30, 441-473.

Garcia, S. M. and Grainger, R. J. R. (2005) Gloom and doom? The future of marine capture fisheries. *Phil. Trans. R. Soc. B.*, 360, 21-46.

Grafton, R. Q., Arnason, R. Bjørndal, T., Campbell, D., Campbell, H. F., Clark, C. W., Connor, R., Dupont, D. P., Hannesson, R., Hillborn, R., Kirkley, J. E., Kompas, T., Lane, D. E., Munro, G. R., Pascoe, S., Squires, D., Steinshamn, S. I., Turris, B. R. and Weninger, Q. (2006) Incentive-based approaches to sustainable fisheries. *Canadian Journal of Fisheries and Aquatic Sciences*, 63, 699-710.

Gutierrez, N. L., Hilborn, R. and Defeo, O. (2011) Leadership, social capital and incentives promote successful fisheries. *Nature*, 470, 386-389.

Hilborn, R. (2007) Moving to sustainability by learning from successful fisheries. *Ambio*, 36, 296-303.

Hilborn, R., Orensanz, J. M. and Parma, A. M. (2005) Institutions, incentives and the future of fisheries. *Phil. Trans. R. Soc. B*, 360, 47-57.

Jacquet, J. L. and Pauly, D. (2007) The rise of seafood awareness campaigns in an era of collapsing fisheries. *Marine Policy*, 31, 308-313.

Jentoft, S. (2006) Beyond fisheries management: the *Phronetic* dimension. *Marine Policy*, 30, 671-680.

Jessop, B. (2003) Governance and meta-governance: on reflexivity, requisite variety and requisite irony. In Bang, H. P., ed., *Governance as Social and Political Communication*. Manchester: Manchester University Press, pp. 101-116.

Khan, A. and Neis, B. (2010) The rebuilding imperative in fisheries: clumsy solutions for a wicked problem? *Progress in Oceanography*, 87(1-4), 347-356.

Kooiman, J., Bavinck, M., Jentoft, S. and Pullin, R., eds. (2005) *Fish for Life: Interactive Governance for Fisheries*. Amsterdam: Amsterdam University Press.

Kurien, J. (2004) Responsible fish trade and food security: towards understanding the relationship between international fish trade and food security. Report of the Study on the Impact of International Trade in Fishery Products on Food Security. Food and Agricultural Organization of the United Nations and the Royal Norwegian Ministry of Foreign Affairs. Rome: FAO.

Larkin, P. A. (1988) The future of fisheries management – managing the fishermen. *Fisheries*, 13(1), 3-9.

Ludwig, D., Mangel, M. and Haddad, B. (2001) Ecology, conservation and public policy. *Annual Review of Ecology and Systematics*, 32, 481-517.

Malakoff, D. (1998) Papers posit grave impact of trawling. *Science*, 282, 2168.

Mascia, M. B., Brosius, J. P, Dobson, T. A., Forbes, B. C., Horowitz, L., McKean, M. A. and Turner, N. J. (2003) Conservation and the social sciences. *Conservation Biology*, 17(3), 649-650.

Offe, C. (2009) Governance: an "empty signifier"? *Constellations*, 16(4), 550-562.

Ommer, R., Perry, R. and Neis, B. (2008) Bridging the gap between social and natural fisheries science: why is this necessary and how can it be done? In Nielsen, J., Dodson, J. J., Friedland, K., Hamon, T. R., Musick, J. and Verspoor, E., eds., *Reconciling Fisheries with Conservation. Proceedings of the 4th World Fisheries Congress*, Vancouver, May 3-6, 2004. Bethesda: American Fisheries Society, pp. 177-185.

Ostrom, E. (1990) *Governing the Commons: The Evolution of Institutions for Collective Action*. Cambridge: Cambridge University Press.

Pauly, D. (1997) Small-scale fisheries in the tropics: marginality, marginalization and some implication for fisheries management. In Pikitch, E. K., Huppert, D. D. and Sissenwine, M. P., eds., *Global Trends: Fisheries Management*. Bethesda, Maryland: American Fisheries Society Symposium 20, pp. 40-49.

Pauly, D. (2006) Major trends in small-scale marine fisheries, with emphasis on developing countries, and some implications for the social sciences. *Maritime Studies (MAST)*, 4(2), 7-22.

Pauly, D. (2009). Aquacalypse now: the end of fish. *The New Republic*, 240, 24-27.

Pauly, D., Christensen, V., Dalsgaard, J., Froese, R. and Torres Jr., F. (1998) Fishing down marine food webs. *Science*, 279, 860-863.

Pitcher, T. J. and Lam, M. E. (2010) Fishful thinking: rhetoric, reality, and the sea before us. *Ecology and Society*, 15(2), 12.

Pollnac, R. and Poggie, J. J. (1978) Economic gratification orientations amongst small-scale fishermen in Panama and Puerto Rico. *Human Organization*, 37, 355-367.

Salomon, A. K., Gaichas, S. K., Jensen, O. P., Agostini, V. N., Sloan, N. A., Rice, J., McClanahan, T. R., Ruckelshaus, M. H., Levin, P. S., Dulvy, N. K. and Babcock, E. A. (2011) Bridging the divide between fisheries and marine conservation science. *Bulletin of Marine Science*, 87(2), 251-274.

Sinclair, P. (1985) *From Traps to Draggers: Domestic Commodity Production in Northwest Newfoundland 1850-1982*. St. John's: ISER Books.

Symes, D. and Hoefnagel, E. (2010) Fisheries policy, research and the social sciences in Europe: challenges for the 21st century. *Marine Policy*, 34, 268-275.

Urquhart, J., Acott, T., Reed, M. and Courtney, P. (2011) Setting an agenda for social science research in fisheries policy in Northern Europe. *Fisheries Research*, 108, 240-247.

Watling, L. and Norse, E. A. (1998) Disturbance of the seabed by mobile fishing gear: a comparison to forest clearcutting. *Conservation Biology*, 12(6), 1180-1197.

Worm, B., Barbier, E. B., Beaumont, N., Duffy, J. E., Folke, C., Halpern, B. S., Jackson, J. B. C., Lotze, H. K., Micheli, F., Palumbi, S. R., Sala, E., Selkoe, K. A., Stachowicz, J. J. and Watson, R. (2006) Impacts of biodiversity loss on ocean ecosystem services. *Science*, 314, 787-790.

Worm, B., Hilborn, R., Baum, J. K., Branch, T. A., Collie, J. S., Costello, C., Fogarty, M. J., Fulton, E. A., Hutchings, J. A., Jennings, S., Jensen, O. P., Lotze, H. K., Mace, P. M., McClanahan, T. R., Minto, C., Palumbi, S. R., Parma, A. M., Ricard, D., Rosenberg, A. A., Watson, R. and Zeller, D. (2009) Rebuilding global fisheries. *Science*, 325, 578-585.

CHAPTER 23

What is this thing called 'community' good for?

Henrik Angerbrandt[1], *Lars Lindström and Maricela de la Torre-Castro*

The community-based and community-driven approach as key to successful small-scale fisheries management conceptually situates the terms 'local' and 'community'. We argue for a relational understanding of scale and for a conception of community in line with Tönnies' distinction between Community (Gemeinschaft) and Society (Gesellschaft), highlighting suppressive as well as affirmative aspects, which the predominant notion – conceptualizing community as a homogeneous place with shared interests – leaves out. Community may provide shared norms and affinity, but also non-democratic elements, making ideas of community-based and community-driven management less straightforward, suggesting a politicization of small-scale fisheries management.

Introduction

The term 'community' figures prominently in the discourse on natural resource management, not the least within small-scale fisheries. It claims that if management takes place within institutional contexts where the community has some substantial say, or where management is left to the 'community' altogether, many of the problems with sustainability of the resource itself, and also of the resource users, will be solved. 'Communities' are also predominantly articulated with the 'local' geographical 'scale'. This seems to be part and parcel of dominant theories of decentralization which emphasize the 'local'. By moving the state closer to the people, the argument goes, local preferences will be met, and accountability and participation will increase (World Bank, 1997). The ideas also fit well within the dominant governance discourse, with its emphasis on a change from hierarchical modes of coordination – govern*ment* – to more horizontal modes – govern*ance*, often ex-

1 Corresponding author: Stockholm University, Sweden, henrik.angerbrandt@statsvet.su.se

pressed in normative terms: governance ought to be. The discourse is articulated with concepts such as governance and management with adjectives. The World Bank's "good governance," "global governance," (FAO, 2006), "earth governance," (Biermann, 2007), "adaptive governance," (e.g., Folke, Hahn, Olsson and Norberg, 2005) "interactive governance," (e.g., Kooiman, Bavink, Jentoft and Pullin, 2005) and "sustainable governance," (e.g., Agrawal, 2001) are cases in point. At the level of management, "co-management," (Jentoft, McCay and Wilson, 1998; Jentoft, 2000b; Carlsson and Berkes, 2005; Jentoft, 2005) "adaptive management," (e.g., Walters, 1986) and "adaptive co-management" (e.g., Olsson, Folke, Galaz, Hahn and Schultz, 2007) have all flooded the conceptual market. Lately, the role of the community in small-scale fisheries management has been particularly emphasized in ideas of community-based or community-driven management (Khan, Mikkola, and Brummett, 2004), and even portrayed as the missing link of fisheries management (Jentoft, 2000a).

But what is this thing called 'community'? What are the links between 'community' and 'the local'? What good do 'communities' do? What is their track record? Empirical analyses have rarely lent support to the ideas of the positive effects of community-driven or community-based management. Results are inconclusive (Agrawal and Gibson 2001; Mansuri and Rao, 2004). Why is that?

We argue that there is a weak conceptualization of 'community' and of 'local', and that there are a host of assumptions tied to the terms that have not really been assessed. The 'community' concept is generally seen as unproblematic, connoting shared norms or a certain area or place (Agrawal and Gibson, 2001; Mansuri and Rao, 2004; Blaikie, 2006). However, communities may be characterized by conflicts constituted in asymmetrical power relationships (Jentoft, 2000a) as well as stretching beyond a certain place. We suggest a relational conceptualization of scale which questions the inherent positive effects of community-based and community-driven management, making it less straightforward. We also propose a view on 'scale' which does not equate community as 'the most local scale'.

We offer a twofold conceptualization of 'community'. On the one hand, community can be empirically analyzed in terms of actors, interactions and institutions, as suggested by Agrawal and Gibson (1999). This means that communities are not confined to a certain scale but need to be contextualized, and considered on the merits of the projects pursued by the different, more or less influential, actors. On the other hand, there is a theoretical aspect of the concept that relates to Tönnies distinction between Community (*Gemeinschaft*) and Society (*Gesellschaft*). This distinction highlights issues of authority, logics of practice, and suppressive as well as affirmative aspects in different social contexts. But, disconnecting community from scale and instead conceiving it as a social system based on culture, religion and tradition (*Gemeinschaft*), in contrast to Society (*Gesellschaft*) that is connected to the state, public opinion and the market, also makes the emphasis on community-based or community-driven management problematic. Even though ethnicity and religion are fundamental community values, they are in turn structured by and articulated with a variety of cultures and traditions. In small-scale fisheries, for example, this may be illustrated with the ways in which communities become structured according to the gears used and species targeted.

Local 'place' and 'scale'

There is a tendency to imagine 'place' as a territorial unit of *local* relations, in opposition to spaces of *global* relations (Massey, 2004). The distinction between good and benign relations within localities, in contrast to bad and external global relations, stems from an ontological separation of place – seen as 'here' and intimate – and space – seen as 'out there' and intrusive. For community-based or community-driven management to be effective, the rationale includes a view of the local as a place isolated from other relations. Instead of seeing localities as spatially circumscribed places beyond which actors have no real influence, we may think of localities as bringing together different scales of practices and processes. That is, place may be thought of as a site of situated practices rather than as a context or setting (Amin, 2002, p. 391). Fisheries management articulates local, national and global scalar dimensions (Mansfield 2001, 2005).

The idea of space and place as separate geographical sites or realms can accordingly be contested and allow for "the ontological presence of both the proximate and the remote at the same geographical level" (Amin, 2002, p. 389). Taylor (1999) defines space as an abstract realm which is politically disenabling, in contrast to place as a lived realm which is politically enabling; a 'place-space tension', which to a large extent coincides with Tönnies' (2001) distinction between *Gemeinschaft* and *Gesellschaft*. In Tönnies' account, however, there are different logics determining practices in the different forms of relationships, thus making them both enabling and disenabling, but for different issues. We will return to this discussion below, as we argue that this distinction is implicitly present in the discourse of the 'local' although its implications are not assessed.

There are problems with regarding place as 'real', 'grounded', 'lived' and more 'meaningful' than space, seen as abstract. With a relational conception of the world, Massey (2004) argues that our lived reality, which is ascribed to place, is in fact not confined to the local but dispersed in its sources and its repercussions. Space and place should not be understood as antagonistic. Instead, places should be regarded as sites of interaction which are formed by particular articulations of social relations – be they local or connections that stretch beyond the local – and with all connections embedded in complex histories, making places open, porous and hybrid (Massey, 1999; Amin, 2002). What is considered local is the product of diverse spatial practices. In all, this perspective challenges the idea of the local place as isolated, possessing some inherent (good) values and as a homogenous sphere with shared interests, attained through deliberation and partnership. We suggest a topological notion of place politics that connects place to space. This is not to say that changes in the institutional organization of the state do not have implications for the abilities of different actors to pursue their political projects. However, it would be a mistake to conclude that an ongoing 'politics of scale', where actors mobilize around a particular spatial representation, means that the *actual* spatial forms of governance are also contained within the boundaries of that spatial representation (Allen and Cochrane, 2007). Spatial restructurings need to be recognized as political processes in which various actors are empowered and disempowered, and it is the character of the political projects pursued by these actors that is of importance.

Moore (2008) argues that mistaking scalar categories for ontological objects stems from confusing the categories used by social scientists as actually existing entities, making ontology of epistemology, such that geographical scales become tools that actors use for different reasons as well as categories used by social scientists to analyze (the same) processes. Thus, it is crucial to make the distinction between scale as a category of practice and as a category of analysis. Rather than treating scale as a starting point, the ways in which scale shapes social life is an open question that needs to be empirically addressed.

Community

Although at the centre of contemporary development thinking, the concept of community has been given little reflection. Communities are predominantly conceptualized as territorially fixed, small, and culturally and politically homogenous (Agrawal and Gibson, 1999; Mansuri and Rao, 2004). These characteristics are supposed to be favorable for certain outcomes. The dominant natural resource management discourse connects community to desirable resource use and conservation and to ideas about poverty reduction; community is associated with enhanced sustainability, inclusiveness, empowerment, building social capital, etc. The propositions of a community having all these effects are, however, not much studied. On the contrary, there seems to be little evidence of the assumed features of community actually leading to the desired outcomes. As Agrawal and Gibson (1999, p. 636) write: "In fact, some community characteristics considered important to collective action may actually thwart conservation efforts". There are problems with all three ascribed features of community: territoriality, size and homogeneity. The notion that a certain territorially fixed area would have some intrinsic significance is reasonably dismissed with a developed idea about place as argued above. Further, small sized groups may have limited ability to defend their resources or to influence processes that shape their conditions, and the focus on homogeneity tends to obscure asymmetrical social relations, as well their scalar dimensions.

A study of boat captains in three small-scale fishing 'communities' on the east coast of Zanzibar, Tanzania, showed that they were far from homogeneous in their opinions and attitudes, either within or between 'communities', and that their differences correlated strongly with fishing gear (de la Torre-Castro and Lindström, 2008). Conflicts over gears and traditional fishing grounds have even resulted in fatalities (de la Torre-Castro and Lindström, 2010). Similarly, Berkes (1985) observes that small communities may be unpleasant places for those who violate local rules and norms.

We argue that there is a mix-up of empirical and theoretical uses of the term community in the current discourse. The conceptualization of community as an empirical notion is merged with the analytical concept of community as an ideal social system, in line with Tönnies' distinction between *Gemeinschaft* (community) and *Gesellschaft* (society). A distinction between empirical and theoretical uses of the term can help us understand different aspects of community rather than making assumptions about its beneficial qualities.

Actors, interactions and institutions

An empirical understanding of community seldom gives a clear indication as to how boundaries are drawn. Ultimately, what is labelled as a 'community' is "often an endogenous construct defined by the parameters of a project, by project facilitators, or by the nature of administrative or identity boundaries" (Mansuri and Rao, 2004, p. 8). It is typically seen as being on par with 'the local'. An alternative is to focus on actors, interactions and institutions in communities (Agrawal and Gibson, 1999). A focus on actors and interests clearly challenges the notion of community as homogenous, obscuring local structures of economic and social power. Consequently, there is need to direct the attention to the politics of the local. Decentralization, for example, does not imply that 'the community' gains power at the expense of the central government, but that certain actors and projects may be empowered to the disadvantage of competing ones. In line with the scalar discussion above, local interactions need to be seen as essentially multi-scalar. Different initiatives and processes have different scalar configurations, which may also be contested through the interactions. Institutions articulate power relations that define interactions among actors at the same time as they structure the interactions that take place around different kinds of resources. These are constantly renegotiated, sometimes explicitly, but more often implicitly.

A focus on actors and institutions challenges the essentialism of community that is prevalent in the dominant management discourse, and makes it possible to concentrate on actual practices and projects that are more or less inclined to fulfil the objectives of sustainability. Community in itself gives little help in predicting the outcomes of community-based or community-driven management projects. It is necessary to recognize the actual exercise of authority and modes of governance in local politics, which include not only formal state institutions but also various other actors such as so-called traditional institutions, vigilante groups, religious organizations etc. (Lund, 2006). Associations claiming to represent a community are not necessarily inclined to increase the inclusiveness of those living in poverty, and can instead function as a means of exclusion. They also tend to lack local legitimacy and to depend on non-local actors for resources (Dill, 2009). The critical point is to recognize that local politics, far from being characterized by consensus agreements in a 'community spirit', is fraught with contestations over local as well as non-local power and resources, both within and among communities, and within which communities are being (re-)produced and defined as categories of practice.

Gemeinschaft or gesellschaft

Turning to the analytical dimension of the concept, Tönnies' distinction between Community (*Gemeinschaft*) and Society (*Gesellschaft*) as contrasting systems of collective social order provides a framework for analyzing different kinds of social formations, forms of authority and of actions. While Community refers to an ideal type of group with intrinsic and non-logical values which rest on affective and emotional elements, Society stands for deliberately formed associations in support of rational achievement of mutual goals (Tönnies, 2001). There are two different logics guiding agency, generating two different kinds of social systems. According to Tönnies, practice is guided by 'natural will' (*Wesenwille*) in Community relations. He identifies three different forms of natural will: preference, habit and memory, which are all related to previous experiences and, thus, the natural will establishes a close connection between Community and traditionalism. The natural will resembles Bourdieu's (1990) notion of *habitus*, which denotes the internalized norms and understandings which make people act and respond in certain ways. Just as *habitus* guides a person on how to act in different situations, and generates the individual's common sense behaviour, natural will makes thought and action one and indivisible. As pointed out by Tönnies (2001), decisions tend to be unreflective.

In Society, practice is guided by 'calculative will' (*Kürwille*), which involves analytically informed decision-making, unfettered freedom and rational choice. Thus, calculative will needs to be understood with reference to developments in the future. It is concerned with rational calculation and is the product of thought. Tönnies also specifies three different forms of calculative will: deliberation, unfettered choice and conception. These are all forward-looking and concern issues such as weighing opportunities and unpleasant experiences against each other.

As stated above, there is a tendency to associate the local with community and the global with society, though seldom explicitly. This trend has been interpreted in accordance with a Society logic only, as is the case with the 'space-place tension' (Taylor, 1999), referred to above. The different logics guiding practice in the different types of relationships have not been recognized, and neither have the tensions these different types of relationships give rise to. The institution of *bwana diko*, beach recorder, found in Zanzibar, Tanzania, illustrates this by recognizing both community and society, as it is constituted in both the natural will (through tradition and custom) and calculative will (through reason and rational choice), and by articulating different scales, as it is regulated both at local and national levels (de la Torre-Castro, 2006). The persons functioning as *bwana dikos* find themselves entangled in a web of social relations with community and society articulated at different scales. The *bwana diko* is caught in a dilemma between reporting members of his/her own 'community' in which norms and traditions legitimize the use of nets but which, from the viewpoint of society, have been declared illegal (for similar arguments, see Barrett 1991).

We find that many of the values advocated in decentralization theories and in notions of the local are attached to community, while issues relating to politics and decision-making are part of society. Mutual understanding and consensus is the norm in community, and acts are more expressions of attitudes than means to an end. Society, in contrast, is instrumental and based on convention, in the form of political legislation. People are driven by ambition and consciousness.

A central aspect of the distinction between Community and Society is the contradictory forms of authority. Patriarchy is the foundation of authority in Community, and the idea of a paternal authority is transferred to community leaders and forms the basis for claims of a historically rooted authority (Tönnies 2001).

The so-called traditional institutions in African countries may not always be so traditional, as they are constantly negotiated and reconstructed over time. But even if communities are 'fictive' or 'imagined', it is the way in which people perceive their relations that is of importance. Conversely, authority in Society rests in public institutions with specialized functions, where the state is the most general and universal embodiment of Society (Tönnies, 2001). This follows from the fact that, in Society, control belongs to the individual and is transferred to the state by convention. Studies have come to the conclusion that the power of so-called traditional authorities has been strengthened by decentralization. They have been given greater freedom to pursue their agendas, and decentralized governments tend to depend on traditional authorities to implement regulations. Studies have also shown that traditional authorities mediate conflicts and control access to resources (Bene, Belal, Baba, Ovie, Raji, *et al.*, 2009; Andrae, 2010). Accordingly, increased focus on community and decentralization seems to strengthen Community authority at the expense of public institutions and parliamentary politics. However, it would be misleading to regard the relationship between community and society as a zero-sum game where, for example, increased modernization automatically means fading traditionalism or vice versa. On the contrary, in many cases we can see traditional values being strengthened in processes of economic and political modernization. The two types of socialities, community and society coexist, and a social group for some can be seen as community and for others society, depending on perspective. The ideal types are two aspects of all human group life, not two different kinds of group life (Gusfield, 1975)

Deliberation is the style of thought in Society as it is connected to the calculative will. In Community, conscience takes the corresponding role. Shame or modesty is, furthermore, the most profound form of conscience; it restricts people and makes them reluctant to voice certain thoughts or ideas (Tönnies, 2001). Community participation may, thus, make it difficult to voice demands. In addition, the logic of homogeneous communities entails little incentives for participation, since the whole community is seen as sharing interests vis-à-vis 'the outside' or 'higher levels'. This fits well with community leaders as the embodiment of the community and the (more or less) legitimate representatives of the group. In society, however, people's acts are means to an end, fulfilling certain objectives in line with individual ambitions and goals. Thus, participating in and trying to influence development projects and natural resource management may be a way to achieve this. How to get people to do so is a question that directs our attention to interactions and institutions; that is, the empirical perspective of community. In that endeavor, community norms may constrain popular participation (Dill, 2010). There is, in short, a need to recognize the multidimensional aspects of the community concept, as Ostrom (2001, p. xi) writes concerning gendered community relations: When community is seen as imposing traditional values and ways of life on members, then the presence of community can be viewed as itself oppressive for women. If, on the other hand, community is viewed as a protection against the inequality fostered by market relations, then community may be a mechanism for shielding women against unfair practices and exploitation.

Politicising small-scale fisheries management: embedding 'community' in 'society'

The assumption that communities are homogeneous and characterised by harmony through common traditions and customs seems difficult to uphold. Studies of community-driven and community-based management are inconclusive as to their positive outcome in terms of sustainability of small-scale fisheries. As argued above, communities may be emancipating as well as retrenching, homogeneous and heterogeneous, and may be characterized by harmony as well as bereft with conflicts. Devolution of management to the community level may thus run the risk of reproducing and even reinforcing asymmetrical power relations (Mohan and Stokke, 2000), also in small-scale fisheries (Bailey and Davis, 1995). This is particularly risky as communities, unlike societies, normally lack institutions of accountability, transparency, and inclusiveness, and representation is not straightforward. The institutional setup at the community level may resemble Village Republics (Wade, 1994), as the traditional authority may be illegitimate and undemocratic (Jentoft 2005). Power over management decisions may be in the hands of an elite, or a 'community within the community' which may, for example, favor the use of certain fishing gears; some of which may be destructive as they are non-discriminatory in terms of fish species and size, hence threatening sustainability.

Most of the countries in the developing world fare low on the indicators of democracy, many of them being authoritarian (Economist Intelligence Unit, 2008) with a corporatist, rather than integrative, mode of inclusion (Törnquist 2004). The power relations at play with its politics of inclusion and exclusion jeopardize citizenship and unequally structure access to and influence over management decisions, with 'communities' and 'the local' being instances at which asymmetrical social relations are expressed. Non-democratic and authoritarian practices may thus be entrenched. The apparent risk of devolution as depoliticization has to be removed by the politicization of small-scale fisheries management through the establishment of institutions of democracy based on integrative modes and on citizenship: community needs to be embedded in society.

So what is this thing called community good for? We're not dismissing the idea that community-based or –driven small scale fisheries management may promote sustainability. We just want to warn against the sometime idealizing and even idolizing conceptualizations of community and with them normative claims which are not always empirically substantiated. Communities need to be embedded in political institutions of democracy which makes it possible to redress asymmetrical community relationships in order to secure equity and sustainability.

References

Agrawal, A. (2001) Common property institutions and sustainable governance of resources. *World Development*, 29(10), 1649-1672.

Agrawal, A. and Gibson, C. C. (1999) Enchantment and disenchantment: the role of community in natural resource conservation. *World Development*, 27(4), 629-649.

Agrawal, A. and Gibson, C. C. (2001) *Communities and the environment: ethnicity, gender, and the state in community-based conservation*. New Brunswick, N.J.: Rutgers University Press.

Allen, J. and Cochrane, A. (2007) Beyond the territorial fix: regional assemblages, politics and power. *Regional Studies,* 41(9), 1161-1175.

Amin, A. (2002) Spatialities of globalisation. *Environment and Planning A,* 34(3), 385-399.

Andrae, G. (2010) De-scaling without democratic substance: water provisioning in peripheral Kano, Nigeria. In Hagberg, S., ed., *Inventing and Mobilising the Local.* APAD Bulletin 31-32. Niamey, Niger: APAD.

Bailey, C. and Davis, A. (1995) Common in custom, uncommon in advantage: common property, local elites, and alternative approaches to fisheries management. *Society and Natural Resources,* 9(3), 251-265.

Baland, J. M. and Platteau, J. P. (1999). The ambiguous impact of inequality on local resource management. *World Development,* 27(5), 773-788.

Barrett, G. (1991) The fish pot pan: reef overfishing and state management in Bermuda. *Maritime Anthropological Studies,* 4(2), 17-39.

Bene, C., Belal, E., Baba, M. O., Ovie, S., Raji, A., Malasha, I., Njaya, F., Andi, M. N., Russell, A. and Neiland, A. (2009) Power struggle, dispute and alliance over local resources: analyzing 'democratic' decentralization of natural resources through the lenses of Africa inland fisheries. *World Development,* 37(12), 1935-1950.

Berkes, F. (1985) Cooperation from the perspective of human ecology. In Berkes F., ed., *Common Property Resources.* New Haven: Belhaven Press.

Biermann, F. (2007) 'Earth system governance' as a crosscutting theme of global change research. *Global Environmental Change,* 17(3-4), 326-337.

Blaikie, P. (2006) Is small really beautiful? Community-based natural resource management in Malawi and Botswana. *World Development,* 34(11), 1942-1957.

Bourdieu, P. (1990) *The logic of practice.* Stanford, CA: Stanford University Press.

Carlsson, L. and Berkes, F. (2005) Co-management: concepts and methodological implications, *Journal of Environmental Management,* 75(1), 65-76.

Cox, K. R. (1998) Spaces of dependence, spaces of engagement and the politics of scale, or: looking for local politics. *Political Geography,* 17(1), 1-23.

de la Torre-Castro, M. (2006) Beyond regulations in fisheries management: the dilemmas of the "beach recorders" Bwana Dikos in Zanzibar, Tanzania. *Ecology and Society,* 11(2), 35.

de la Torre-Castro, M. and Lindström, L. (2008) Heterogeneity, complexity and natural resource management: a coastal East African example. Human Dimensions of Global Environmental Change, Berlin, Germany, February 22-23, 2008.

de la Torre-Castro, M. and Lindström, L. (2010) Fishing institutions: addressing regulative, normative and cultural-cognitive elements to enhance fisheries management. *Marine Policy*, 34(1), 77-84.

Dill, B. (2009) The paradoxes of community-based participation in Dar es Salaam. *Development and Change*, 40(4), 717-743.

Dill, B. (2010) Community-based organizations (CBOs) and norms of participation in Tanzania: working against the grain. *African Studies Review*, 53(2), 23-48.

Economist Intelligence Unit (2008) Democracy Index 2008. http://graphics.eiu.com/PDF/Democracy%20Index%202008.pdf

FAO (2006) Making global governance work for small-scale fisheries. New directions in fisheries – A series of policy briefs on development issues, No. 07. Rome: FAO.

Folke, C., Hahn, T., Olsson, P. and Norberg, J. (2005) Adaptive governance of social-ecological systems. *Annual Review of Environment and Resources*, 30, 441-473.

Gusfield, J. R. (1975) *Community: a critical response (key concepts in the social sciences)*. Oxford: Blackwell.

Jentoft, S. (2000a) The community: a missing link of fisheries management. *Marine Policy*, 24, 53-59.

Jentoft, S. (2000b) Co-managing the coastal zone: is the task too complex? *Ocean and Coastal Management*, 43(6), 527-535.

Jentoft, S. (2004) The community in fisheries management. In Hersoug, B., Jentoft, S. and Degnbol, P., eds., *Fisheries development: the institutional challenge*. Delft, The Netherlands: Eburon Academic Publishers.

Jentoft, S. (2005) Fisheries co-management as empowerment. *Marine Policy*, 29(1), 1-7.

Jentoft, S., McCay, B. and Wilson, D. C. (1998) Social theory and fisheries co-management. *Marine Policy*, 22(4-5), 423-436.

Khan, A.S., Mikkola, H. and Brummett, R. (2004) Feasibility of fisheries co-management in

Africa. *NAGA, WorldFish Center Quarterly*, 27(1), 60-64.

Kooiman, J., ed. (1993) *Modern governance. New government-society interactions*. London: Sage.

Kooiman, J., Bavink, M. T., Jentoft, S. and Pullin, R., eds. (2005) *Fish for Life. Interactive Governance for Fisheries*. Amsterdam: Amsterdam University Press.

Lund, C. (2006) Twilight institutions: public authority and local politics in Africa. *Development and Change*, 37(4), 685-705.

Mansfield, B. (2001) Thinking through scale: the role of state governance in globalizing North Pacific fisheries. *Environment and Planning A*, 33(10), 1807-1827.

Mansfield, B. (2005) Beyond rescaling: reintegrating the 'national' as a dimension of scalar relations. *Progress in Human Geography*, 29(4), 458-473.

Mansuri, G. and Rao, V. (2004) Community-based and -driven development: a critical review. *World Bank Research Observer*, 19(1), 1-39.

Massey, D. (1999) Power-geometries and the politics of space-time: Hettner-Lecture 1998, Hettner-Lectures, 2. Heidelberg: University of Heidelberg.

Massey, D. (2004) Geographies of responsibility. *Geografiska Annaler Series B: Human Geography*, 86(1), 5-18.

Mohan, G. and Stokke, K. (2000) Participatory development and empowerment: the dangers of localism. *Third World Quarterly*, (21)2, 247-268.

Moore, A. (2008) Rethinking scale as a geographical category: from analysis to practice. *Progress in Human Geography*, 32(2), 203-225.

Olsson, P., Folke, C., Galaz, V., Hahn, T. and Schultz, L. (2007) Enhancing the fit through adaptive co-management: creating and maintaining bridging functions for matching scales in the Kristianstads Vattenrike Biosphere Reserve, Sweden. *Ecology and Society*, 12(1), 28.

Ostrom, E. (2001) Foreword. In Agrawal, A. and Gibson, C. C., eds., *Communities and the environment: ethnicity, gender, and the state in community-based conservation*. New Brunswick, N.J.: Rutgers University Press.

Taylor, P. J. (1999) Places, spaces and macy's: place-space tensions in the political geography of modernities. *Progress in Human Geography*, 23(1), 7-26.

Tönnies, F. (2001) *Community and civil society. Cambridge texts in the history of political thought*. Cambridge: Cambridge University Press.

Törnquist, O. (2004) The political deficit of substantial democratisation. In Harris, J., Stokke, K. and Törnquist, O., eds., *Politicising Democracy: The New Local Politics of Democratisation*. Basingstoke: Palgrave Macmillan.

Wade, R. (1994) *Village Republics: Economic Conditions for Collective Action in South India*. San Francisco: ICS Press.

Walters, C. (1986) *Adaptive Management of Renewable Resources*. New York: McGraw Hill.

World Bank. (1997) *World Development Report 1997: The State in a Changing World*. New York: Oxford University Press.

CHAPTER 24

Fisheries development as freedom

Svein Jentoft[1]

In this essay I present two contrasted images of what causes fishing people to be poor and what is needed to help them secure a better livelihood. These images are represented by Garrett Hardin's idea of the Tragedy of the Commons and by Amartya Sen, as elaborated in his seminal book "Development as Freedom." Whereas Hardin perceives the freedom of resource users as the problem underlying resource degradation and poverty, Sen regards freedom as an essential part of the solution to poverty. Could it be that they are both right? To illustrate this dichotomy, and to demonstrate some of the policy implications that can be drawn, some historical events in Norwegian fisheries will be discussed.

Introduction[2]

In this essay I shall compare and contrast Garrett Hardin's (1968) and Amartya Sen's (1999) concepts of freedom and development. In particular, I discuss the implications their different ideas may have for resource management and poverty alleviation in small-scale fisheries. To provide context and illustration, I shall tell the story of how my own country, Norway, transformed from among Europe's poorest to become one of the world's wealthiest. This extraordinary shift is not only about oil and gas, as one would be led to believe when looking at Norway today. It is also about how Norway's second most important export industry – the fishing industry – changed. Small-scale fishers in Norway used to be the poorest of the poor. But some important legislation introduced in the 1930s drastically altered the situation in their favour. Norwegian small-scale fishers used to think of themselves as

---

1 Corresponding author: Norwegian College of Fishery Science University of Tromsø, Norway, svein.jentoft@uit.no

2 A shorter version of this essay was presented as a plenary lecture at the World Small-Scale Fisheries Congress, Bangkok, Thailand. October 19-22, 2010.

poor. They may still consider themselves as disadvantaged and threatened on several fronts, but they are no longer associated with poverty. Neither do they think of themselves as such.

The Norwegian fisheries development story is thus by and large a positive example. It does not illustrate Garrett Hardin's "tragedy of the commons." Most fish stocks are now in good shape, which is relatively rare in the global context. The Norwegian case, I argue, is more demonstrable of Amartya Sen's "development as freedom." Many of the concepts that Sen champions have been part of Norwegian government fishery policy since the 1930s. Still, there is more to the story than meets the eye. Not everything is as one would expect if Norway were to follow Sen's development model. In fact, from the perspective of a viable small-scale fishery, the picture now seems rather bleak, at least considering what has happened in recent years with fisheries policies guided much more by Hardin's recipe than by Sen's. The fishery commons, for the most part, is now closed and essentially privatized. Largely as a consequence of that policy, Norway has witnessed a dramatic loss of fishers, a loss that is still ongoing. Since 1990, when a new individual vessel quota system was introduced as a response to a crisis in the cod fishery at that time (cf. Jentoft, 1993), the number of registered fishers has declined by more than half – from 27,000 to 12,000 in 2009.

The positive effect of this policy is that the economic outlook has improved for those remaining in small-scale fisheries. For those of the younger generation who aim to establish themselves in this sector, the entrance costs have increased dramatically. Recruitment has therefore halted. For many small-scale fishing communities, particularly in the north, this trend has indeed been tragic. The down-scaling has meant that people have lost their jobs and young people leave home, never to return. It has been a gradual and scattered process, not very noticeable outside the local community. It certainly does not receive similar attention from the media as when a big industrial enterprise in the central part of Norway is closed, and people are laid off. Therefore it does not create much political clamour. One needs to check the official statistics to see how the numbers add up. However, if the government had announced that it intended to reduce the number of fishers by fifty percent in 1990, it would have provoked uproar. This drastic outcome has not necessarily been a deliberate policy. I do believe that the government, at least in hindsight, considers the extreme decrease more as "collateral damage" – unfortunate but unavoidable. Many would also describe this, by and large, as a positive trend; not something to feel nostalgic about. The total quota is, after all, still being fished but with less manpower. This, therefore, is just sound economic policy. And small-scale peripheral fishing communities in the north are just a charge on the public purse anyway. Those arguing the opposite are therefore portrayed as romanticizing the past. The fact that so many are leaving the fishery is understood as a confirmation of the idea of fishing as "an occupation of last resort." Thus, freeing them from this burdensome job in the small-scale fishery can only be positive. In many coastal communities, people have lost faith of having a future in the small-scale fishing industry and are therefore hopeful that the oil and gas industry will save them.

Perceptions of freedom

A reader familiar with Amartya Sen's famous book, "Development as Freedom," will already have recognized the title of this essay: I have just added "fisheries." Sen is the 1998 Nobel Prize winner in economics. He has written on a lot of issues, most recently on social justice, and he is a philosopher as much as an economist. To my knowledge he has not focused on fisheries. Still, since his work is relevant to fisheries development, he is frequently referred to by people interested in small-scale fisheries, poverty and development. As for fisheries, it is Garrett Hardin, more than Sen, who has set the fisheries management agenda in recent years, in Norway (Hersoug, 2005) and worldwide. People who know Hardin's work will also recognize his statement:

> Therein is the tragedy. Each man is locked into a system that compels him to increase his herd without limit – in a world that is limited. Ruin is the destination toward which all men rush, each pursuing his own best interest in a society that believes in the freedom of the commons. Freedom in a commons brings ruin to all. (Hardin, 1968, p. 1244)

Then, compare this statement with what Amartya Sen says:

> Freedoms are not only the primary ends of development, they are also among its principal means. (Sen, 1999, p. 10)

I find it intriguing that Hardin and Sen are both talking about poverty and freedom. Hardin considers freedom as a problem, indeed as poverty's root cause; he therefore wants less of it. Sen, on the other hand, wants more freedom; but not only as a desirable goal of development to be pursued for its own value. For Sen, freedom is also a necessary condition: Without freedom, there is no development. The question is: Are Hardin and Sen talking about different things, and could it be that they both are right?

It should be noted that Hardin was not the first to identify the commons problem. In Scandinavia we would claim that Jens Warming deserves the credit, but he wrote in Danish and is therefore not as well known. Håkan Eggert at University of Gothenburg has recently translated Warming's 1931 article on the Danish eel fishery (see Eggert, 2010) into English. But let us stick to Hardin, as he talks explicitly about poverty and freedom.

For Hardin, the idea is that in a commons, resource users such as fishers are responsible for their own misery; they are both villain and victim. Therefore, if resource users do not take the initiative to restrain themselves, government must step in to help. Governments would therefore do people a service if they took their freedom away: Less freedom, more resources, less poverty, more well-being. However, it is often ignored that Hardin's main argument in his famous *Science* paper is that the tragedy of the commons and commoners belongs to the kind of problems that do not have a technical or a scientific solution. Instead solutions depend, as he says, on "our values and ideas of morality." Thus, the tragedy of the commons and poverty qualify as what Rittel and Webber (1973) called "wicked problems" for which there are no right or wrong solutions, only good and bad ones; that need a definition before they can be resolved; and for which our idea of a solution often precedes the definition. In fact, poverty is their prime example – where defining the problem is a problem, as years of debate among academics and policymakers would suggest. Thus the sort of solutions that we envisage for wicked problems, such as the tragedy of the commons, depend on what our initial image of the problem is – on what the German philosopher Hans-Georg Gadamer (1989) called pre-judgment. Indeed, the image of a tragedy is in itself an example of such pre-judgement. Once mentally internalized, any resource crisis is automatically taken for a Hardin tragedy. As Ostrom (1990) warned against, metaphors converted into policies can do more harm than good. There is always a danger that simplified problem images may lead to ineffective solutions. Therefore, a reality check is always recommended (Chuenpagdee and Jentoft, 2009; Jentoft *et al.*, 2010). This is particularly relevant with regard to small-scale fisheries, which are always embedded in local communities and cultures of great diversity and complexity (McGoodwin, 1990; Kooiman *et al.*, 2005). There is basically no single formula to poverty alleviation in small-scale fisheries.

It is also useful to employ more than one image. There are those – for instance the economist Kenneth Boulding (1977) and the cultural anthropologist Bonnie McCay (1995) – who have argued that if Hardin had used a different Greek theatrical plot as metaphor and defined the problem as a *comedy* rather than as a tragedy, he would have prescribed different remedies other than closing the commons, removing the freedom of the resource users. For poor small-scale fishers, perhaps the classic theatrical *romance* would have been even more appropriate, as that plot is about the struggle between good (like a loving young couple) and evil (like a fierce dragon), where good prevails in the end. We would for example recognize the romance in the writings of Karl Marx (see White, 1975), and in much of the fisheries discourse about fisheries management. Cast as a romance, the spotlight would then be on the *conflict* between small-scale, artisanal fisheries existing in harmony with nature on the one hand and, on the other, the large-scale, industrial vessels exhausting the resource and invading the fishing grounds traditionally occupied by small-scale fishers, thus limiting whatever freedom they may enjoy in the fishery commons. In the romance plot, the idea of small-scale fishers as both the villain and the victim, as in the Hardin narrative, would be misleading: Victim yes, but villain no.

Instrumental freedoms

Amartya Sen is not into such lofty metaphors. Instead, he starts from concrete experience, like famine in India and the lives that poor people are actually leading in that part of the world. He contends that it is not necessarily lack of food *per se* that causes people to starve, but their lack of access to it; they might be shut off from the stream of supply for various reasons. Poverty must therefore, he argues, be understood as a relationship between people and the food resource, which is as much a social arrangement and a justice issue as a natural phenomenon.

For Sen, freedom of choice is essential, but in order for people to enjoy it, they need to possess certain entitlements (like property rights, political rights, human rights) and capabilities (like those that come from having an education and being in good health). Sen (1999, p. 154) says: "Political rights, including freedom of expression and discussion, are not only pivotal in inducing social responses to economic needs, they are also central to the conceptualization of economic needs themselves." He also includes access to credit and commodity markets among his entitlements. Surely, as any fisheries development expert would know, entitlements represent external obstacles which, if removed, can enhance small-scale fishers' room to maneuver and, hence, the freedom that is necessary for investment and growth.

These are all what Sen calls "instrumental freedoms" – freedoms that equip people with the assets they need in order to live the life they have reasons to value, given the opportunities that actually exist or could potentially be created. This concept includes the notion of "protective security," the need to be safe, which is also what the philosopher Isaiah Berlin (1970 [1958]) was thinking of with his distinction between "freedom to" and "freedom from." To pursue the former (described as "positive freedom") one also needs the latter ("negative freedom"). One cannot, for instance, expect people to take risks, like venturing into alternative livelihoods to fishing, or to assume management roles and responsibilities, if they do not feel secure. There are certain things, like your family's well-being and your children's future, that you do not gamble with. Small-scale fisherfolk may be poor but they also have much at stake. The poor are already vulnerable and cannot allow themselves to become more so. Moving poor people out of the fishery may be coined by some as an act of liberation, but as Kris Kristofferson wrote and both he and Janis Joplin sang: "Freedom's just another word for nothing left to lose." And this is a situation no one wishes to be in. There must therefore be alternative ways to reduce overfishing and combat poverty and vulnerability that do not run the risk of ruining fishing people's lives and destroying the social fabric of their communities. On this note, Sen's message is far more optimistic than Hardin's.

The role of institutions

I do not believe that lessons can easily be transferred from one country to another, and certainly not from the north to south (or vice versa). Each country, each fishery, each community has circumstances and features that are unique and that count. Still, comparisons are interesting and useful, but perhaps not as much for the answers that they give as for the questions they inspire.

In Norway, no group was as hard hit by the economic depression in the late 1920s and 30s as the fishing population. What caused the problem in the fishery during that time was not a resource crisis but a market crisis that was felt most severely by fishers. As described by Abraham Hallenstvedt (1982), when export prices soared, fishers saw nothing of the profits made. Instead they helped to build a wealthy class of fish traders. When export prices plummeted, fishers had to shoulder the loss. When the buyer lowered the price, fishers had no other option than to increase their fishing effort to compensate for the loss. But by bringing more fish ashore, they induced the fish buyer to lower the price even more. And prices often changed during the day, from when fishers went out to harvest the catch and when they returned.

Some events radically changed this situation. First, fishers decided not to tolerate the situation any longer. This was all they talked about, and they were angry. Then they organized themselves and demanded the government to support them, which it did. Fishers converted from price-takers to price-setters in 1938, when a new law (the so-called Raw Fish Act) granted fishers' sales organizations the monopoly right to fix minimum landing prices. The second paragraph of the Raw Fish Act reads:

> The King may decide that the processing, sale or export of raw fish ... or products thereof shall be prohibited regardless of where the fish is caught if first-sale of the raw fish has not taken place through, or with the approval of a fishermen's sales organisation whose statues have been approved by the Ministry concerned. Sale by an approved sales organisation is regarded as first-sale.

By this law, the subordinate position of fishers relative to fish buyers was instantly changed, the bondage broken. Because fishers controlled the price at the dockside, processors and exporters had to become more effective in foreign markets, which would benefit the whole industry and the national economy. Thus the law was not only aimed at income distribution but also at expanding the total income made from fish exports.

The Raw Fish Act, which still stands, bears the popular name "The Fisher's Constitution." It is a good example of an entitlement that Amartya Sen talks about. Norwegian fisheries history is full of similar legislative initiatives aimed at strengthening protective security, like the 1908 ban on trawlers inside Norway's territorial borders, credit lines (a state fisher's bank, established 1919), and bargaining power of small-scales fishers (cf. Holm, 1995). But they were introduced at a time when fishers were much more numerous and, hence, politically powerful, than they are today. The government was also instrumental in facilitating the National Fisher's Association in 1926, and thus laid the foundation of a fruitful cooperative relationship between the state and the fishing industry, which is still working.

There is nothing to suggest that governments in other countries should not initiate similar measures to strengthen the protective security of small-scale fisheries – not to me at least. You do not have to be a wealthy country to do it, as the Norwegian case illustrates. Even so, the Norwegian model is not easily transferred to other countries, as Norway discovered in the 1950s when it tried to provide development assistance to Kerala, India. The Norwegian style fishers' sales organizations had very limited success there, to say the least. As pointed out by John Kurien (1985) in the article he published about this development program, there is a major difference between forming such organizations *by* fishers, as in Norway, and *for* fishers, as they tried to do in India. In Norway these sales organizations enjoyed overwhelming support among fishers, whereas in Kerala people appeared to be much more reluctant.

New institutions such as the Raw Fish Act and the fishers' sales organizations that it mandated can do a lot to empower the poor. Government can also help to facilitate such institutions. Because of their collective organization at local, regional and national levels, Norwegian small-scale fishers have historically been a powerful force in fishery development, in the building of strong coastal communities, and indeed in the creation of the welfare state as a whole – and through all this, in the eradication of poverty in Norway. (That the privatization process triggered by the individual quota system seems to have weakened their traditional role in recent decades, is another matter that I will not go into here, but see Mikalsen *et al.*, 2007). But, as the Norway/India example shows, institutions cannot just be imposed on people and then expected to work, regardless of how good in principle and well-intended they are. In Norway, the Raw Fish Act resulted from grassroots mobilisation. When instituted, it was "by the people" as much as "for the people." In 1938, the government initiated a referendum to gauge support for the new law. It turned out that 88 percent of the 21,000 fishers who cast their vote favoured it, which gave parliament a clear marching order (Christensen and Hallenstvedt, 1990). This also explains the legitimacy of this law among the fishers until this day. They consider the law as their own. No wonder the fish buyers and exporters never liked it and have tried to remove it, but so far the government has not been willing to listen to them. Some changes have been made since the law was enacted. For instance the rule that the sales organisation can determine who is allowed to buy fish has been removed. But the main principles, still stand.

Institutions and social justice

Garrett Hardin and Amartya Sen think differently about the role of institutions in poverty alleviation. Hardin emphasizes the regulatory and restrictive role of institutions; he talks about "mutual coercion mutually agreed upon." Sen, on the other hand, highlights their empowering, liberating and therefore enabling potential. I believe that institutions can do all of this (Jentoft, 2004). Institutions may well limit the freedom in the fishery commons and create social order and accountability. But they may also enhance the freedoms that fishers need in order to become effective stewards. As common resources are concerned, this includes their freedoms to enforce "un-freedoms," as when Ulysses needed to bind himself to the mast. Overfishing cannot be ruled out. People break rules – at least sometimes. Sustainability is therefore never guaranteed. Rather, it is a wicked problem that is not solved once and for all but needs to be *re*solved on a continuous basis (Jentoft and Chuenpagdee, 2009).

It is not easy to tell how Hardin and Sen would agree if they were to compare notes. Hardin and the tragedy of the commons are not among Sen's references. But Hardin and Sen do share the view that social values and norms have a role to play in economic development and environmental management, in as far as it involves the freedom of resource users in fundamental ways (see Sen, 1999, p. 269). If people are not allowed or are unable to pursue what they value in life, they can hardly be said to be free. What they actually value and want to realize are intrinsic to the community and culture they grow up and live in. Thus values are social through and through.

This is also why communities must be part of the equation, and why losing community may be detrimental to human well-being. But it is also true that, as far as freedom is concerned, communities come with a catch. To be member of a community involves commitments and responsibilities that may hold you back. If needed, other members may be of assistance, but they will be concerned about what you are doing. As sociologist Zygmund Bauman (2001, p. 4) so poignantly puts it, while community provides protective security, it also limits freedom: "Whatever you choose, you gain some and lose some. Missing community means missing security; gaining community, if it happens, would soon mean missing freedom." Still, we need both security and community. We cannot stay healthy in splendid isolation. We are social animals and need to fit in. A sense of belonging gives us identity. This is also how our values are nurtured. But in addition to all this, we need to be free to be creative. This is also the fundamental message in Thomas Friedman's (1999) bestseller "The Lexus and the Olive Tree." Modernity is a blessing, but only insofar as it does not destroy community.

I assume that Sen would agree with Hardin's lifeboat ethics, that "injustice is preferable to total ruin." Who wouldn't? But I do not think that Sen, without reservation, would support the idea that injustice is necessary to avoid the tragedy of the commons. Neither do I believe that Sen would dispute the notion of community which, as Kenneth Boulding (1977) observes, is needed to bring the lifeboat safely ashore. Community is not among the concepts that figure prominently in Sen's thinking, as it would have if he were an anthropologist or a sociologist, instead of an economist. He does, however, talk about social capital and trust as essential entitlements, which are qualities that we typically attribute to community. After reading Sen's (2009) book "The Idea of Justice," I believe he would maintain that neither justice nor community can be sacrificed. Like freedom, in addition to being intrinsically important, they are also necessary for avoiding the tragedy of the commons and alleviating poverty. Healthy marine ecosystems require healthy social communities, with governance systems that work. The same applies to poverty alleviation.

On certain accounts, Hardin and Sen complement each other. They each have solutions that together point in the same direction. Hardin regards poverty as an outcome of common resource over-exploitation, and concludes that harvest limitations must therefore be enforced. Sen has something to teach us about what might bring people out of this situation. Poverty may also lead people to over-exploit, and poverty alleviation may help relieve the pressure on scarce resources. Their perceptions of freedom are not identical and have contradictory policy implications. Whereas Hardin sees a *vicious* circle of poverty that can only be broken by external intervention, Sen's image is a *virtuous* circle involving capable and empowered actors. Small-scale fisheries on the brink of over-exhausting their resource commons need remedies based on Hardin's philosophy. But they also need those of Sen to avoid descending into poverty. Government and communities can play coordinating roles, preferably together, as the co-management concept aims to facilitate.

Conclusion

Since fisheries are an industry that relies on common pool, renewable natural resources, development must happen within safe ecological limits. Initiatives to promote economic development can easily exceed the limits of nature. The capacity of the resource to reproduce may be ruined. Economic growth can therefore be counterproductive. What is gained in the short run may be lost in the long run. Efforts to reduce poverty can end up aggravating it. Fisheries development is therefore a balancing act.

Since we cannot always be sure when we are about to transgress these limits, we need to be cautious. If not, there is a possibility that nature will strike back. Indeed, this has happened in many parts of the world where overfishing has brought fish stocks to the verge of collapse, and sometimes beyond. The Northern Atlantic Cod in Canada is a prime example. Norway was close to ending up in a similar cod crisis in the early 1990s, and for some years the picture was pretty bleak (Apostle *et al.*, 1998). The vessel quota system that was introduced to prevent this from happening again has, however, had consequences that, in many instances, seems to have killed rather than cured the patient. The fishing industry has been radically transformed, much to the expense of small-scale fishers. This is a negative lesson that other countries should learn from.

There are ways of dealing with resource crisis and poverty that are more sensitive to the concerns of small-scale fishers and local communities than others. But if limiting people's freedom in the commons in Hardin's way is the only remedy, more hardship to fishing people, who are already poor and vulnerable, is guaranteed. This is particularly the case if they have no other options. In Norway fishers had alternatives. They did not end up unemployed, but were absorbed by the growth in other sectors. They could also benefit from social security. But it did bring many fishing communities beyond the tipping point, with their cultural heritage lost, and a population base too small to sustain them.

Hardin can hardly be held responsible for what his followers in Norway and elsewhere have done when putting his ideas into regulatory policy and practice. He made a good point; his analysis should not be ignored. Despite the simple lifeboat ethics that he outlined, his argument is much more nuanced and intriguing than one is often led to believe when reading those who criticize him. As far as small-scale fisherfolk are concerned, however, there is reason to think that policies, practices and, therefore, outcomes, would have been very different than they are today if it had been Sen's idea of freedom, rather than Hardin's, that inspired them.

References

Apostle, R., Barrett, G., Jentoft. S., Holm, P., Mazany, L., McCay, B. and Mikalsen K. (1998) *Community State and Market on the North Atlantic Rim. Challenges to Modernity in Fisheries.* Toronto: University of Toronto Press.

Bauman, Z. (2001) *Community: Seeking Safety in an Insecure World.* Cambridge: Polity Press.

Berlin, I. (1970) *Four Essays on Liberty.* Oxford: Oxford University Press.

Boulding, K. (1977) Commons and community: the idea of a public. In Hardin, G. and Baden, J., eds., *Managing the Commons.* San Francisco: W.H. Freeman and Co.

Christensen, P. and Hallenstvedt. A. (1990) *På første hånd. Norges Råfisklag gjennom 50 år.* Tromsø: Norges Råfisklag.

Chuenpagdee, R. and Jentoft, S. (2009) Governability assessment for fisheries and coastal systems: a reality check. *Human Ecology,* 37, 109-120.

Eggert, H. (2010) Jens Warming on open access, the Pigovian tax, and property rights. *History of Political Economy,* 42 (3), 469-481.

Friedman, T. L. (1999) *The Lexus and the Olive Tree.* New York: Ferrar, Straus and Giroux.

Gadamer, H. G., Weinsheimer, J. and Marshall, D. G., trans. (1989) *Truth and Method.* New York: Crossroad.

Hallenstvedt, A. (1982) *Med lov og organisasjon. Interesser og markeder i norsk fiskerinæring.* Oslo: Universitetsforlaget.

Hardin, G. (1968) The tragedy of the commons. *Science,* 168, 1243-1248.

Hersoug, B. (2005) *The Closing of the Commons. Norwegian Fisheries from Open Access to Private Property.* Delft: Eburon Academic Publishers.

Holm, P. (1995) The dynamics of institutionalization. *Administrative Science Quarterly,* 40, 398-422.

Jentoft, S. (1993) *Dangling Lines. The Fisheries Crisis and the Future of Coastal Communities. The Norwegian Experience.* St. John's: ISER Press.

Jentoft, S. (2004) Institutions in fisheries: what they are, what they do, and how they change. *Marine Policy,* 28 (2), 137-149.

Jentoft, S. and Chuenpagdee, R. (2009) Fisheries and coastal governance as a wicked problem. *Marine Policy,* 33, 553-560.

Jentoft, S., Chuenpagdee, R., Bundy, A. and Mahon, R. (2010) Pyramids and roses: alternative images for the governance of fisheries systems. *Marine Policy,* 34, 1315-1321.

Kooiman, J., Bavinck, M., Jentoft, S. and Pullin, R., eds. (2005) *Fish for Life. Interactive Governance for Fisheries.* Amsterdam: Amsterdam University Press.

Kurien, J. (1985) Technical assistance projects and socio-economic change. *Economic and Political Weekly,* 20 (25/26).

McCay, B. J. (1995) Commons and private concerns. *Advances in Human Ecology,* 4, 89-116.

McGoodwin, J. R. (1990) *Crisis in the World's Fisheries. People, Problems, and Practices.* Stanford: Stanford University Press.

Mikalsen, K. H., Hernes, H.-K. and Jentoft, S. (2007) Leaning on user-groups: the role of civil society in fisheries. *Marine Policy,* 31, 201-209.

Ostrom, E. (1990) *Governing the Commons. The Evolution of Institutions for Collective Action.* Cambridge: Cambridge University Press.

Rittel, H. W. J and Webber, M. M. (1973) Dilemmas in a general theory of planning. *Policy Sciences,* 4, 155-169.

Sen, A. (1999) *Development as Freedom.* Oxford: Oxford University Press.

Sen, A. (2009) *The Idea of Justice.* London: Allen Lane.

White, H. (1975) *Metahistory: The Historical Imagination in Nineteenth-century Europe.* Baltimore: The Johns Hopkins University Press.

Section 7:

Way forward

CHAPTER 25

Too big to ignore: global research network for the future of small-scale fisheries

Ratana Chuenpagdee[1]

According to the Food and Agriculture Organization of the United Nations (FAO), fisheries support the livelihoods of about 540 million people, or about 8% of the world's population (FAO, 2010). More than 90% of them are related to the small-scale fishing sector. While these figures suggest that small-scale fisheries are "too big to ignore," the reality is quite the opposite. Information and knowledge about small-scale fisheries remains scattered and scarce. Their actual and potential contributions to food security, livelihoods, well-being, and resource conservation are generally undermined and unaccounted for in policy and decision-making. Rectifying this disparity was the main rationale behind the organization of the first World Small-Scale Fisheries Congress in 2010. This essay describes the congress background and focus, key outcomes, and a proposed global research network, *Too Big To Ignore*, arising from it.

Introduction

As the essays in this volume show, small-scale fisheries are important at local, regional, national and global levels. They are the main source of food and employment for millions of people. Current environmental, social, market and institutional changes put small-scale fishing communities at risk and in a vulnerable position, unless their assets, including capital, social network and cultural identity, are protected. At the onset, management decisions and policies need to be sensitive to the conditions and needs of this sector of the population, acknowledging, for instance, their roles and contributions to the society and global sustainability. This requires, among other things,

1 Corresponding author: Memorial University of Newfoundland, Canada, ratanac@mun.ca

comprehensive knowledge and fine-scale information about small-scale fisheries.

The local complexity and dynamic reality of this fishing sector poses major challenges to researchers, managers, and policymakers alike. Social science research aimed at understanding small-scale fisheries in certain locations faces difficulties in elevating issues to the national policy agenda. Managers lack suitable tools to assess the context-specific situation of small-scale fisheries. Decisions about fisheries are often influenced by powerful industrialized fishing enterprises, and rarely have the interests of small-scale fishing communities at the core.

In many areas of the world, issues and concerns important to small-scale fisheries are underrepresented in national and international policy forums. The situation has improved with efforts by the FAO, research institutions, and organizations working for the interests of small-scale fishers like the International Collective in Support of Fishworkers (ICSF). Nevertheless, a lot more is required to correct the imbalance in terms of research, funding, and policy attention between small- and large-scale fisheries.

From the research perspective, sustained and substantial financial support for the continuation of social and natural science disciplinary research, interdisciplinary research, and participatory research is necessary. However, other avenues are also possible and need to be explored. This essay recounts one of the latest attempts to raise the profile of small-scale fisheries research through a scientific gathering titled the World Small-Scale Fisheries Congress (WSFC). It was held for the first time in Bangkok, Thailand in October 2010. In the following, I provide the rationale for the meeting and present some highlights and the main outcomes. I conclude with a call for the establishment of a global research network for the future of small-scale fisheries, describing its key tenets deliberated at the WSFC.

Research on small-scale fisheries

Several forum for dialogue about fisheries research exist, one of which is the World Fisheries Congress (WFC). The first WFC was held in 1992 in Athens, Greece. Subsequent meetings have been held every four years in Brisbane, Australia; Beijing, China; Vancouver, Canada; and Yokohama, Japan, consecutively. The central theme of each congress is a telling story about how research and policy interests in the world's fisheries have evolved. The Athens conference focused on the state of the resources, while the second one in Brisbane revolved around developing and sustaining fisheries.

The central theme at the Beijing meeting was how to increase production to feed the world, whereas in Vancouver, the focus shifted to reconciling use and conservation. The latest conference held in Yokohama in 2008 addressed global aquatic issues and sustainability such as fish habitats and marine and freshwater ecosystems and effective utilization of resources. In 2012, the sixth conference in the series will be held in Edinburgh, Scotland, with 'sustainability in the changing world' as the central theme.

How prominent are small-scale fisheries at the WFCs? The only systematic assessment of this was conducted for the fourth WFC held in Vancouver in 2004 by a group of fisheries scholars working together under the auspices of AQUALINK.[2] They reported that about 38% of the 223 papers presented focused on artisanal and small-scale fisheries, which was a comparable proportion to those related to the industrial fishing sector (Chuenpagdee and Bundy, 2005). Although many social issues, such as resource allocation, stakeholder conflicts, and property rights, were raised, the majority of papers were concerned with ecological aspects of fisheries, and little emphasis was placed on knowledge integration and policy discussion about small-scale fisheries.

An 'issue scan'[3] was conducted to examine the breadth of research on small-scale fisheries and to determine the need for a scientific congress focusing exclusively on this sector. The results of this scoping exercise, coupled with a literature search of peer-reviewed journal articles written in English about small-scale fisheries published between 1917 and 2009 (see Song and Khan, this volume for details on the method), revealed eight main themes most frequently discussed (Table 1).[4] The level of coverage presented in this table reflects the proportion of issues (from the total of 570) addressed in the 1,717 papers analyzed. It should be noted that since many more papers have been written in other languages, the findings may represent only a fraction of research in small-scale fisheries.

---

2 www.fisheries.ubc.ca/aqualink/index.htm

3 A short open-ended questionnaire posted on the website and circulated through research networks.

4 Based on the research conducted by Andrew Song in 2010 as part of the preparation for the first WSFC.

Table 1 Coverage of small-scale fisheries issues in published research through time

	Level of coverage*			
Theme	*Prior to 1980*	*1980-1989*	*1990-1999*	*2000-2009*
Fisheries information and knowledge	2	3	2	1
Ecosystem complexity, dynamics and interactions	2	4	2	1
Governance challenges	1	1	2	2
Overexploitation, gear conflict, resource/space competition	3	3	2	3
Fisheries development, livelihoods and well-being	1	2	3	3
Access, rights and responsibility	3	4	4	4
Poverty, marginalization, and vulnerability	4	3	4	4
Post-harvest process and globalized markets	4	4	4	4

* 1= more than 20% of the papers; 2= between 16-20%; 3= between 11-15%; 4= less than or equal 10%

Table 1 shows that topics that have long captured research interest on small-scale fisheries are related to governance; mostly concerning governance effectiveness, arrangements and processes. With heightened interest in stakeholder participation and inclusion in decision-making and resource governance, research has focused on limitations and opportunities that inhibit or enable this transition. Some of the current discourses on governance emphasize the complex linkages between ecological and social small-scale fisheries systems that call for flexible, adaptive (Armitage *et al.*, 2009) and interactive governance (Kooiman *et al.*, 2005) approaches, multi-level governance and networks (Fanning *et al.*, 2007), direct attention on occupational pluralism in small-scale fisheries (Allison and Ellis, 2001), and legal aspects (Jentoft *et al.*, 2009).

Another topic that is high on the research agenda is ecosystem complexity, dynamics and interactions. Research on fishery ecology and effects of fishing on ecosystems has traditionally been done in the context of the commercial, industrialized fishing sector. The ecosystem approach to fisheries management, however, requires comprehensive understanding of how system components function and how they interact. Impacts of small-scale fishing on fisheries resources, and bycatch and habitat degradation issues need to be examined, as some studies argue (see Peckham *et al.*, 2007). Similarly, the role of small-scale fisheries in conservation and restoration is emphasized. This topic is linked to concerns of over-exploitation, gear conflict and resource/space competition, all of which have been surprisingly less covered in scholarly research. This by no means suggests lack of interest or low importance of these issues. Conflicts between small- and large-scale fishing sectors, and in some instances with recreational fishing, as well as competition for resources and space, also with aquaculture, are rife in most fisheries and have been the subject of discussion at local, national and global levels. The low coverage simply signifies the lack of scientific information on the issue, and likely the lack of scientific engagement in policy discourse on these topics.

Similar observation can be made on issues that are commonly associated with small-scale fisheries, such as livelihoods and well-being, access and rights, and poverty and vulnerability. The low number of published articles addressing these important issues may suggest an unequal proportion of investment (e.g., funding and training programs) on natural and social science research. The recent debate on catch-share systems and wealth- and welfare-based fisheries (Costello *et al.*, 2008; Bromley, 2009; Béné *et al.*, 2010) will certainly change this analysis in the next couple of years. A new book on poverty in small-scale fisheries (Jentoft and Eide, 2011) based on fifteen case studies around the world is another important contribution that will provide astute understanding about how small-scale fishing communities perceive, negotiate and cope with poverty, thus offering a sound basis for poverty alleviation policy.

The relatively low scientific coverage of the post-harvest process and globalized markets in the past decades will likely change with increased public interest on environmental issues and food safety, as well as with the surge in product-certification schemes. While small-scale fisheries should be subjected to the same scrutiny as large-scale fisheries for their impacts on environmental sustainability, the existing standards, such as those applied by the Marine Stewardship Council, are problematic in their application to the small-scale fishing sector (Ponte, 2008). More inputs from social science research are required to ensure that these certification processes do not result in marginalization of small-scale fisheries in the globalized world.

The final research theme, fisheries information and knowledge, is an ongoing priority in the context of both large- and small-scale fisheries. Numerous studies aim to provide information about various aspects of fisheries to address the knowledge gaps on issues such as biology and ecology of fish, archaeological characterization of the fishery, effort and landing data, and socio-economic conditions. Integration of knowledge from different disciplines and sources, including local and traditional knowledge, for the purpose of fisheries management has been emphasized (Wilson *et al.*,

2006). Finally, it is widely recognized that specific sampling and monitoring protocol, as well as assessment tools and techniques sensitive to small-scale fisheries contexts are required (Salas *et al.*, 2007).

The 'World Small-Scale Fisheries Congress'

While there seems to be some attention on small-scale fisheries at the WFCs, findings from the literature search, as well as those contributed through the issue-scan exercise, suggest that a scientific conference devoted to this sector is fitting. Other conferences like the 'People and the Sea' biennial conference series organized by the Centre for Maritime Research based in Amsterdam, and meetings focusing on small-scale fisheries that the FAO frequently convenes, serve similar functions. The main goal of the World Small-Scale Fisheries Congress (WSFC), however, is to provide opportunities for interactive discussion between researchers from the north and the south, which can help build the knowledge base and strengthen the arguments in support of small-scale fisheries.

The first WSFC[5] was held in Bangkok, Thailand in October 2010, and was attended by about 220 people from 39 countries. It was hosted by three local institutions, the Coastal Development Centre, Thailand; the Department of Fisheries, Government of Thailand; and the Southeast Asian Fisheries Development Center (SEAFDEC), and supported by several other institutions from around the world. The congress themes mirrored the issues identified through the above research, and known experts on the topics were invited to present their work on these themes and participated in panel discussions. Except for two hours in the afternoon, all presentations were conducted in plenary, which made it possible for participants from natural and social science disciplines, and interdisciplinary backgrounds to interact and cross-fertilize their ideas. Students constituted about 25% of the participants. Because of the scientific focus and limited funding, few members of environmental and community organizations were present at the meeting.

On the last day of the congress, participants split into five groups to discuss research and policy priorities for different regions of the world; the outcomes of which are summarized below.

5 http://www.seafdec.or.th/wsfc2010

Africa

The priorities proposed by the participants discussing small-scale fisheries in Africa were prefaced by a list of principles, serving as foundation for any research activities. They suggested that research needs to have policy relevance, be gender sensitive, and responsive to the local context. They also called for participatory and interdisciplinary research, as well as partnership approaches to understanding fisheries. Research priorities for this region include an understanding of impacts and implications of global change processes on livelihoods and food security; not only climate-related change, but also others induced by urbanization, migration and information technology. They emphasized the need to examine the role and contribution of small-scale fishers, including women and children, to development processes. Documenting success and failure of different governance approaches was also listed as a priority. They referred, for instance, to the need to bridge the gap between nationally-managed fisheries by government departments and community-based approaches.

Latin America and the Caribbean

Discussion among participants interested in small-scale fisheries in this region led to a wide range of issues that require research, considering all aquatic systems at every scale. These include diverse dependency on fisheries resources, market and non-market valuation, user perspectives on management, and power and conflicts of interest in policy-making. Better use of existing data, improving information about small-scale fisheries and sharing information with resource users and policymakers at national and regional levels, were specified as important. The group also stressed the need for developing synthesis and analytical tools within integrated frameworks, such as ecosystem-based management and environmental impact assessment, and called for research about small-scale fisheries in the region to be better represented in the literature.

North America

One of the key issues identified by this group was the knowledge gap in small-scale fisheries and the lack of appropriate tools for management. Bridging the knowledge gaps and developing alternative management tools were thus listed as some of the priorities for this region. The group also emphasized the need to situate small-scale fisheries in the broader context, such as in ecosystem-based management and integrated coastal management. The importance of fisheries for aboriginal communities, as well as their rights, was recognized, especially in the context of climate change. Sharing stories, experience and lessons among fisheries stakeholder groups, and cross-fertilization of knowledge with other regions, through network and partnership building, topped the list of priorities for this region.

Asia-Pacific

For Asia-Pacific, several research topics were presented as priorities. For instance, group members highlighted issues related to mangroves, aquaculture, marine protected areas, other spatial and transboundary management, gender relations, migration and mobility of fishing communities, along with governance research to address corruption, zoning of coastal activities, and funding for local government. Action-oriented research that responds to the problems and can be transformed into practice, with better connection to the communities in terms of their needs and their contribution to knowledge, was also emphasized. Similar to the other groups, they called for an integration and recognition of local knowledge in fisheries management, partnership building and communication to increase the profile of small-scale fisheries in national and regional policies, as well as sharing of best practices. In addition to research into the concept of ecosystem-based fisheries management and its implication for small-scale fisheries, the group suggested that interdisciplinary and multi-sectoral approaches to fisheries, such as linkages with forestry, climate change and global seafood consumption, should be promoted. Finally, they noted the importance of simplifying messages for policymakers and creating space for policy change.

Europe

The group discussing research and policy priorities in Europe commented on the relative neglect and under-reporting of small-scale fisheries, and the lack of recognition of their diversity in the current Common Fisheries Policy (CFP). Some doubts were expressed about what this fisheries reform and the future policy would hold. One of the first priorities is thus to examine the impact of CFP and the new policy on small-scale fisheries. Further, they called for an investigation of the impact of decentralization and reorientation of fisheries policy for small-scale fisheries, and of the history of funding and research. Issues such as inshore and offshore marine protected areas and sea tenure arrangements, and their impacts on fishing communities, as well as the role of fisher institutions, were raised. Recognizing the changing position of small-scale fisheries in public discourse vis-à-vis the environment, and the important role of women in fisheries, the group highlighted the need for a comprehensive data collection and information system on small-scale fisheries, to enable analyses such as economic performance.

Global research network for the future of small-scale fisheries

Drawing from the key regional priorities deliberated by the participants, the first WSFC provided an impetus to create a global network of researchers interested in the future of the world's small-scale fisheries. The *Too Big to Ignore* network aims to address current gaps in knowledge and information availability about small-scale fisheries, and lack of appropriate governance mechanisms and policy initiatives to support this sector, by conducting the following activities.

Local studies and global analysis of small-scale fisheries
The network proposes to examine local and global issues facing small-scale fisheries, including global change processes. Knowledge about these issues will be assessed and gaps identified, serving as a framework to conduct evidence-based, problem-driven case studies. A general methodological framework and standardized data template will be developed and used by network members to enable comparative analysis and regional synthesis. Drawing from general discourse in resource governance, these include, but are not limited to, the analysis of fishing impacts, commodity and value chains, gender roles, well-being and livelihoods, social capital, employment mobility, career patterns, recruitment, and governance performance.

Small-Scale Fisheries Information System
A Small-Scale Fisheries Information System (SFIS) will be developed as a means to improve understanding about this sector and for knowledge sharing. It will capture the key characteristics of small-scale fisheries in any given location across the 'fish chain' (pre-harvest, harvest and post-harvest), including nature and type of fisheries and fishing activities, fishers' relationships with fish traders and moneylenders, and formal and informal rules and norms. Local fishers, practitioners and community members will be trained to assist with secondary data collection and systematic documentation of knowledge in the information system. In addition to facilitating information sharing, the online SFIS will provide a solid foundation for policy development and decision-making for the secure future of small-scale fisheries.

Global interdisciplinary training program in fisheries
Another key activity of the *Too Big to Ignore* network is the development of global interdisciplinary training modules that can be offered in various ways, such as distance learning, online self-taught, field course training, or as part of the curriculum for degree and non-degree programs. The

training consists of professional development as much as lifelong learning strategies. Each module contains theories drawn largely from the synthesized knowledge obtained through the research in the first component, and the analyses conducted in the second component, and through collaboration with other fisheries experts. The practicum will contain both region-specific and globally applicable case studies. Trainees will be able to interact and communicate with 'resident experts,' served by the partnership members and other invited scholars, and among themselves to encourage exchange and learning. Institutions and community groups interested in using the courses will be encouraged to contribute to the case studies to enrich the knowledge about small-scale fisheries of the world.

This global partnership is necessary and timely because of the heightened value of small-scale fisheries to the world's economy and food security. The viability of this partnership requires commitment from researchers, governments, environmental and community organizations to share knowledge about fisheries, and contribute data to the information system and the training modules. It is expected that the network will mobilize knowledge about small-scale fisheries and influence the development of policies linking economic viability, environmental sustainability and fisheries governance with an increased sensitivity to the small-scale fisheries way of life.

Moving forward

A case has been made in this essay, and all others in this volume, for the recognition of small-scale fisheries' contribution to food security, livelihoods, poverty alleviation and the sustainable future of the world's fisheries. Scientific, interdisciplinary, participatory, action-oriented and applied research is called for to address current and emerging issues facing small-scale fishing people and for the development of the sector. This research needs to be situated in the local context and have purposeful goals to contribute to improving the livelihoods of local communities, but should also have direct policy implications and put small-scale fisheries issues on the policy agenda at all levels. Through database systems like SFIS, information and knowledge about this sector can be brought on par with that of large-scale fisheries. Tools and approaches specific to the small-scale fisheries context can be developed, exchanged and shared through the *Too Big to Ignore network*. Scientific Interdisciplinary fisheries research, participation by fishers and communities, supported by environmental groups and funding agencies, and ongoing dialogue with policy- and decision-makers for knowledge-based management and innovative governance are key ingredients to moving forward.

Acknowledgments

I thank Andrew Song of Memorial University for his help in the issue-scan exercise and the literature search that took place prior to the conference, as well as his assistance with the conference organization. Further, I thank Anthony Charles (Saint Mary's University), Maarten Bavinck (University of Amsterdam), Patrick McConney (University of West Indies), Merle Sowman (University of Cape Town), and Nireka Weeratunge (WorldFish Center) for their help in leading the regional group discussion. Finally, the idea about the proposed network for the future of small-scale fisheries presented in this essay was developed and submitted for a Partnership Grant application to the Social Sciences and Humanities Research Council of Canada. Numerous colleagues from around the world, many of whom contributed the essays to this volume, are participants and partners in the proposal. Thanks especially to Reade Davis (Memorial University), Svein Jentoft (University of Tromso), and Derek Johnson (University of Manitoba), for their help in drafting the proposal.

References

Allison, E. H., and Ellis, F. (2011) The livelihoods approach and management of small-scale fisheries. *Marine Policy*, 25(5), 377-388.

Armitage, D. R., Plummer, R., Berkes, F., Arthur, R. I., Charles, A. T., Davidson-Hunt, I. J., Diduck, A. P., Doubleday, N. C., Johnson, D. J., Marschke, M., McConney, P., Pinkerton, E. W. And Wollenberg, E. K. (2009) Adaptive co-management for social-ecological complexity. *Frontiers in Ecology and the Environment*, 7, 95-102.

Béné, C., Hersoug, B., and Allison, E. H. (2010) Not by rent alone: Analysing the pro-poor functions of small-scale fisheries in developing countries. *Development Policy Review*, 28(3), 325-358.

Bromley, D. (2009) Abdicating responsibility: the deceits of fisheries policy. *Fisheries*, 34(6), 280-302.

Chuenpagdee, R. and Bundy A. eds. (2005). *Innovation and Outlook in Fisheries: An Assessment of Research Presented at the 4th World Fisheries*

Congress. Fisheries Centre Research Report, 13(2). Vancouver, Canada: Fisheries Centre, University of British Columbia.

Costello, C., Gaines, S. D., and Lynham, J. (2008) Can catch shares prevent fisheries collapse? *Science*, 321, 1678-1681.

Fanning, L., Mahon, R., McConney, P., Angulo, J., Burrows, F., Chakalall, B., Gil, D., Haughton, M., Heileman, S., Martinez, S., Ostine, L., Oviedo, A., Parsons, S., Phillips, T., Santizo Arroya, C., Simmons, B. and Toro, V. (2007) A large marine ecosystem governance framework. *Marine Policy*, 31, 434-443.

FAO (2010) The State of World Fisheries and Aquaculture 2010. Rome: FAO Publication.

Jentoft, S., Bavinck, M., Johnson, D. S. and Kaleekal T. (2009) Co-management and legal pluralism: how an analytical problem becomes an institutional one. *Human Organization*, 68, 27-38.

Jentoft, S. and Eide, A., eds. (2011) *Poverty Mosaics: Realities and Prospects in Small-Scale Fisheries*. Amsterdam: Springer.

Kooiman, J., Bavinck, M., Jentoft, S. and Pullin, R. S. V., eds. (2005) *Fish for Life: Interactive Governance for Fisheries*. Amsterdam University Press, Amsterdam.

Peckham, S. H., Diaz, D. M., Walli, A., Ruiz, G., Crowder, L. B. and Wallace, J. N. (2007) Small-scale fisheries bycatch jeopardizes endangered Pacific Loggerhead turtles. *PLoS ONE*, 2(10): e1041. doi:10.1371/journal.pone.0001041.

Ponte, S. (2008) Greener than thou: the political economy of fish ecolabeling and its local manifestations in South Africa. *World Development*, 36(1), 159-175.

Salas, S., Chuenpagdee, R., Seijo, J. C., and Charles, A. (2007) Challenges in the assessment and management of small-scale fisheries in Latin America and the Caribbean. *Fisheries Research*, 87(1), 5-16.

Wilson, D. C., Raakjær, J., and Degnbol, P. (2006) Local ecological knowledge and practical fisheries management in the tropics: a policy brief. *Marine Policy*, 30(6), 794-801.

LIST OF CONTRIBUTORS

Henrik Angerbrandt is a PhD candidate at the Department of Political Science, Stockholm University, Sweden. His research focuses on decentralization, identities and conflict, with particular emphasis on issues of scale and community. He has recently published on these issues in a Nigerian context in *Journal of Contemporary African Studies*, 29(1).

Maarten Bavinck is Associate Professor at the University of Amsterdam and Director of the social science Centre for Maritime Research (MARE). He has done extensive ethnographic research on small-scale fisheries, particularly in South Asia. He is also a key figure within the Fisheries Governance Network, and is currently President of the Commission on Legal Pluralism.

Ben Belton is a post-doctoral fellow at the WorldFish Center, Bangladesh office. His doctoral research dealt with the origins and consequences of aquaculture development across South and Southeast Asia. His research interests also include private environmental governance, sustainable development and natural resource use, and agrarian change.

Anthony Charles is Professor of Environmental Studies at Saint Mary's University, Halifax, Canada, a Pew Fellow in Marine Conservation, and a widely-recognized expert on fishery governance, management and economics, including sustainability and resilience, human dimensions of ecosystem-based management, and community-based fishery and coastal management.

Ratana Chuenpagdee is Canada Research Chair in Natural Resource Sustainability and Community Development at Memorial University, Canada. Her research emphasizes interdisciplinary approaches to fisheries and coastal ecosystems management, focusing particularly on small-scale fisheries, marine protected areas, community-based management, food security and fisheries governance. In addition to conducting research in Thailand and Canada, she has worked in other countries including Cambodia, Malawi, Mexico and Spain.

Shelly-Ann Cox is a doctoral candidate at the Centre for Resource Management and Environmental Studies (CERMES), University of the West Indies, Cave Hill Campus, Barbados. As a member of the MarGov project, her research emphasizes social-ecological systems in fisheries, adaptive co-management and social network analysis.

Raquel De la Cruz Modino is a social researcher at the Institute of Social and Political Sciences, University of La Laguna. Her research combines fisheries and tourism studies, with the North-Atlantic and the Mediterranean Spanish coast as study areas. She is currently a post-doctoral research fellow at Memorial University of Newfoundland and the University of Tromso, working on a project about governability of multiple-use MPAs.

Maricela de la Torre-Castro is Assistant Professor in Natural Resource Management at Stockholm University, Sweden, and has published extensively in the field of social-ecological systems, particularly on the links between humans, seagrasses and fisheries in poverty contexts.

Steve Donda is a fisheries socio-economist who started his fisheries career as a fisheries biologist. He is currently working as Deputy Director of Fisheries with the Department of Fisheries in Malawi. His field of interest is fisheries management and governance.

Larry Felt has worked with commercial fishers as well as the Labrador Inuit on collaborative research including aboriginal governance, resource harvesting and the use of traditional knowledge in resource management. He is Professor of Sociology and Chair of the Research Ethics Board (REB) at Memorial University in St. John's, Newfoundland and Labrador, Canada.

Katia Frangoudes is a senior researcher at AMURE-Center for Law and Economics of the Sea, a joint research unit of the University of Brest and the French Research Institute for the Exploitation of the Sea (IFREMER). She specializes in women in fisheries and aquaculture, and institutional issues in fisheries management. She is the facilitator of AKTEA, the European Network of Women Organizations in Fisheries, and editor of the AKTEA Newsletter.

Maria A. Gasalla is Associate Professor at the University of Sao Paulo Oceanographic Institute, head of the Fisheries Ecosystems Laboratory (LabPesq), and researcher at the Brazilian National Scientific Research

Council (CNPq). She has been studying and developing innovative and integrative approaches to better understand fisheries social-ecological systems in the search for sustainable pathways.

Sapta Putra Ginting is Deputy Director for Coastal Community Empowerment, Ministry of Marine Affairs and Fisheries, Indonesia. He has been involved in many major coastal management projects in Indonesia since 1994. He currently chairs a Coastal Community Development Project dealing with small-scale fisheries livelihoods in 11 provinces in Indonesia.

Moenieba Isaacs is a senior lecturer and researcher with the Institute for Poverty Land and Agrarian Studies (PLAAS), University of the Western Cape. Her research focuses on understanding poverty, HIV/AIDS impacts on fish communities, and fisheries policy reforms in South Africa as well as MPAs and co-management. She conducts development work and field research in fishing communities on the west, south and east coasts of South Africa.

Svein Jentoft is a sociologist and Professor of fisheries management and organization at the Norwegian College of Fishery Science, University of Tromsø, Norway. He specializes in communities and coastal development and has published extensively on co-management, fisheries co-operatives, marine protected areas, indigenous peoples and interactive governance.

Guy Jobbins is a Senior Program Officer with the Climate Change Adaptation in Africa Program at the International Development Research Center, based in Cairo. His work is focused on adaptation to rising sea levels, increased freshwater scarcity, and the impacts of climate change on human health.

Kungwan Juntarashote is Associate Professor in Fisheries, Faculty of Fisheries, Kasetsart University, Bangkok, Thailand, and director of the Coastal Development Centre. He has extensive experience in small-scale fisheries, community-based fisheries management and fisheries development and economics in Thailand and Southeast Asia.

Ahmed Khan is a PhD candidate in the Department of Geography, Memorial University of Newfoundland. He has worked on global fisheries management and policy issues, and is familiar with Sierra Leonean, Gambian, and Canadian fisheries. His interests include fisheries policy and governance, integrated management and regional economic development.

Abdellatif Khattabi is Professor of Natural Resource Management at the Ecole Nationale Forestière d'Ingénieurs, Salé, Morocco. His research interests are focused on rural livelihoods and the integrated management of natural resources, mainly in the context of climate change.

John Kurien is an activist-researcher who has lived in and worked with fishing communities in several parts of Asia. He was Professor at the Centre for Development Studies, Trivandrum, India and a founding member of the International Collective in Support of Fishworkers (ICSF). For a decade he was Vice-Chair of the FAO/UN Advisory Committee on Fisheries Research.

Lars Lindström is Associate Professor in Political Science at Stockholm University, Sweden, and has published in the areas of globalization, labor, and democratization, and the management of marine resources in poverty contexts.

David Little is Professor of Aquatic Resource Development at the Institute of Aquaculture, University of Stirling, UK. He has worked extensively in Asia on the role of aquatic resource management in sustainable livelihoods. He currently heads a major EU funded project which combines value chain and lifecycle analysis to explore the sustainability of the trade in aquaculture products from Bangladesh, China, Thailand and Vietnam to the European Union.

Gayathri Lokuge is a researcher at the Centre for Poverty Analysis, and is reading for her Masters in Development Studies at University of Colombo, Sri Lanka. She has done extensive research and advisory work on poverty and conflict, especially in North and Eastern Sri Lanka. She also works as a trainer in conflict sensitivity, impact monitoring and understanding poverty.

Kristen Lowitt is a doctoral candidate in Interdisciplinary Studies at Memorial University of Newfoundland. Her PhD research focuses on community food security and fisheries in the coastal region of Bonne Bay on Newfoundland's west coast. Her research shows a long-standing interest in food systems, including her Master's research about farmers' markets in Nova Scotia.

Robin Mahon is Professor of Marine Affairs and Director of the Centre for Resource Management and Environmental Studies (CERMES), University of the West Indies, Cave Hill Campus, Barbados. His research is on institutional arrangements for marine resource governance and how to effect organizational change in complex systems, particularly in the Wider Caribbean Region.

Patrick McConney is Professor at the Centre for Resource Management and Environmental Studies (CERMES), University of the West Indies, Cave Hill Campus, Barbados. He is leader of a research team investigating Marine Resource Governance in the Eastern Caribbean (MarGov), funded mainly by the International Development Research Centre (IDRC) of Ottawa, Canada.

Mohamed Munas is a researcher at the Centre for Poverty Analysis, and holds a Masters in Economics from the University of Colombo, Sri Lanka. He has researched poverty, conflict, agriculture and livelihoods, and is presently working towards new thematic research on migration and poverty.

David C. Natcher has done collaborative harvesting research with aboriginal partners in Alaska and Canada. Formerly a Canada Research Chair in Aboriginal Studies at Memorial University, he is currently Assistant Dean, Aboriginal Programs and Research, College of Agriculture and Bioresources at the University of Saskatchewan in Saskatoon, Saskatchewan, Canada.

Kemraj Parsram is a doctoral candidate at the Centre for Resource Management and Environmental Studies (CERMES), University of the West Indies, Cave Hill Campus, Barbados. His research focuses on fisheries social-ecological systems, adaptive co-management and social networks. He is a member of the MarGov project team.

José J. Pascual-Fernández is Professor of social anthropology and Director of the Institute of Social and Political Sciences at the University of La Laguna, Tenerife, Spain. Since the 1980s, his research has been on fisheries, tourism and governability, with recent focus on MPAs in Spain. He research has resulted in consolidating a line of social science research on coastal population and fisheries in Spain.

Gabriela Sabau is Associate Professor of Economics/Environmental Studies at Memorial University of Newfoundland, Grenfell Campus, Canada. She also teaches the Master's course Economics of Coastal and Marine Areas at the University Center of the Westfjords, Iceland. Her research interests are knowledge-based economy, sustainable development, values and valuation.

Jyothis Sathyapalan is Associate Professor at the Centre for Economic and Social Studies, Hyderabad, India. He has also been a Visiting Scholar at the Amsterdam Institute of Metropolitan Studies, University of Amsterdam, and at Memorial University of Newfoundland, Canada.

Joeri Scholtens graduated cum laude in Environment and Resource Management at the VU University Amsterdam. He is currently a PhD candidate at the Amsterdam Institute of Social Science Research, University of Amsterdam, and affiliated with the Center for Maritime Research, Amsterdam.

Andrew Song is a PhD candidate in the Department of Geography, Memorial University of Newfoundland. He is interested in fisheries governance, particularly institutional arrangements and meta-level governance, and holds some familiarity with several fishery settings including Malawi, South Korea and Philippines.

Jeena T. Srinivasan is Associate Professor at the Centre for Economic and Social Studies, Hyderabad, India. She has also been a Visiting Scholar at the University of Namur, Belgium, and Brock University, Canada.

Deatra Walsh is a postdoctoral research fellow at York University, Toronto, Canada, specializing in gender, work, labor mobility and rural communities. She holds a doctorate in Sociology from Memorial University, and a Master's of Rural Development from Brandon University. She has worked extensively on coastal community issues, fisheries, and rural development.